Advances in Intelligent Systems and Computing

Volume 963

AF181707

The series "Advances in Intelligent Systems and Computing" contains publications on theory, applications, and design methods of Intelligent Systems and Intelligent Computing. Virtually all disciplines such as engineering, natural sciences, computer and information science, ICT, economics, business, e-commerce, environment, healthcare, life science are covered. The list of topics spans all the areas of modern intelligent systems and computing such as: computational intelligence, soft computing including neural networks, fuzzy systems, evolutionary computing and the fusion of these paradigms, social intelligence, ambient intelligence, computational neuroscience, artificial life, virtual worlds and society, cognitive science and systems, Perception and Vision, DNA and immune based systems, self-organizing and adaptive systems, e-Learning and teaching, human-centered and human-centric computing, recommender systems, intelligent control, robotics and mechatronics including human-machine teaming, knowledge-based paradigms, learning paradigms, machine ethics, intelligent data analysis, knowledge management, intelligent agents, intelligent decision making and support, intelligent network security, trust management, interactive entertainment, Web intelligence and multimedia.

The publications within "Advances in Intelligent Systems and Computing" are primarily proceedings of important conferences, symposia and congresses. They cover significant recent developments in the field, both of a foundational and applicable character. An important characteristic feature of the series is the short publication time and world-wide distribution. This permits a rapid and broad dissemination of research results.

**** Indexing: The books of this series are submitted to ISI Proceedings, EI-Compendex, DBLP, SCOPUS, Google Scholar and Springerlink ****

More information about this series at http://www.springer.com/series/11156

Waldemar Karwowski ·
Tareq Ahram · Salman Nazir
Editors

Advances in Human Factors in Training, Education, and Learning Sciences

Proceedings of the AHFE 2019 International Conference on Human Factors in Training, Education, and Learning Sciences, July 24–28, 2019, Washington D.C., USA

Springer

Editors
Waldemar Karwowski
Department of Industrial Engineering
and Management Systems
University of Central Florida
Orlando, FL, USA

Tareq Ahram
Institute for Advanced Systems
Engineering
University of Central Florida
Orlando, FL, USA

Salman Nazir
Institute of Maritime Operations
University College of Southeast Norway
Borre, Norway

ISSN 2194-5357 ISSN 2194-5365 (electronic)
Advances in Intelligent Systems and Computing
ISBN 978-3-030-20134-0 ISBN 978-3-030-20135-7 (eBook)
https://doi.org/10.1007/978-3-030-20135-7

This Springer imprint is published by the registered company Springer Nature Switzerland AG
The registered company address is: Gewerbestrasse 11, 6330 Cham, Switzerland

Advances in Human Factors and Ergonomics 2019

AHFE 2019 Series Editors

Tareq Ahram, Florida, USA
Waldemar Karwowski, Florida, USA

10th International Conference on Applied Human Factors and Ergonomics and the Affiliated Conferences

Proceedings of the AHFE 2019 International Conference on Human Factors in Training, Education, and Learning Sciences, held on July 24–28, 2019, in Washington D.C., USA

Advances in Affective and Pleasurable Design	Shuichi Fukuda
Advances in Neuroergonomics and Cognitive Engineering	Hasan Ayaz
Advances in Design for Inclusion	Giuseppe Di Bucchianico
Advances in Ergonomics in Design	Francisco Rebelo and Marcelo M. Soares
Advances in Human Error, Reliability, Resilience, and Performance	Ronald L. Boring
Advances in Human Factors and Ergonomics in Healthcare and Medical Devices	Nancy J. Lightner and Jay Kalra
Advances in Human Factors and Simulation	Daniel N. Cassenti
Advances in Human Factors and Systems Interaction	Isabel L. Nunes
Advances in Human Factors in Cybersecurity	Tareq Ahram and Waldemar Karwowski
Advances in Human Factors, Business Management and Leadership	Jussi Ilari Kantola and Salman Nazir
Advances in Human Factors in Robots and Unmanned Systems	Jessie Chen
Advances in Human Factors in Training, Education, and Learning Sciences	Waldemar Karwowski, Tareq Ahram and Salman Nazir
Advances in Human Factors of Transportation	Neville Stanton

(continued)

(continued)

Advances in Artificial Intelligence, Software and Systems Engineering	Tareq Ahram
Advances in Human Factors in Architecture, Sustainable Urban Planning and Infrastructure	Jerzy Charytonowicz and Christianne Falcão
Advances in Physical Ergonomics and Human Factors	Ravindra S. Goonetilleke and Waldemar Karwowski
Advances in Interdisciplinary Practice in Industrial Design	Cliff Sungsoo Shin
Advances in Safety Management and Human Factors	Pedro M. Arezes
Advances in Social and Occupational Ergonomics	Richard H. M. Goossens and Atsuo Murata
Advances in Manufacturing, Production Management and Process Control	Waldemar Karwowski, Stefan Trzcielinski and Beata Mrugalska
Advances in Usability and User Experience	Tareq Ahram and Christianne Falcão
Advances in Human Factors in Wearable Technologies and Game Design	Tareq Ahram
Advances in Human Factors in Communication of Design	Amic G. Ho
Advances in Additive Manufacturing, Modeling Systems and 3D Prototyping	Massimo Di Nicolantonio, Emilio Rossi and Thomas Alexander

Preface

This book provides researchers and practitioners a forum to share research and best practices in the application of human factors to training, education, and learning sciences. Just as human factors discipline has been applied to hardware, software, and the built environment, there is now a growing interest in the optimal design of training, education, and learning experiences. Principles of behavioral and cognitive science are extremely relevant to the design of instructional content and the effective application of technology to deliver the appropriate learning experience. These principles and best practices are important in corporate, higher education, and military training environments.

The book also aims to share and transfer not just knowledge, learning experiences, and best training approaches that are of real value in practical terms; a value that can help leaders ensure their organizations stay ahead of the competition through continued innovation, strong competitive advantage, and inspired leadership.

This book is organized into nine sections that contain the following subject areas:

Section 1 Internationalization in Learning Science
Section 2 Behavioral and Cognitive Science
Section 3 Future of Training, Education, and Learning Sciences
Section 4 Education, Learning, and Flipped Classroom
Section 5 Human Factors in Training, Education, and Learning Sciences
Section 6 Education in Medicine and Rehabilitation
Section 7 Educational Strategies
Section 8 Instructional Design and Learning Strategies
Section 9 Usability of Learning Technology

Each section contains research papers that have been reviewed by members of the International Editorial Board. Each section contains research papers that have been reviewed by members of the International Editorial Board. Our sincere thanks and appreciation to the board members as listed below:

Thomas Alexander, Germany
Tibor Barath, Hungary
Joana Bartnicka, Poland
Juan Crisótbal Castro-Alonso, Chile
Jo Elliott, USA
Michael Freeman, USA
Anders Gronstedt, USA
Charles Madinger, USA
Steven Mallam, Norway
Bilal Mansoor, Qatar
Chris McClernon, USA
Kevin Moore, USA
Kara Orvis, USA
Bob Pokorny, USA
Roy D. Roper, USA
Demetrios Sampson, Greece
Jennifer Smith, Canada
Jon Lars Syversen, Norway

July 2019

Waldemar Karwowski
Tareq Ahram
Salman Nazir

Contents

Educational Strategies

Instructional Design and Learning Strategies

Usability of Learning Technology

Internationalization in Learning Science

A Query Generation Technique for Measuring Comprehension of Statistical Graphics

Mark A. Livingston(✉), Derek Brock, Jonathan W. Decker, Dennis J. Perzanowski, Christopher Van Dolson, Joseph Mathews, and Alexander S. Lulushi

Naval Research Laboratory, 4555 Overlook Ave SW, Washington, DC 20735, USA
{mark.livingston, derek.brock, jonathan.decker, dennis.perzanowski, joseph.mathews}@nrl.navy.mil, christopher.vandolso@navy.mil, alulushi657@gmail.com

Abstract. In our information-driven society, there is increasing use of statistical graphics to convey information in a variety of settings, including industry, mass media, government operations, and health care. Current methods for assessing a reader's ability to comprehend statistical graphics are custom-written, not widely accepted, usable only once, and/or reliant on subjective interpretations and inferences. We have developed a method for generating queries suitable for evaluating graph comprehension capability. Our method is based on the Sentence Verification Technique (SVT), an empirically validated framework for measuring an individual's comprehension of prose material. Compared to ad hoc methods for testing graph comprehension, our technique is less subjective, requires less manual effort and subject matter expertise, and addresses the essential features of a given graph: values and relationships depicted, frames of reference, and style attributes. The SVT, and therefore our method, combat superficial comprehension by testing what the reader has encoded, as opposed to testing the reader's ability at visual recall or ability to look up data without reaching real comprehension. We motivate and describe our query generation method and report on a pilot study using queries generated with it.

Keywords: Graph comprehension · Sentence Verification Technique (SVT) · Statistical graphics · Quantitative evaluation

1 Introduction

Statistical graphics have become ubiquitous in modern mass media, scientific and technical publications, and government reports. Thus, some consider the abilities to read, write, and perhaps design statistical graphics important for visual or even general literacy [1–3]. An essential component of literacy is an individual's ability to comprehend information; to know whether a person has achieved comprehension (or literacy), we must have a reliable and robust test of comprehension. According to Kintsch [4], "[w]e comprehend a text, understand something, by building a mental model." Comprehension research first focused on how this model was structured, progressed to

W. Karwowski et al. (Eds.): AHFE 2019, AISC 963, pp. 3–14, 2020.
https://doi.org/10.1007/978-3-030-20135-7_1

consider how it was constructed, and then focused on iteration and interaction between the construction process and the resulting model. Testing methods for reading comprehension are well-established (albeit with strenuous disagreements).

There are multiple tests of graph literacy or interpretation in the literature, but none seem to be widely-used (although some were introduced recently, cf. Sect. 2). Standard practice is subjective development of test items by experts in relevant fields, which is a time-consuming process that tends to produce a single test. The effort required to generate suitable test queries from visual communication was noted as a concern long ago [5]. We overcome this challenge with a more algorithmic (but not automated) approach.

Given the extensive use of graphs in modern communications and the interest in developing comprehension tests for graphs, an algorithmic method of constructing tests of graph comprehension would be of great value. Covering the range of forms for statistical graphics requires a large corpus of questions [6]. A single test enables graph authors to determine whether a particular graph or set of graphs is understandable by a target population of users (via testing with representative readers). But a battery of tests (requiring an even larger corpus of queries) could determine the parameters of a class of graphs that make an instance harder or easier to read. A series of tests could help an educator identify whether a particular individual has learned the skills necessary to read a particular type of graph. With a large base of results from such a test battery, a general level of skill required to successfully read a particular graph (akin to reading level or grade level of prose) could be assessed through the graph properties. A precise test battery could even help ascribe the resulting difficulty level to individual properties. For all these reasons, we desire a reliable and robust method of generating not just a single test of graph comprehension, but a large corpus of graph comprehension queries. Further, even test questions custom-written by experts in accordance with standard test procedures may not truly measure comprehension. Our approach is based on a reading comprehension assessment methodology designed to overcome this challenge as well.

Our primary goal is to develop an algorithmic method of generating queries to measure comprehension of statistical graphics. The technique for generating queries described in this paper is adapted from a validated test construction method for prose reading comprehension known as the Sentence Verification Technique (SVT) [7] and is built on its principles applied to graphs. For brevity, applicable features of the SVT are described below (Sect. 3) as they become relevant to our presentation.

2 Related Work

Most test development strategies for graph comprehension focus on the type of tasks graph readers are asked to do, rather than the effort required to develop queries or the definition of comprehension implicit in queries. Bertin [8] introduced a task taxonomy of elementary (e.g. data extraction), intermediate (e.g. understanding trends), and overall (e.g. comparing trends) query tasks. This is a common choice [6, 9–13] for distribution of graph tasks, although it does not and cannot lay claim on its own to testing comprehension. In cognitive science, comprehension requires the construction of a mental model [14]; comprehension can thus only be tested by querying this mental model, which in turn requires removal of source material during queries.

The Test of Graphing in Science (TOGS) [9] was designed for science students in grades seven through twelve. Its development and use demonstrate several challenges for test development. Test items were validated by a review panel and a validation study (strategies which have been used for other tests [10, 13] as well). These reviews often resulted in items being removed or re-written. Multiple tests (including [11]) reuse TOGS questions rather than develop new items, decreasing the independence of tests and offering some evidence of the difficulty of writing questions.

Curcio [10] found that scores on her custom-designed graph comprehension test significantly correlated with measures of reading achievement, mathematics achievement, and prior knowledge of the topic, mathematical content, and graphical forms (all collected at the same time). However, our examination of her test material leads us to believe that some questions may have been answered through general knowledge rather than comprehension of the graph. To us, this argues for building a graph comprehension test that controls for general knowledge, which the SVT does by verifying agreement of query probes with source material.

Boy et al. [15] employed a test development method based on evaluation of manually-constructed test items through item response theory [16]. They found that a first test of line graphs provided more information about below-average examinees. A second test found discrepancies in the ability of questions to discern differences in examinees. Half the questions on a bar graph test were either too easy or too hard. To us, this argues for building a graph comprehension test that controls for general knowledge. The SVT limits application of general knowledge by asking readers to verify agreement of query probes with source material rather than asking for the truth value of query probes or for repetition of statements of facts presented in source material.

The Visualization Literacy Assessment Test [13] was developed according to the established procedure of test creation in psychological and educational measurement. The authors developed several types of graphs and maps and a series of three to seven questions for each graph. Of 61 questions developed, only 54 were deemed by a panel of five experts to measure the ability to read and interpret visually represented data. One further item was dropped due to low discriminability found after administration of the test to 191 volunteers. While VLAT is likely to be a useful tool, we note that the authors reported taking a month to develop these 61 test items from twelve source graphs, which were only then given to the expert panel for review and subsequently tested with volunteers. Our examination of their test materials leads us to believe that some questions also may have been answered from general knowledge the SVT framework mitigates this challenge through a four-fold structure for query probes (Sect. 3.2).

We believe these contributions and results with them show several challenges for writing tests of graph comprehension. It requires many queries to adequately test many aspects of graph comprehension, emphasizing the need for a better way to generate test questions. Thus, the process becomes quite labor-intensive. Even experts, writing subjective questions, may not realize the difficulty of a query and it may have to be removed from the test. We thus devised a more rigid, algorithmic query generation methodology for graph comprehension, based on the SVT.

3 New Technique for Generating Graph Queries

The arguments in favor of the SVT for reading comprehension tests all apply well to visual representations of information. As noted, Royer and Cunningham [5] long ago foresaw the possibility of adapting the SVT to visual forms, but argued the difficulty of generating test material was considerable with tools then in existence. We noted this difficulty in developing comprehension questions related to a node-link diagram using subjective development techniques [17]. We saw a way to overcome this difficulty with a graph specification language, converting the challenge from one of image manipulation into a set of rules to alter a (textual) graph specification. We developed rules for governing changes to graph specifications; these changes generate paraphrase, meaning change, and distractor query probes that are central to the SVT.

3.1 Graph Specification

Our clients make information dashboards for their customers. They use, and thus we adopted, HighCharts <http://www.highcharts.com/> to build graphs. HighCharts is a JavaScript library intended to ease the addition of interactive graphs to web applications. Options for graph configuration are given in JavaScript Object Notation (JSON). This forms a hierarchical set of keys and values (Table 1), which lends itself to our need to manipulate graphical elements (Fig. 1) systematically.

Table 1. A JavaScript Object Notation (JSON) specification for a graph in HighCharts. See Fig. 1 for the visual form of this graph.

```
{ chart: { type:“bar”,
            width:800,
            height:600,
          },
  exporting: { scale:1, },
  credits: { enabled: false, },
  legend: { enabled: false, },
  colors: [ ‘rgb(153,255,153)’, ‘rgb(51,153,51)’, ‘rgb(0,102,0)’, ],
  series: [{ data:[67,58,54],
              name: “Landfill”,
              colorByPoint: true,
              maxPointWidth:75,
              pointPadding:0,
          }],
  title:  { style: {color: “#000000”, font-size: “x-large”, fontWeight: “bold” },
             text: ”Percentage of Garbage going to Landfills”,
          },
  xAxis: [{ categories: “1990”, “2000”, “2010”],
             labels: { style:
                         { color: “#000000”, font-size: “20px”, fontWeight: “bold” }, },
          }],
  yAxis: [{ linewidth:1, gridLineWidth:0, max:100, tickInterval:20,
             title: { style: { color: “#000000”,font-size: “20px”, fontWeight: “bold” },
             text: “Percentage”, },
             labels: { style:
                         { color: “#000000”, font-size: “16px”, fontWeight: “bold” }, },
          }],
  tooltip: { enabled: false, },
}
```

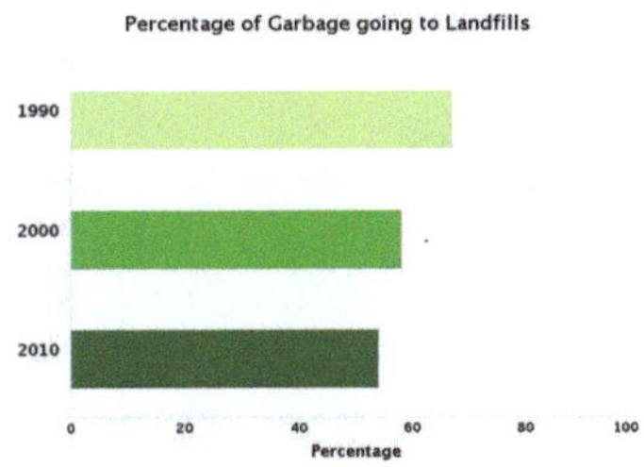

Fig. 1. The bar graph corresponding to the specification in Table 1.

3.2 Graph Query Definitions and Rules

Kosslyn [18] decomposed graphs into five components, of which three are important for our work (our *background* component thus far is a solid white field, and we do not use *captions*). The *framework* for most statistical graphs consists of the axes. The *content* is the representation of the data: points, lines, or bars. *Labels* name the variables, give titles to the graph or axes, or create a legend.

The SVT defines transformations of prose sentences into four types of query probes. Readers are asked to identify whether a probe gives information that was "stated" or "not stated" in the source prose. However, Kosslyn's graph components do not convey complete thoughts; rather, they function akin to words in a sentence. On the other hand, "sentences" in graphs are the meaningful informational statements or assertions that are coordinated, collectively, by the graph's components. A lone bar, divorced from a graph, is not an informational statement, but it becomes one when shown together with (at a minimum) a framework and labels. Two bars from the same graph convey an abstract relationship, but fail to make a meaningful informational statement – unless their display is coordinated by a framework and labels. By analogy, points and lines on line graphs require a framework and labels to join them in a construct equivalent to a sentence. When constructing a query, we need not include all the data in the source graph; this is analogous to the SVT using a single sentence at a time for a query. We may opt to use one data point or multiple data points, to reflect the various information statements that are shown in a graph.

With the above analysis of what constitutes simple sentence-level information in a graph, we need rules that define alterations to these information statements that come from graphs. This completes the analogy to the sentence transformations defined by Royer et al. [7]. However, there are numerous subtle features of graphs that may be altered without changing the meaning of the graph. Navigating these features is a key contribution to applying the SVT to graphs. We now use two source graphs (Fig. 2) for examples of applying (some) rules for transformations from source to query graphs.

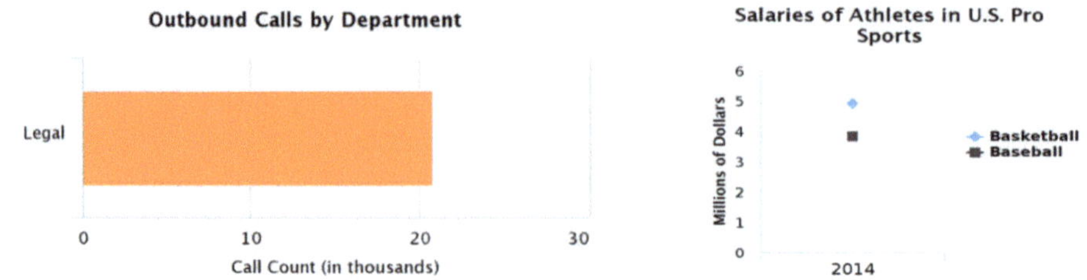

Fig. 2. An example bar graph and line graph used in the tutorial instructions for the study and used here to demonstrate the query variations as we adapted them from sentences to graphs.

3.3 Original Query Type

In the SVT, an *original* query type is defined as a verbatim copy of a sentence in the reading passage. Here we take some license with the definition of "verbatim." We assert that style features in a graph that do not alter the meaning of the underlying data are not fundamental to the graph. *Content* may have different colors, fill, shapes, et al. *Labels* may be drawn in different font family, size, or style and be centered differently. We note also that the *framework* could theoretically be changed without altering the meaning, but this would necessarily change the syntax of the *content*, and Royer et al. [7] recommended avoiding such "gray areas" in queries. Figure 3 shows examples of how some of these considerations are manifested for *original* queries.

Fig. 3. *Original* query probes for the source information graphs shown in Fig. 2.

3.4 Paraphrase Query Type

An SVT *paraphrase* query type calls for "as many words as possible to be changed without altering the meaning or the syntactical structure of" the source sentence. All style changes permitted in an *original* query are also permitted in a *paraphrase* query (Fig. 4); we argued above that these changes would not change the meaning, so they fit both definitions. Thus, style changes to *content* are the same as for original queries. We also deem rounding to be acceptable (so long as it moves the content by amounts that do not confuse the value); we argued similarly about smoothing data, but with few data points per graph, we did not adopt this. In retrospect, this is challenging and we recommend not adopting this change in combination with others. *Labels* may still have different style; however, a *paraphrase* should also change the wording of *labels* when

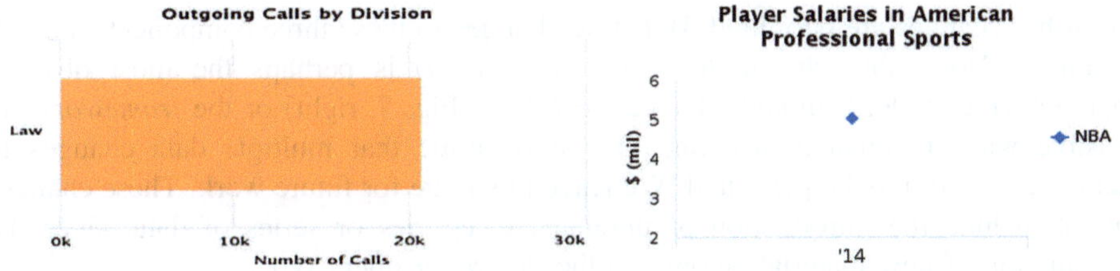

Fig. 4. *Paraphrase* query probes for the source information graphs shown in Fig. 2.

possible, using synonyms or different units for numbers (e.g. converting to scientific notation, or giving numbers in thousands). On this, we must accept subjective judgments about equivalence of the words substituted into *labels*. As with the application of the SVT to prose, a thesaurus may mitigate this challenge, although the jargon associated with the domain of a graph could create additional complexity (and perhaps limit the applicability of the resulting test to those who can be expected to know the domain). However, with the wide use of statistical graphics, we feel that domain-specific issues are easily avoided without limiting the range of style attributes explored in a test. In the *framework*, we allow changes to major and minor units (denoted by gridlines and/or tick marks). As for *original* query types, we choose not to transpose axes, change the range of an axis or convert to logarithmic. We assert that such changes alter the syntax of the graph. If we decide in the future to relax adherence to Royer's definition, then we may study whether such framework changes could be permitted.

3.5 Meaning Change Query Type

The SVT rule for a prose *meaning change* is to "alter one word in an original sentence such that the meaning of the sentence is changed." Since we adopt the paradigm that the "words" of a graph are the constituents in the *content*, *labels*, and *framework*, it follows that we should change one constituent in a way that alters the meaning, and that

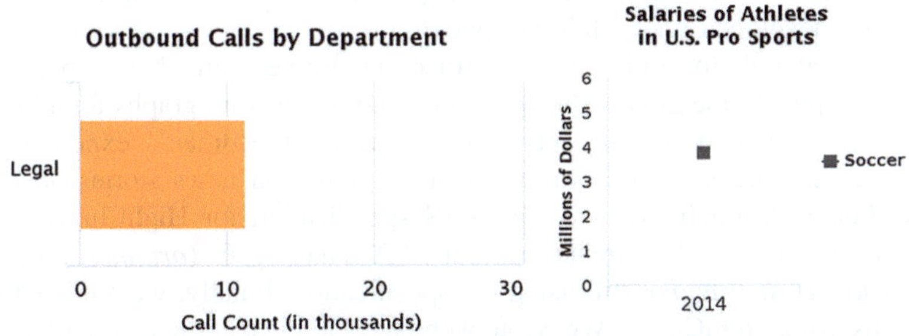

Fig. 5. *Meaning change* query probes for the source information graphs shown in Fig. 2. Note that the datum on the graph at right corresponds to the bottom point in the source graph, rather than the top point, which was used in the original and paraphrase query probes.

no further changes are permitted. But style changes to these three components are still permitted. Noticeable change to a datum (*content*) is perhaps the most obvious approach (Fig. 5, left), though changes to *labels* (Fig. 5, right) or the *framework* are possible ways to change meaning. One may argue that multiple data changes to maintain a trend may be permitted. We leave this issue for future work. These changes cannot include the introduction of unrelated categories or series of data, since the introduction of new material belongs to the *distractor* query type.

3.6 Distractor Query Type

The SVT definition of a prose *distractor* query is "a sentence that is consistent with the general theme of the source material but is unrelated to any original sentence; it should also have the same length, syntactical structure, and conceptual complexity as sentences in the source material." This tells us that we may make multiple changes of the type we may make for a *meaning change*, or introduce new material (Fig. 6). However, we must limit ourselves to changes that stay within the topic of the source graph.

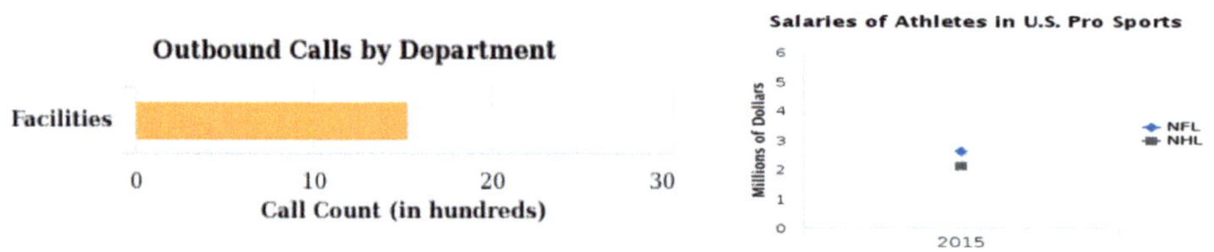

Fig. 6. *Distractor* query probes for the source information graphs shown in Fig. 2.

4 Pilot Test of Queries

To validate test items constructed using our method will require field testing them in a population with known graph comprehension abilities. Since one of the motivations for our work is the lack of a widely-validated test, we cannot yet undertake this test. With recently available tests, such as VLAT [13], perhaps future work can test the consistency of tests developed under different paradigms.

To build materials for a pilot test, we constructed nine source bar graphs and nine source line graphs. Some graphs showed data pared down from graphs found in media sources; two were reduced data sets from Shah and Freedman's experiment [19]. Others were constructed from a variety of ideas based on news stories or technical literature. For each graph, we wrote a JSON specification for HighCharts. We then applied the rules (Sect. 3) to create the four SVT query types (*original*, *paraphrase*, *meaning change*, *distractor*), still using the specification. Finally, we rendered images of all graphs using HighCharts. We wrote web pages to present the instructions, source graphs, and queries, as well as two diversionary tasks, described next. Of the nine graphs of each type (bar and line), one was embedded in the instructions, two were used for practice (described below), and six were used for testing.

To reduce reliance on visual memory, we added two diversionary tasks. We showed participants two images in sequence, each for three seconds. These were intended to interrupt visual pattern memory and were taken from a public database for eye tracking data [20]; they showed a variety of natural and urban imagery, with a few close-up images of common items. Participants also read brief, successive excerpts (about 200 words) from a novella.

For each trial, participants were asked to study a graph and a prose excerpt (as sources) and to answer corresponding queries; they were asked simply to look at the diversionary images for whatever they found interesting. The prose also gave us a baseline for comparison against the graph comprehension task. Thus, the complete sequence of a data trial was

- show a source graph (minimum time: 30 s, maximum time: 3 min),
- show a diversion image (3 s),
- show a blank screen (1 s),
- show a second diversion image (3 s),
- show a blank screen (1 s),
- show a source prose excerpt (also 30 s to 3 min),
- show a graph query and ask the participant whether the information in this graph query was "stated" or "not stated" in the previous source graph, and
- show a prose query and ask the participant whether the information in this prose query was "stated" or "not stated" in the previous source prose.

All material and layout from the study may be requested from the contact author. Participants completed a pre-study questionnaire with demographic and background information. They next read four pages with instructions for the task: (1) examples of the SVT on prose, (2) our adaptation with a bar graph example, (3) our adaptation with a line graph example, and (4) a brief summary of the procedure. They next completed four practice trials of the above sequence. During this practice, the above sequence was followed by two screens: one for giving the correct answer for the graph query (confirming that the participant was correct or informing the participant of the correct answer), and one for giving the correct answer for the prose query (again, with confirmation or correction). After the practice, a short break was permitted and the participant was asked if he or she had any questions about the procedure. (Participants did not generally ask questions; one asked to clarify what was to be done during the display of the diversion images and was told to simply look at them for whatever may be of interest.) Then the twelve trials were conducted, grouped by graph type (bar or line). Half the participants saw the six bar graph trials as their first group; the other half saw the line graphs first. Within each group, a Latin square ordered the graphs and another Latin square ordered the SVT query types. After the first group of queries, another break was permitted; no participants took a break for more than a few seconds.

Control software was implemented in web pages viewed with Google Chrome (version 49 for some data, version 54 for some data – with no effect expected of the version), under Windows 8.1. The volunteer sat at a standard desktop environment and viewed the stimuli on a 28-inch Dell U2412M running at 1920 $\times$ 1200@60 Hz.

Twenty-four participants (20 male, 4 female) completed the study; they ranged in age from 19 to 58 (mean and median age were both 38). All self-reported having normal or corrected-to-normal visual acuity and normal color vision. All but one of our participants also reported being heavy computer users; ten reported that they closely read bar graphs or line graphs for work or personal reasons on at least a weekly basis. Thirteen said that they create such graphs for work or personal projects. Our participants came from the research and clerical staff at our laboratory; fourteen held a graduate degree. For the procedure as described above, participants took an average of 54 min (minimum 31 min, maximum 98 min).

Overall, participants got 92.0% correct on graph queries; they got 82.6% correct on prose queries. We conducted a series of one-way analysis of variance (ANOVA) tests with Greenhouse-Geisser correction to look for statistically significant differences. We found a main effect of SVT query type on response time, for both the graph queries and the prose queries (Table 2(a)). For graph queries: $F(3,69) = 7.978$, $p < 0.001$, $\eta^2 = 0.100$ and for prose queries: $F(3,69) = 5.638$, $p = 0.010$, $\eta^2 = 0.081$. Royer et al. [7] previously noted that *paraphrase* and *meaning change* queries could be expected to be harder than *original* and *distractor* queries; this effect on response time gives some evidence of this being the case for our *paraphrase* queries (but not *meaning change* queries). Participants spent more time studying source graphs that had more data points on them, summed over all series (Table 2(b)), $F(3,69) = 10.604$, $p < 0.001$, $\eta^2 = 0.112$, so we feel confident that our participants focused on the task they were attempting to complete. However, the number of points on the source graph did not show a main effect on accuracy, $F(3,69) = 1.442$, $p = 0.238$, $\eta^2 = 0.048$.

While our graph sources had between three and six data values, our graph queries contained one, two, or three data values. (One query showed all three of the source data values.) We noticed a slight tendency for participants to be more accurate as queries showed more data values, $F(2,46) = 2.712$, $p = 0.093$, $\eta^2 = 0.069$ (Table 2(c)). This gives rise to a hypothesis for future studies that more context on the graph query (in the form of more of the source graph being shown) may help participants recall the information content of a graph. There was no significant main effect of sequence number on error (Pearson $r = -0.3789083$, but $t(10) = 1.2948$, $p = 0.2245$). So, we did not find that the length of the study session limited the performance of our participants. (Note that negative correlation would imply improvement on successive queries.)

Table 2. (a) SVT query type had a main effect on response time (shown in seconds) for both graph and prose queries. (b) The number of data points on a graph source had a main effect on the study time. We enforced a minimum study time of 30 s. (c) The number of data points on a query showed a tendency to yield more accuracy with more data points. RT = response time

(a)	Graph Queries		Prose Queries	
Query Type	RT (sec)	Std. Dev.	RT (sec)	Std. Dev.
Original	15.7	10.6	10.3	5.3
Paraphrase	18.4	11.3	13.8	9.5
Meaning Change	14.5	8.9	10.1	6.7
Distractor	11.7	7.2	10.4	7.5

(b) Number of Source Data	Study Time (sec)	Std. Dev	(c) Number of Query Data	Error (pct)	Std. Dev.
Three	36.3	0.7	One	0.125	0.334
Four	40.7	1.3	Two	0.089	0.286
Five	39.9	1.6	Three	0.028	0.107
Six	45.4	1.7			

5 Discussion and Conclusion

We believe that our adaptation of the SVT provides a foundation for developing reliable and robust graph comprehension tests. By combining the SVT structure with graph specification languages and a taxonomy of graph components, we can systematically vary graphs within the boundaries defined by the SVT. The SVT's foundation, grounded in cognitive theory, thus applies to our adaptation. The SVT query types were designed to defeat a solution of relying on rote memory. The taxonomy for graph components enables our adaptation to provide a mostly objective construction (Sect. 3) for a comprehension query. The specification language enables us to transform a text language rather than a graph image. We believe that the combination of the taxonomy and the SVT structure also will eventually enable us to compare the difficulty (level of comprehension in a given population) of varied attributes and styles of graphs.

As stated above, our primary goal in this work was to develop an algorithmic method for generating tests of graph comprehension. To that end, we adapted the methodology of the SVT, selected a graph specification that fit our purposes and our clients, and developed rules for generating queries of each type mandated by the SVT. Furthermore, we conducted a pilot study, with the goal of showing that the visual form of the SVT was functional (that participants understood the task and that queries were generally found to be reasonable). Subjectively, we found that readers generally believed that they understood the task in the resulting graph comprehension test, and they objectively demonstrated comprehension of the graphs. A far larger study will be needed to fully assess the validity of our approach, however, and this must be left for future work.

We also collected eye tracking data in the pilot study; we noted [21] that the pattern of fixations does not match the patterns that are typical for natural imagery. This leads to a hypothesis that people have distinctive patterns for reading statistical graphs; this has been noted in other work [22] and is an area for further study.

We seek ultimately to develop objective, extensible metrics by which we can measure how difficult graphs are to comprehend. As a first step, we have a reliable and algorithmic method through which we can generate tests of comprehension of statistical graphics. There are numerous obvious extensions to our first effort. We began with bar, column, and line graphs because they are frequently used by our clients, but we plan to include other types of statistical graphics (e.g. pie graphs and scatterplots). As we have previously demonstrated [17], the SVT may be adapted for more general visual representations of relational information. Eventually, we expect to include more complex graphs, interfaces composed of multiple graphs, and animated and interactive graphs in our research.

Acknowledgements. The authors wish to thank Mike Royer, Joseph Coyne, Priti Shah, Michael Svec, and the pilot study volunteers. This research was supported by the Naval Research Laboratory Base Program.

References

1. Roth, W.-M.: Reading graphs: contributions to an integrative concept of literacy. J. Curric. Stud. **34**(1), 1–24 (2002)
2. Galesic, M., Garcia-Retamero, R.: Graph literacy: a cross-cultural comparison. Med. Decis. Mak. **31**(3), 444–457 (2011)
3. Börner, K., Maltese, A., Balliet, R.N., Heimlich, J.: Investigating aspects of data visualization literacy using 20 information visualizations and 273 science museum visitors. Inf. Vis. **15**(3), 193–213 (2016)
4. Kintsch, W.: Comprehension: A Paradigm for Cognition. Cambridge University Press(1998)
5. Royer, J.M., Cunningham, D.J.: On the theory and measurement of reading comprehension. Technical report no. 91, University of Illinois at Urbana-Champaign (1978)
6. Wainer, H.: A test of graphicacy in children. Appl. Psychol. Meas. **4**(3), 331–340 (1980)
7. Royer, J.M., Hastings, C.N., Hook, C.: A sentence verification technique for measuring reading comprehension. J. Read. Behav. **11**(4), 355–363 (1979)
8. Bertin, J.: Sémiologie Graphique, 2nd edn., Gauthier-Villars (1973). English translation: Berg, W.J.: Semiology of Graphics. University of Wisconsin Press (1983)
9. McKenzie, D.L., Padilla, M.J.: The construction and validation of the Test of Graphing in Science (TOGS). J. Res. Sci. Teach. **23**(7), 571–579 (1986)
10. Curcio, F.R.: Comprehension of mathematical relationships expressed in graphs. J. Res. Math. Educ. **18**(5), 382–393 (1987)
11. Svec, M.: Improving graphing interpretation skills and understanding of motion using microcomputer based laboratories. Electron. J. Sci. Educ. **3**(4) (1999)
12. Lai, K., Cabrera, J., Vitale, J.M., Madhok, J., Thinker, R., Linn, M.C.: Measuring graph comprehension, critique, and construction in science. J. Sci. Educ. Technol. **25**(4), 665–681 (2016)
13. Lee, S., Kim, S.-H., Kwon, B.C.: VLAT: development of a visualization literacy assessment test. IEEE Trans. Vis. Comput. Graph. **23**(1), 551–560 (2017)
14. Van Dijk, T.A., Kintsch, W.: Strategies of Discourse Comprehension. Academic Press (1983)
15. Boy, J., Rensink, R.A., Bertini, E., Fekete, J.-D.: A principled way of assessing visualization literacy. IEEE Trans. Vis. Comput. Graph. **20**(12), 1963–1972 (2014)
16. Baker, F.B.: The Basics of Item Response Theory, 2nd ed. ERIC Clearinghouse on Assessment and Evaluation (2001)
17. Livingston, M.A., Brock, D., Maney, T., Perzanowski, D.: Extending the sentence verification technique to tables and node-link diagrams. In: Proceedings of Applied Human Factors and Ergonomics (2018)
18. Kosslyn, S.M.: Graph Design for the Eye and Mind. Oxford University Press (2006)
19. Shah, P., Freedman, E.G.: Bar and line graph comprehension: an interaction of top-down and bottom-up processes. Top. Cogn. Sci. **3**(3), 560–578 (2011)
20. Judd, T., Ehinger, K., Durand, F., Torralba, A.: Learning to predict where humans look. In: IEEE International Conference on Computer Vision, pp. 2106–2113 (2009)
21. Harrison, A., Livingston, M.A., Brock, D., Decker, J., Perzanowski, D., Van Dolson, C., Mathews, J. Lulushi, A., Raglin, A.: The analysis and prediction of eye gaze when viewing statistical graphs. In: Proceedings of Augmented Cognition. Neurocognition and Machine Learning. LNCS, vol. 10284, pp. 148–165. Springer (2017)
22. Matzen, L.E., Haass, M.J., Divis, K.M., Stites, M.C.: Patterns of attention: how data visualizations are read. In: Proceedings of Augmented Cognition. Neurocognition and Machine Learning. LNCS, vol. 10284, pp. 176–191. Springer (2017)

Comprehensive Internationalization at HAN University of Applied Sciences. Administrative Leadership, Structure, and Staffing

Florentin Popescu(✉)

HAN University of Applied Sciences, Ruitenberglaan 31, 6826 CC Arnhem, The Netherlands
florentin.popescu@han.nl

Abstract. A research on institutional policy regarding internationalization has been headed by the Netherlands Universities Foundation for International Cooperation (Nuffic) [1, 2], including many of the Dutch Higher Education Institutions (HEIs). From these higher education institutions, 27 of them representing 59% had a central-plan level, 8 of them (meaning 17%) were developing this type of plan, while for 7 of these institutions, representing 15%, such a central-level plan was missing. Just 4 institutions of the HEIs in this research were without a central-level policy on internationalization. Consequently, it can be stated that internationalization policies are adopted at a high level in the Netherlands. This paper assigns the new institutional progress as part of the framework for the strategic planning and internationalization policies as part of a comprehensive research project developed at HAN University of Applied Sciences. Mainly, this paper discusses with the "Administrative Leadership, Structure, and Staffing" as a dimension in the Comprehensive Internationalization CIGE model [3].

Keywords: Comprehensive Internationalization · Internationalization policies · Higher education · Internationally connected university

1 Introduction

A percentage of 89% of the universities around the globe claim to have applied internationalization as part of their global strategy or to benefit from an institutional policy, and the remaining 22% of the institutions claim that they are developing a strategy focused on internationalization, according to the 4th Global Survey of International Association of Universities [4].

Universities are often directed into unexplored waters since they must apply a more strategic approach for internationalization in the context of an increasingly fierce competition. Many universities are not operating in the ideal circumstances with implementing a strategic approach for internationalization. They do not have the option to make a free choice into this direction, but they are following it as an imperative direction externally required. The success of many internationalization strategies that are introduced based on a trial-and-error principle is very fluctuating. In order to help both governments and universities to discover and develop a sustainable and valuable

W. Karwowski et al. (Eds.): AHFE 2019, AISC 963, pp. 15–26, 2020.
https://doi.org/10.1007/978-3-030-20135-7_2

international dimension, many lessons may be learned from the experiences across both national and institutional contexts.

The rapid growth of the internationalization procedure of higher education institutions represents one of the core factors that may support the adaptability of university's policies regarding the feedback received from the external environment, according to many authors. The increase of the overall university competitiveness on the academic market is another consequence of this rapid growth of the internationalization procedure [5–7].

The Center for Internationalization and Global Engagement's (CIGE) model representing the foundation for extensive internationalization is used within this paper. This extensive internationalization is a process that tries to align university strategic approaches with different policies, initiatives and programs in order to become internationally oriented and globally connected [3]. This CIGE model includes 6 target domains, as follows: (1) Articulated Institutional Commitment, (2) Administrative Leadership, Structure, and Staffing, (3) Curriculum, Co-curriculum, and Learning Outcomes, (4) Faculty Policies and Practices, (5) Student Mobility, and (6) Collaboration and Partnerships. Moreover, this model will support internationalization at university level by identifying implementation strategies. Mainly, this paper discusses with the "Administrative Leadership, Structure, and Staffing" as a dimension in the Comprehensive Internationalization CIGE model [3] (Fig. 1).

Fig. 1. Center for Internationalization and Global Engagement's Model for Comprehensive Internationalization [3]

This article attempts to point out the manner in which the above mentioned aspects of internationalization are identified by the Faculty of Business, Management and Law (FEM) at HAN University of Applied Sciences, the Netherlands. Moreover, this research tries to understand how these dimensions of internationalization are helpful for a better management as policies and strategic planning. An assessment of certain university practices and policies linked to the model and the theme utilized is planned to be included by the researcher. Moreover, the researcher is planning to include also an assessment of institutional responses of Arnhem Business School (ABS) regarding a large variety of strategies, policies and issues associated with the internationalization.

The major contribution of this study is represented by the attempt to develop faculty management in the strategic planning of the internationalization procedure by establishing the best practices. At the same time, this internationalization process will enhance the global competitiveness of the university. The responses to the internationalization process of the studied university regarding the policy framework that the university sets is also discussed within this research. This paper analyzes the numerous levels of internationalization into a theoretical framework trying to relate them to the general framework of the institutional strategic plan developed by the studied university.

2 Problem Definition and Research Design

The recent development regarding the institutional internationalization strategic planning of the studied university is analyzed in this article. This paper investigates how university administrators perceive the different dimensions and characteristic of internationalization. Moreover, some of the primary problems related to the institutional responses regarding the effect of internationalization in the context of duties that vary from being local to global in essence, are highlighted within this paper. Mainly, this article analyzes the strategic planning of internationalization since the literature points out that it is a crucial aspect in identifying the arguments for institutional feedback offered by universities as complex organizations.

According to the "Administrative Leadership, Structure, and Staffing" element of the model elaborated by CIGE, the key stakeholders are implied in the strategic planning of internationalization. These stakeholders contribute with a guideline for implementation by expressing an institution's engagement regarding internationalization. This commitment is reinforced by formal evaluation mechanisms that determine the institution to accomplish them.

An indispensable context for the implementation of internationalization is represented by the involvement of top leaders, relevant administrative and reporting structures. The following steps will be utilized in order to structure this study [3]:

- *Step 1: Senior leadership.* The CAO and president are involved from the beginning in the process dedicated to internationalization.
- *Step 2: International office.* Campus-wide internationalization activities are coordinated by an office or more offices. President or CAO get reports regarding internationalization from the faculty or staff member who are responsible for internationalization activities.

In order to align with these suggestions, researchers have adopted a mixture of approaches such as observations, archives and interviews, putting more emphasis on the last two methods. The aim of the interviews was to understand the point of view obtained from the interviewee regarding the research theme and to obtain an explanation of this personal perspective. This goal of the research is aligning with the explorative aspect of this study.

During the primary and secondary data collection on site, supplementary material like different documentation and policy for the higher education institution was gathered. These sources of information were utilized in order to scale the status of the chosen university at local, national and international levels. Different governments' websites were used as sources to obtain positional and policy data. Moreover, diverse search engines and digital databases referenced in numerous research studies were utilized as online sources. The documentation from university was gathered in order to obtain an insight regarding the policies and strategies of the higher education institution. Among the documentation collected from the university it could be mentioned annual reports, management structure graphics, internationalization policy reports and strategic plans. These kinds of documents were used first in the research.

3 Results and Outcomes

It is a commonly accepted fact that a university will fulfill its internationalization objective if it determines targets, for medium and long terms [8], with regard to creative tools used for autonomy and academic freedom [9–12] but only if they favor one specific profile (training, services, research, business etc.), when applying for an accessible and participatory management [13] and administration oriented towards results [14].

Step 1: Senior leadership. The CAO and president are involved in the process from the beginning dedicated to internationalization.

Precise points of action related to the internationalization policy were set by senior management at the HAN University of Applied Sciences following the Strategical Internationalization Agenda 2016–2020 [15]:

For a long time internationalization is more than only a central point or a theme in economic education at HAN. Nowadays, internationalization is a norm. It is quite right, therefore, that HAN's institutional plan demonstrates strong ambition with regard to internationalization. The HAN Institutional Plan 2016–2020 includes the following strategic priorities:

"People with different backgrounds are getting in touch with our students. By internationalizing research and education, we achieve the target on diversity. Enhancing and simplifying internationalization at home and supporting international work placements and exchanges can represent examples of how internationalization may be strengthening" [15].

"Internationalization is categorically included in all our activities and goals. The curriculum includes intercultural competences and foreign language skills. Moreover, students have the opportunity to join international projects with a high social relevance and to take part in joint research and education programs developed by our institutions in order to promote diversity in the classroom" [16].

Han developed activities that are by their nature international. The internationalization of Han has been improved based on several factors, such as the increased mobility of information, people and resources beyond national borders. Due to the increased cooperation at global level of higher education institutions, the necessity for a global view on the management of higher education institutions is established. Consequently, all higher education institutions must integrate the international dimension in their strategy at the national level. Moreover, higher education institutions should include the international dimension into the core activities.

According to Morais and Ogden [17] one of the transformative effects of mobility is represented by the global citizenship, while Leask and Bridge [18] consider that the internationalization of the curriculum represents a strategy to train global citizens. This idea may be obvious based on the university's internationalization policy and university's mission [19]

The mission of Han's Faculty of Economics and Management (FEM) is to give its students a solid grounding for future professional practice. Because that will be important in a globalized world, we want the future professional to have intercultural skills, knowledge of the international aspects of professional practice and a sense of being a global citizen. Internationalization is no longer merely a theme but is the norm [15].

"Internationalization is an important means of improving the intensity, quality and influence of the connection we make between research, education and the professional field. This will aid us in our efforts to produce well-qualified professionals with a strong moral compass, who are socially engaged, not afraid to explore new horizons and keen to continue developing. In particular, HAN seeks to train its students in the international and intercultural competencies increasingly demanded by the professional field" [15].

Step 2: International office. Campus-wide internationalization activities are coordinated by an office or more offices. President or CAO get reports regarding internationalization from the faculty or staff member who are responsible for internationalization activities.

FEM has a specific office called International Relations that is designated to coordinate internationalization daily activities, which fall under the six core tasks and described as follows (Fig. 2):

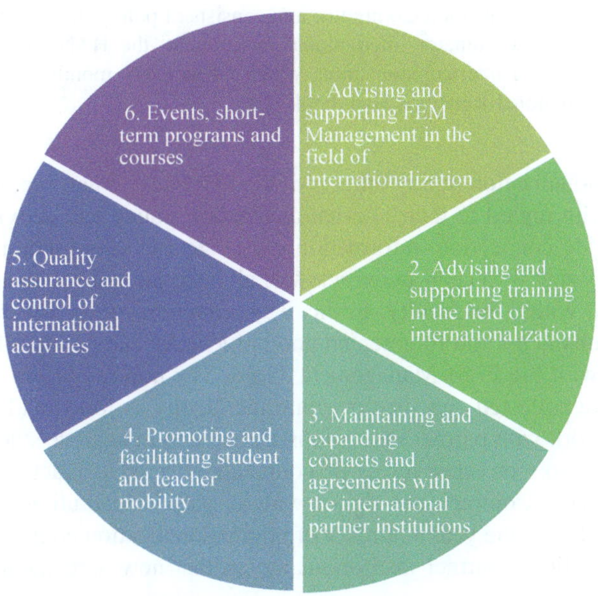

Fig. 2. FEM International Relations internationalization activities core tasks

The FEM International Relations forms an important link between the internal context of the FEM and the external context such as HAN, the Netherlands, the EU and beyond. It is by facilitating these connections that FEM can anticipate external developments and incorporate these in internal decision-making regarding internationalization. The FEM International Relations has an important role in formulating the policy objectives with respect to internationalization as specified hereunder (Table 1):

Table 1. Advising and supporting FEM Management in the field of internationalization

Category	Activity	Target audience
HAN policy	- Participation HAN Internationalization consultation (HINT)* - HAN participation in broad working groups	Policy officers internationalization
Advice FEM Management	- Feedback developments HINT and vice versa - Advice on internationalization of the curriculum - Advice on partner policy and potential new partners - Evaluation of policy goals - Annual report	FEM Management Team
Policy plan internationalization FEM	- Formulate objectives - Progress report	FEM Management Team and educational programs
Advice and consultation staff department FEM	- Advising and supporting Role - Switch between FEM/HAN	FEM support team

*HINT is HAN Internationalization Committee and consists of policy staff Internationalization of the four faculties, the coordinator international recruitment, the HAN subsidy desk and the senior policy coordinator internationalization, which meets every month.
Source: FEM International Relations Handbook [20]

The internationalization policy officers of the FEM advise and support the institutes and programs with regard to their internationalization policy and activities. In the first place they provide connection, information exchange and coordination. An important task of the policy staff is to keep an overview of all internationalization activities that are being developed within the faculty. In this way, communication towards international partner institutions is not too fragmented and bottlenecks can be identified in time and communicated with the FEM Management Team. The policy officers also ensure that best practices are collected and shared within the faculty as follows (Table 2):

In order to work with partner institutions abroad and to achieve student and teacher mobility, benchmarks or other forms of cooperation, good contacts and clear agreements are necessary. Contracts are signed with all partner institutions in which details are stated. Also during the year a number of internationalization conferences are visited where our international partner institutions are so that new agreements can be made.

Table 2. Advising and supporting training in the field of internationalization

Category	Activity	Target audience
Advising per institute/educational program	- Advising on vision and mission - Advise on medium-term planning and activities - Keep an overview of all internationalization activities that have been developed - Feedback to the FEM Management Team	Institutes/educational programs
Internationalization of the curriculum	- Supporting courses development - Advising courses development	Institutes/educational programs
Preferred partnerships	- Identifying potential preferred partners - Make, maintain and expand contact with preferred partners - Facilitating cooperation at educational level - Monitor the collaboration - Evaluate the collaboration, activities and outcomes	Institutes/educational programs
International networks of educational programs	- Identifying potential networks - Make contact, maintain and expand with the networks - Support in forming consortia - Monitor the collaboration - Evaluate the collaboration	Institutes/educational programs
Teacher mobility (incoming/outgoing)	- Information to training coordinators and teachers - Contact partner institutions and facilitate the right connections - Preparing and evaluating teachers	Institutes/educational programs Teachers
Management reports	- Collect data internationalization activities	FEM Management Team institutes and educational programs
Monitoring internationalization training/institute	- Monitoring progress implementation - Signaling of bottlenecks - Advising on improvement processes	FEM Management Team institutes and educational programs
Knowledge dissemination	- Collect, document and disseminate 'best practices' - Knowledge dissemination	FEM Management Team institutes and educational programs

Source: FEM International Relations Handbook [20]

Where necessary, partner institutions are visited, preferably with one or more training courses to discuss cooperation in more detail. The policy officers also ensure that best practices are collected and shared within the faculty as follows (Table 3):

Table 3. Maintaining and expanding contacts and agreements with the international partner institutions

Category	Activity	Target audience
Contract management	- Contracts - Keep track of statistics and balances	International partner institutions (e.g. legal affairs)
International conferences	- Visit EAIE, NAFSA, other conferences: discussing collaborations, contracts and networks with contacts	International partner institutions
Partner institutions visits (incoming and outgoing)	- Discussing collaboration and curriculum - Making work agreements	International partner institutions
Acquisition of new partner institutions	- Identifying potential partner institutions - Recruit new partner institutions	International partner institutions
Promotional gifts	- Select and manage business gifts	International partner institutions

Source: FEM International Relations Handbook [20]

The FEM International Relations promotes and facilitates all activities related to student mobility with the international partner institutions of the FEM to offer students of study programs the opportunity to stay abroad for a semester (mobility window).

The entire cycle of informing students, selecting and placing students and guiding them up to departure takes about one academic year. After returning, all administrative processes concerning credits are completed. Incoming student mobility is promoted and It also promotes and facilitates teacher mobility with the aim of professionalizing the teachers and internationalizing the curriculum. Because the FEM International Relations acts as a connecting factor between educational programs and partner institutions abroad, the right matches can be made and contact with the partners about students can be kept with sort lines of command as follows (Table 4):

The FEM International Relations contributes to the measurement and analysis of data concerning mobility, international partner institutions and procedures to organize the administrative process as effectively and efficiently as possible. This benefits the service to students and study programs (Table 5).

In addition to the aforementioned tasks, FEM International Relations also carries out a number of activities and projects to support and promote internationalization. These projects are not always directly traceable to the curriculum of the study program, but explicitly contribute to the further internationalization of the study programs and the faculty (Table 6):

Table 4. Promoting and facilitating student and teacher mobility

Category	Activity	Target audience
Study Abroad (SA)	- Educate students - Coordinate SA procedures - Design and guide the intake and selection process - Communication with partners - Supervising students - Arrangements of administrative processes	Students of the FEM courses and education al programs coordinators and staff
Exchange program	- Promoting activities with partners - Coordinate and supervise the registration process - Monitoring and assisting academic coordinator - Arrangements of administrative processes	Incoming exchange students of partners e.g. partners and academic coordinators
Outgoing teacher mobility	- Information to training coordinators and teachers - Contact partners and facilitate the right connections - Preparing and evaluating teachers - Promotional gifts	FEM teachers and academic coordinators
Incoming teacher mobility	- Contact partners and facilitate the right connections	FEM teachers and academic coordinators

Source: FEM International Relations Handbook [20]

Table 5. Quality assurance of international activities

Category	Activity	Target audience
International partner institutions	- Data collection and data analysis - Advice on quality and potency	Partners, students and educational programs
Statistics mobility	- Study abroad - Exchange Program	Students
Management reports	- Delivery management reporting	Management team
Quality assurance manual	- Setting up and updating the quality manual FEM International Relations	Internal use

Source: FEM International Relations Handbook [20]

Table 6. Events, short-term programs and courses

Category	Activity	Target audience
Summer course	- Promotion with partners - Organization summer course - Support teachers and students during summer course - Administrative arrangements summer course	Teachers ABS and students of partners
ABS talent event	- Promotion with partners - Organization ABS talent event - Support teachers and students during ABS talent event - Administrative arrangements of ABS talent event	Students and teachers ABS and teachers of partners
Buddy program	- Promotion and guidance buddies	ABS and exchange students

Source: FEM International Relations Handbook [20]

4 Conclusions and Further Research

A central role is played by higher education institutions in social development all over the world, no matter if this is the result of new methods and innovations in research or broad general education, and no matter if the social challenges are at local or global scale. In order to overcome crucial global challenges and reach the global sustainable development goals, international cooperation and high quality in education, innovation and research are indispensable. Therefore, HAN must contribute actively to the positive development where worldwide partnerships must bring reciprocal benefits for all collaborative partners and these partnerships should be relevant over the long term. Both academic freedom and social responsibility are defended by the higher education institutions. An essential context for the implementation of internationalization is represented by the involvement of top leaders, relevant administrative and reporting structures:

- As section of their framework of strategic internationalization efforts, higher education institutions must consider their Strategic Internationalization Agenda 2020/2030.
- Together with assistance for third-cycle education and research, higher education in partner countries must prioritize internationalization.
- As part of the framework of development assistance, relationship between higher education strategic partners should be strengthened and facilitated by appropriate administrative and reporting structures.

Occasionally, different kinds of collaboration question adequate practices and regulations by different innovative kinds of partnership. The need to enhance the quality of their research and education should be major conditions for which higher education institutions should act into the internationalization direction. Consequently, eliminating the barriers that limit the higher education to benefit from the optimal conditions to establish operations beyond national borders and act internationally may be seen as a question of priority.

In order to promote the success of other objectives of the strategy it is necessary comprehensive and strategic effort of the institutions at the executive management level. The success of other objectives of the strategy is ensured if the internationalization is part of the operations belonging to the higher education institution and the international perspective becomes part of the institution. The connection between research and education can be promoted by internationalization. Moreover, the relationships between education and research may be improved by internationalization to the local and wider communities, but also beyond its borders. Education and research internationalization may be seen as reciprocally consolidated. Consequently, the internationalization of research reinforces education, while the internationalization of education reinforces research.

As seen at the HAN University of Applied Sciences:

- the development of operations and policy are primarily based on internationalization;
- opportunities to set operations abroad have been improved by the FEM International Relations policies;
- based on the terms prevailing at institutional level, a strategic and documented approach for extensive internationalization of the activities has been analyzed and developed.

Since this paper is a part of a larger and comprehensive research project developed at the HAN University of Applied Sciences by the International Business Centre of Expertise aimed to highlight the new developments at institutional level in the context of international policies and strategic planning, final interpretations and conclusions will be presented after the ending of this research project.

Acknowledgments. I would like to thank my colleagues Ingrid Le Coultre-den Ouden, Humera Saeed, Judith van de Looij and Erna Helsen from the FEM International Relations for their efforts and dedication in their work and providing me with internal documents and tables regarding the International Relations policies and procedures on Internationalization.

References

1. van Gaalen, A., Roodenburg, S., Hobbes, H.J., Huberts, D., Gielesen, R.: Internationalising students in the home country – Part II In Practice. Nuffic, The Hague (2014)
2. van Gaalen, A., Hobbes, H.J., Roodenburg, S., Gielesen, R.: Internationalising students in the home country – Part I. Nuffic, The Hague (2014)
3. American Council on Education (ACE), Center for Internationalization and Global Engagement (CIGE) Model for comprehensive internationalization. www.acenet.edu/news-room/Pages/CIGE-Model-for-Comprehensive-Internationalization.aspx (2011, 2013)
4. Egron-Polak, E., Hudson, R.: Internationalization of higher education: growing expectations, essential values. IAU 4rd Global Survey Report ed., IAU, Paris (2014)
5. Bartell, M.: Internationalization of universities: a university culture-based framework. High. Educ. **45**(1), 43–70 (2003)

6. Marsella, A.J.: Essay: internationalizing the psychology curriculum. Psychol. Int. **12**(2), 7–8 (2001)
7. Sporn, B.: Current issues and future priorities for European higher education systems. In: Altbach, P.G., Peterson, P.M. (eds.) Higher Education in the 21st Century: Global Challenge and National Response, pp. 67–77. Institute of International Education, Report No. 29, New York (1999)
8. Paulsen, M.B., Toutkoushian, R.K.: Economic models and policy analysis in higher education: a diagrammatic exposition. High. Educ.: Handb. Theory Res. **23**, 1–48 (2008)
9. Abdullahi, S.A.: Rethinking global education in the twenty-first century. Glob. Comp. Educ. Policy Res. **12**(1), 23–34 (2010)
10. Altbach, P.G, Reisberg L., Rumbley L.E.: Trends in global higher education: tracking an academic revolution. A report prepared for the UNESCO 2009 World Conference on Higher Education. UNESCO, Paris (2009)
11. Yazmín, C.: Quality and the social responsibility of universities. https://upcommons.upc.edu/bitstream/handle/2099/8461/Quality_socialresponsibility_universities_angl.pdf?sequence=2 (2008)
12. Romo de la Rosa, A.: Institutional autonomy and academic freedom: a perspective from the American continent. High. Educ. Policy **20**, 275–288 (2007)
13. Jones, G.A., Oleksiyenko, A.: The internationalization of Canadian university research: a global higher education matrix analysis of multi-level governance. High. Educ. **60** (2010)
14. Soderqvist, M.: Internationalisation and Its Management at Higher Education Institutions: Applying Conceptual, Content and Discourse Analysis. In Acta Universitatis Oeconomicae Helsingiensis, Helsinki School of Economics (2007)
15. Strategical Internationalization Agenda 2016–2020, HAN University of Applied Sciences
16. Institutional Plan 2016–2020, HAN University of Applied Sciences
17. Morais, D., Ogden, A.: Initial development and validation of the global citizenship scale. J. Stud. Int. Educ. **15**, 445–466 (2011)
18. Leask, B., Bridge, C.: Comparing internationalisation of the curriculum across the disciplines: theoretical and practical perspectives. Comp.: J. Comp. Int. Educ. **3**(1), 79–101 (2013). Special Issue: Contemporary issues on the internationalisation of higher education: critical and comparative perspectives
19. Lilley, K., Barker, M., Harris, N.: Exploring the process of global citizen learning and the student mind-set. J. Stud. Int. Educ. **19**(3), 225–245 (2014)
20. FEM International Relations Handbook, HAN University of Applied Sciences (2017)

Opening Up Higher Education: An E-learning Program on Service-Learning for University Students

Covadonga Lorenzo(✉) and Epifanio Lorenzo

Department of Architecture and Design, Institute of Technology,
CEU University, Madrid, Spain
{clorenzo, fablabceu}@ceu.es

Abstract. Higher education in Europe has been slowly implementing service-learning courses at the university in the last years. It allows students to take part in a community project as part of their course, providing a progressive learning experience while meeting societal needs. Madrid-based CEU University has been implementing service-learning elective courses on-site for the last five years through educational programs carried out in Fab Lab Madrid CEU, the digital fabrication laboratory based at the university. Thanks to it, students use digital fabrication technologies to design and make low-cost devices in cooperation with communities in need. As in the last two years, students have been increasingly taking their elective courses on-line; the number of students demanding e-service-learning courses has increased. To meet this demand, we have started to design an e-service-learning program, providing students remote access to Fab Lab technologies through the NEWTON Project platform and studying the outcomes of such effort.

Keywords: E-learning · Service-Learning · Higher education · Digital fabrication

1 Introduction

With the rapid growth and development of the information and communication technologies, e-learning has seen a fast evolution over the past decade. Being a viable alternative to traditional in-class learning, as well as a cost-effective training solution even in difficult economic circumstances, e-learning is increasingly adopted by educational, corporate and governmental worlds as an important component of information dissemination. The global e-learning market was accounted for $165.21 billion in 2015 and is expected to reach $275.10 billion by 2022, growing at a compound annual growth rate estimated at around of 7.5%. Several world regions appear to have significantly higher growth rates such as Asia at 17.3%, followed by Eastern Europe, Africa, and Latin America at 16.9%, 15.2%, and 14.6%, respectively.

The exponential growth of the e-learning market is making that both, private and public educational sectors, acknowledge the possibilities of novel technologies enhanced learning. These advances offer the chance to extend the learning experience to educational settings outside traditional classrooms and laboratories. Furthermore, these technologies

W. Karwowski et al. (Eds.): AHFE 2019, AISC 963, pp. 27–38, 2020.
https://doi.org/10.1007/978-3-030-20135-7_3

can encourage research skills and interactive learning while considering students' individual learning styles. Most of higher education authorities believe that personalized, adaptive learning could make a positive impact in education. Preliminary research results have shown a link between a reduction in drop-out rates and utilizing e-learning software in education [1].

These technologies are not aimed at replacing teachers in the classrooms, but to empower them to teach at a deeper level, instead of merely trying to get through the curriculum. Several studies look at the benefits of integrating technologies in the learning process, e.g. improvement in attitudes of both, teachers and students and increase in skills [2]. Furthermore, technologies like adaptive intelligent systems can help bridge the gap between low and high achievers [3].

E-service-learning (electronic service-learning) combines on-line learning and service-learning, an educational approach that integrates learning objectives with community service in order to provide a pragmatic, progressive learning experience while meeting societal needs. The combination of both, e-learning and service-learning, enables the delivery of the instruction and the service to occur partially or fully online. E-service-learning allows students anywhere, regardless of geography, physical constraints or other access limitations, to experience service-learning. It reciprocally also equips online learning with a powerful tool for engaging students [4].

In spite of the rapid growth in e-learning in all its forms and the reality that an increasing number of students are pursuing their education online, service-learning has kept pace without exposure to the benefits of the innovative technology enhanced learning methods and tools. Goertzen and Greenleaf [5] highlighted diverse examples of e-service-learning courses representing various levels and academic disciplines. It is also worth to mention Lih-Juan, Hong-Yen and Tze-Han [6] initiatives that integrates service learning into the curriculum in learning projects where college students e-tutoring children in remote areas. However, despite all, nowadays, yet few university students are exposed to service-learning in their online coursework. According to Waldner, McGorry and Widener [7], service-learning must go online to remain relevant. E-service-learning holds massive potential to transform both service-learning and online learning by freeing service-learning from geographical constraints and by equipping online learning with a powerful tool to promote engage.

2 Objectives

The Department of Architecture and Design at Madrid-based CEU University has been implementing service-learning elective courses on-site for the last five years through educational programs carried out in Fab Lab Madrid CEU, the digital fabrication laboratory based at the university [8]. Thanks to it, students use digital fabrication technologies to design and make low-cost devices in cooperation with communities in need [9]. As in the last two years, students have been increasingly taking their elective courses on-line; people demanding e-service-learning courses has also increased. To meet this demand, we have started to design an e-service-learning program.

The program provides remote access to Fab Lab technologies through an innovative platform that allows university students to acquire skills on digital fabrication.

It has been created as part of a large European Union Project called NEWTON. It allows students to design and fabricate low-cost devices to be provided to communities in need, thanks to courses that run between on-site activities and remote access to digital fabrication facilities.

The main objective of the training program has been to raise awareness of the pedagogical potential of the combination of e-learning and service-learning at the university level, freeing service-learning from geographical constraints and providing college students remote access to digital fabrication tools to work with communities in need. To achieve this main objective, the following specific objectives were established. First of all, providing on-line university training on digital fabrication through quality programs thanks to the NEWTON Project platform. Secondly, design and fabricate products and services to solve a need of a community based on technology. Thirdly, reinforcing the social awareness of university students promoting initiatives that involve the cooperation among the university with partners such as communities, NGOs and foundations, and finally, contributing to the development of an integral educational model, which will allow students and faculty to experience the benefits of e-service-learning.

3 Methodologies

In order to achieve the previously stated objectives, a variety of activities has been designed, as e-learning classes on theoretical and practical foundations of service-learning, as well as collaboration with NGOs working in direct cooperation with communities that need support through both, on-site and on-line fieldwork. It also involves an on-line training course on the use of digital fabrication technologies through the NEWTON platform, which includes open access to our pedagogical materials and remote access to Fab Lab machines and designs. This innovative platform is allowing us to move from on-site service-learning courses to on-line classes.

The present paper describes a case study that focuses on the practical outcomes of the use of the NEWTON Project platform to allow university students to acquire skills on digital fabrication technologies and remote access to Fab Lab machines, in order to design and fabricate low-cost devices in cooperation with communities in need. Specifically, we focus on an initiative that involved fifteen students on the design and fabrication of a low-cost incubator for the Makeni Public Hospital (Sierra Leone).

The purpose of the workshop is to compare two methods of fabricating the above-mentioned device: on the one hand, the usual procedures we have been following in previous workshops, in which we use digital fabrication machines on-site and on the other hand, testing the NEWTON Project platform to remotely monitoring Fab Lab machines. The scope of this study is to evaluate the usability of the NEWTON platform, in order to assess whether it is worthwhile to implement its use as an educational tool in the e-learning program that we are trying to implement.

The workshop is organized in two activities: the first one is related to design procedures and focuses on the implementation of an open-source low-cost incubator design in a computer based software; and the second target the digital fabrication of the previous designs, both, on site and remotely through the NEWTON platform.

Design Process. The first activity focused on design procedures. Based on open-source previous designs of a low-cost newborn incubator, students were requested to implement the design in computer software according to the requirements of the Public Hospital of Makeni, in Sierra Leona. The previous design of the incubator was designed by Alejandro Escario, a former student of Fab Lab Madrid CEU, who is still collaborating on service-learning programs carried out at the laboratory. His goal was designing and making an affordable incubator for hospitals in developing countries easily replicable and repairable. In order to do so, designs, sources and documents were uploaded on a public repository under an open source and open hardware license [10] to allow free access to them for those who are interested in collaborating on the project, improving the design or fabricating incubators for communities in need.

In terms of design, the structure of the incubator is robust, but it is also easy to build and repair without having much knowledge about design or electronics. Inexpensive materials such as wood, plastic as well as recycled components were used to reduce the cost. The incubator's bed could be slightly tilted manually and the isolation of the incubator is assured through an easy removable plastic surface, so that the interior could be quickly and easily accessible to medical staff. It was made using 1.75 cm plywood. A press fit method was used to fasten parts, as it is achieved by friction after the parts are pushed together, rather than by any other means of fastening, assuring a quick and easy assembly (Fig. 1).

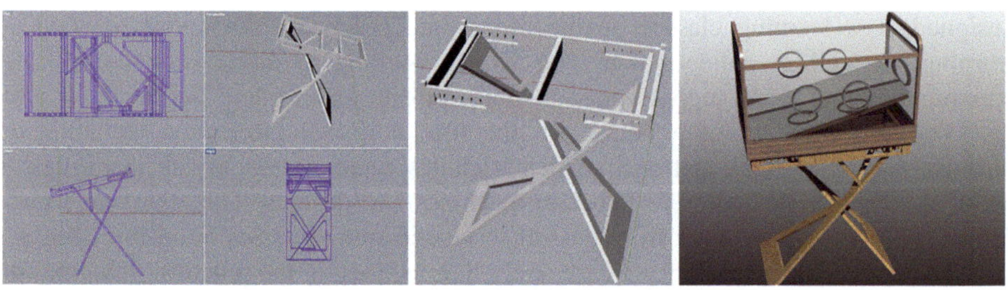

Fig. 1. Original design of the low-cost incubator designed by Alejandro Escario in Rhinoceros.

The methodologies used during the design process combine Self Directed and Independent Learning with Project Based Learning (PBL), two innovative approaches that are possible in the e-learning environment that lead to new models of education as well as new teaching and learning methods.

3.1 Self-Directed and Independent Learning

Flipped Classroom involves creating online multimedia content that is used out-of-class in conjunction with in-class learning activities such as individual and collaborative problem solving, group-work and class discussion. From the published research [11], there is evidence that the Flipped Classroom pedagogy inadvertently follows a constructivist approach towards learning where a student constructs their own learning and promotes self-directed and independent learning outside-of-class. It flips Bloom's taxonomy [12] and entails other thinking activities (creating, evaluating, analyzing) in the classroom. The use of this pedagogy during the first phases of the design process allowed university students to get access to pedagogical content related to the first activity. Students were requested to get access to design files and were trained on the use of a 3D design software called Rhinoceros that allowed them to visualize, modify and design new pieces that could improve the design of the incubator. After that, a Design Thinking approach is followed by the group of students to find out the best way to integrate the requirements of hospital's medical staff.

3.2 Problem-Based Learning

Problem-based learning (PBL) is another modern student-centered teaching method that uses real-world problems as the motivation of a self-directed learning process. Particular for PBL is that the problem comes before the knowledge needed for the problem. Typically, the learner applies new knowledge provided through teaching to solve a problem. This pedagogy requires learners to use the problem to direct them to research and learn the new knowledge needed. A number of educators have explored its potential by designing and delivering modules that make use of this teaching technique [13]. It supports the achievement of higher cognitive levels and develops life-long problem solving skills. Confronting students with the task of solving a design problem for a real client, make them to feel motivated. Receiving feedback from the users makes the final product the result of a collaborative effort in which users and designers propose improvements, taking product design to a higher level.

Digital Fabrication. After finishing the design improvements of the low-cost newborn incubator in Rhinoceros, the second activity targeted the digital fabrication of the previous designs first, using digital fabrication machines on-site and after that, remotely through the NEWTON platform.

3.3 Digital Fabrication Laboratories (Fab Labs)

Fab Labs are small-scale workshops with a set of flexible computer-controlled tools and machines such as 3D printers, 3D scanners, a mid-size laser cutter, CNC machines, a printed circuit board miller and some basic tools. Fab Lab environment is an inspiring and creative surrounding that provides students digital fabrication resources and allows educators to develop hands-on and collaborative teaching practices [14]. For their own intrinsic nature, they can be seen as laboratories to acknowledge and embrace different learning styles and where "students can concretize their ideas and projects with intense personal engagement" [15]. Fab Labs stimulate students to come up with new ideas and especially to fabricate prototypes. It is a perfect illustration of 'learning by doing', as all tools to bring a product to a realization are in reach.

The Fab Lab concept is gaining a worldwide interest and governments are starting to recognize the importance of digital fabrication technologies. Presently, there exists a worldwide network of more than 1.000 Fab Labs located in more than forty countries coordinated by the Fab Foundation. From community-based labs to advanced research centers in universities, Fab Labs are democratizing access to the tools for technical invention. This community is simultaneously a manufacturing network, a distributed technical education campus, and a distributed research laboratory working to digitize fabrication and inventing the next generation of personal fabrication.

As the scope of the case study is to compare digital fabrication using both, Fab Lab machines on-site and remotely through the NEWTON platform, students were requested to fabricate some pieces of the incubator's prototype using the first approach and the rest of the parts using the second. In the first one, students needed to create a dxf (drawing exchange file) starting from Rhinoceros, including all the pieces to be milled on a computer numerical control machine (CNC). After that, students used VCarve, a software solution for cutting parts on a CNC router. The software can import 2D designs providing a full set of drawing and editing tools.

VCarve toolpath options cover all routing operations such as profiling, pocketing, auto-inlays and drilling. Besides, each toolpath includes options to customize the settings and provide a high level of control for different types of operation. Thanks to it, students can easily introduce all needed parameters, such as the size of the material, the cutting depth, the type of profile's toolpath, the number of tabs, the size of the end mill and the type of cut according to the layout.

The work can be previewed to show how the parts will look when they are cut. After that, toolpaths are saved in the proper file format to drive the machine. Finally, students are also trained on the use of the machine on-site, which includes the placement of the material on the table, the way to place the mill on the router and the use of the software that monitor the CNC machine (Fig. 2).

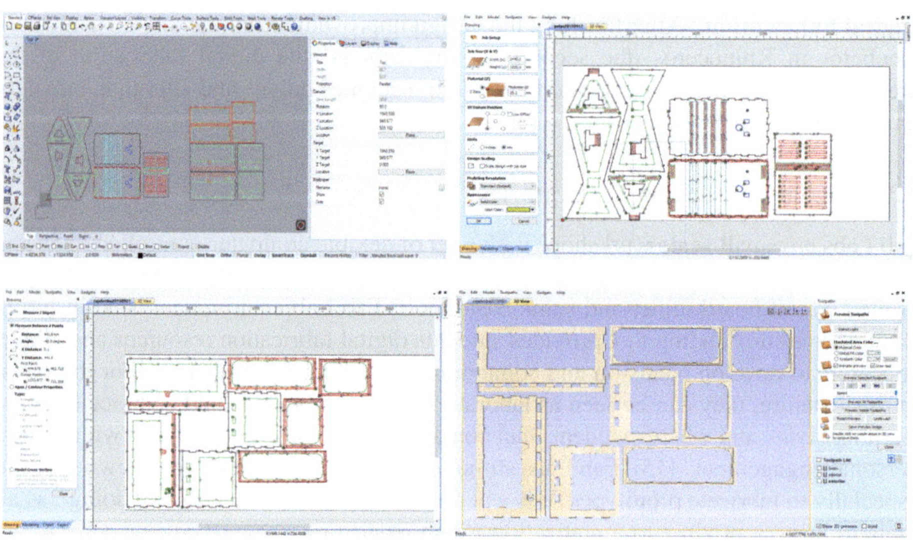

Fig. 2. Screenshots of the incubator designed in Rhinoceros (top left), a dxf file imported in VCarve (top right), toolpaths in VCarve (bottom left) and a preview in VCarve of the work to be milled in plywood (bottom right).

3.4 Remote Access to Fab Lab Facilities Through the NEWTON Platform

The integration of the Fab Lab workflow as part of the e-service learning program is possible thanks to the Cloud Hub application designed for the NEWTON Project [16]. NEWTON is a large European Union project focused on the integration of new solutions for technology-enhanced learning that enables learning content use and supports the generation of new material and content exchange among students to increase learner quality of experience for all. This tool creates new opportunities by providing an integrated approach where students can use remotely Fab Lab machines.

In order to do that, students started creating a portable network graphic (png) file starting from the Rhinoceros file that includes all the incubator's pieces to be milled. After that, students needed to get access to the platform, through which students got a quick and easy access to Fab Lab machines without having any knowledge on how to use the software associated to each machine. Using the application, students selected the Fab Lab machine to be used (a CNC machine), the material to be milled (plywood) and the size of the end mill.

After students sent the png file, it is received on a Raspberry Pi connected to the CNC machine at the Fab Lab. Then, a lab technician is in charge of preparing the material on the Shopbot and finish the process. Parameters needed to fabricate the pieces are assigned automatically according to the information selected by students before sending the file. After that, the file can be milled on the CNC (Fig. 3).

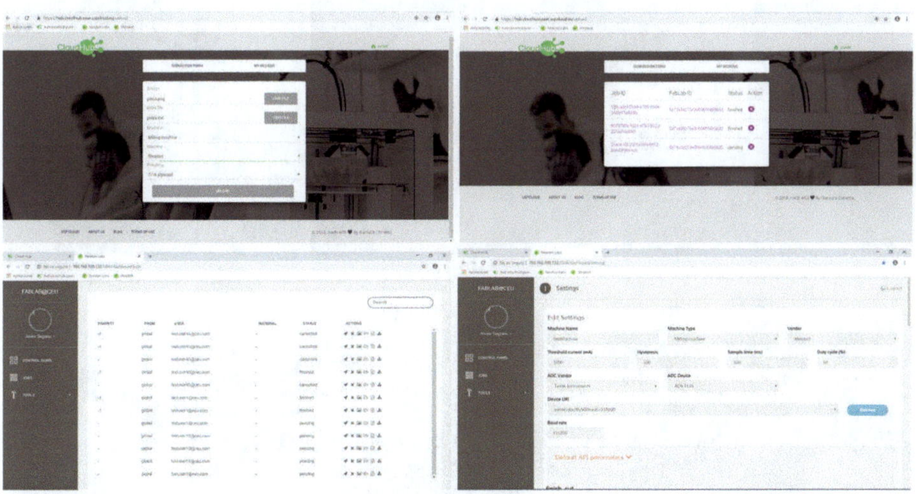

Fig. 3. Cloud Hub app student's interface (top) and Cloud Hub technician's interface (bottom).

Thanks to the NEWTON platform, students with no previous training in Digital Fabrication interested in participating in our e-service-learning program will be able to work on the implementation of the design of these medical devices to be manufactured in our Fab Lab. The Cloud hub app is responsible for transforming a simple png file exported from a 3D design program into a file suitable to be milled on a CNC machine. Students only select the machine to be used, the material and the type of end mill. The ease of use of the application will allow opening Fab Lab technologies to people with special needs or with some type of disability.

After the pieces were milled, the incubator was ready to be sent to the Public Hospital of Makeni in Sierra Leone. Both, the plywood's pieces and the electronic components were placed into a flat box that included some assembly instructions to be easily assembled by medical staff at the destination hospital.

During the last five years, our Fab Lab have been making and donating low-cost incubators through our service-learning programs to various maternities and hospitals on the African continent, such as Nikki's Maternity in Benin, thanks to Alaine Foundation; Magbenteh Hospital in Sierra Leone, thanks to the NGO Swis Sierra Leone Development Foundation; Pouponniere Orphanage in Senegal; St. Joseph Hospital in Ghana and Children's Center of Kakikoka in Benin, thanks to the NGO Solidarity Foundation with Benin.

In this case, the incubator was sent to the Public Hospital in Makeni, Sierra Leone, thanks to the cooperation with the University of Makeni. Clara Abella Dutrús, former student at CEU University, was responsible for the assembly and even organized a training course at Makeni's University, in Sierra Leone, to involve students of Public Health in the project. They are now in charge of collecting data related to the usability of the incubator that will allow testing the medical device (Fig. 4).

Fig. 4. Medical staff assembling a low-cost incubator in the destination hospital (top) training course for students of Public Health at the University of Makeni, Sierra Leone (bottom).

4 Results

The core of the evaluation procedure employed in the case study was developed by the NEWTON project's Pedagogical Assessment Committee (PAC) [17] that provided templates and guidelines for various assessments including various questionnaires. Firstly, a Learner Demographic questionnaire was used to collect information about learners, such as gender, age or background. Secondly, a Learner Motivation questionnaire (pre and post activities) was used to assess learners perception toward service-learning and e-service-learning activities. Finally, a Learner Usability questionnaire was handed-out to assess learners feeling about the usability of Fab Lab machines on-site and remotely through the NEWTON Project platform. Besides, instructors carried out learner observational assessment during activities.

Results obtained after collecting data are discussed below. The Learner Demographic questionnaire revealed that 55% of university students involved on the case study were males and 45% were females. Participants' ages ranged from 19 to 23 years. On the other hand, 30% of students were studying a degree associated to cooperation while 70% of students combined service-learning activities with university studies related to diverse fields. The Learner Motivation questionnaire was used to assess learners perception toward service-learning and e-service-learning activities. When asked about their interest on service learning programs that run in our Fab Lab 80% of students answered that they were "Extremely interesting" while 20% of participants said "Very interesting". When asked about their possible interest into participating in an e-service-learning program, 70% of students answered they were "Extremely interested" while the rest of students said they were "Somewhat interested". The evaluation shows a greater interest in e-service-learning in those students who cannot devote all their time to service learning activities, as they have to combine volunteering with university studies.

Regarding the Learner Usability questionnaire students had to answer three types of questions referred to usefulness, ease of use and ease of learning. In relation to usability and usefulness, students were asked if Fab Lab technologies (on-site and remote) help them to be more effective. As it is shown in the tables below, similar answers were given from students when asked about the effectiveness and usefulness of both, on-site Fab Lab tools and remote access to them through the NEWTON platform. However, the percentages slightly vary when participants are asked if this technologies save them time. 30% of students "Agree" and 30% of students "Strongly agree" when asked about on-site Fab Lab technologies, while 40% "Agree" and 60% "Strongly agree" when asked about their use through the NEWTON platform (Table 1).

Table 1. Effectiveness (left) and usefulness (right) of Fab Lab technologies used on-site and remotely through the NEWTON Project platform.

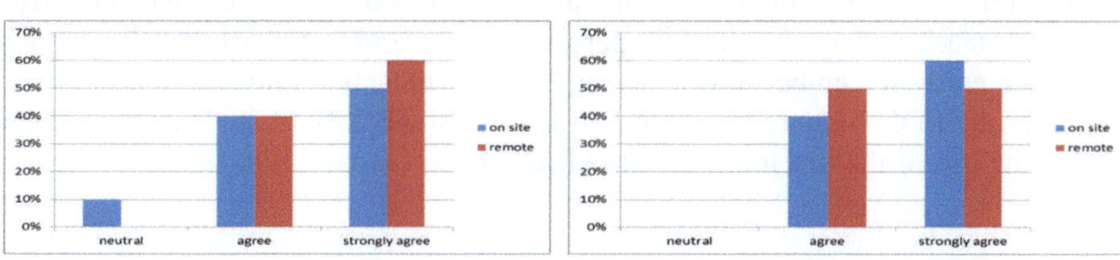

When asked about the ease of use, it is worth to mention that 70% of students "Strongly agree" on the ease of use of Fab Lab machines remotely versus 20% of students that answered the same to this question related to the use of machines on-site. Percentages vary considerably when students were asked about machine's use without instructions on-site and remotely. 60% "Strongly agree" to that questions when asked about the NEWTON platform (Table 2).

Table 2. Ease of use (left) and use without instructions (right) of Fab Lab technologies used on-site and remotely through the NEWTON Project platform.

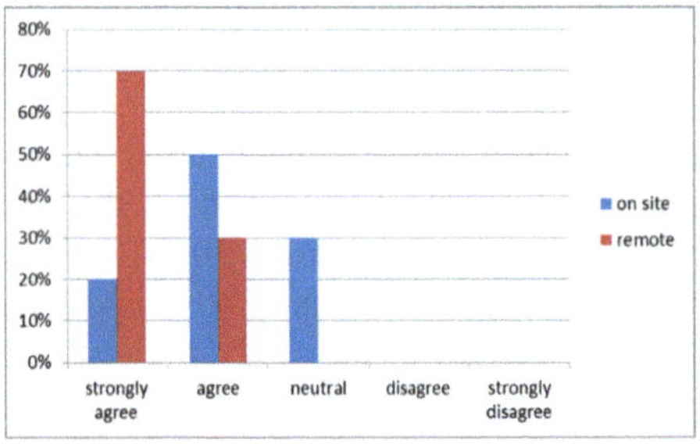

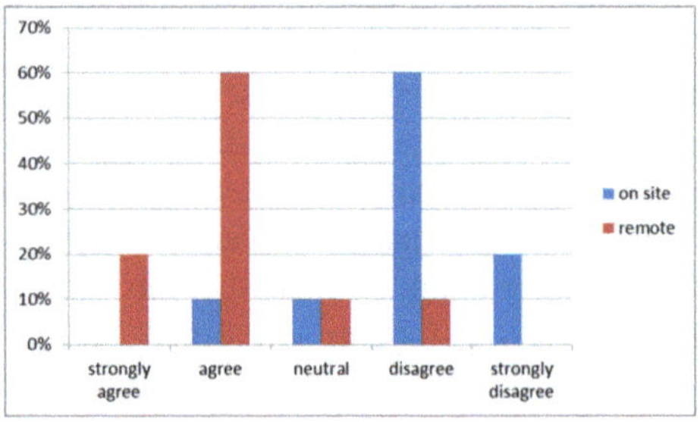

On the other hand percentages of students that said that the technologies requires the less fewest steps possible reach to a 60% when asked about the NEWTON platform. Finally, when asked about the safety in use, percentages reach a 80% of students when refer to the remote use of Fab Lab machines (Table 3).

Table 3. Feeling about steps requirements (left) and ease of remembering steps (right) of Fab Lab technologies used on-site and remotely through the NEWTON Project platform.

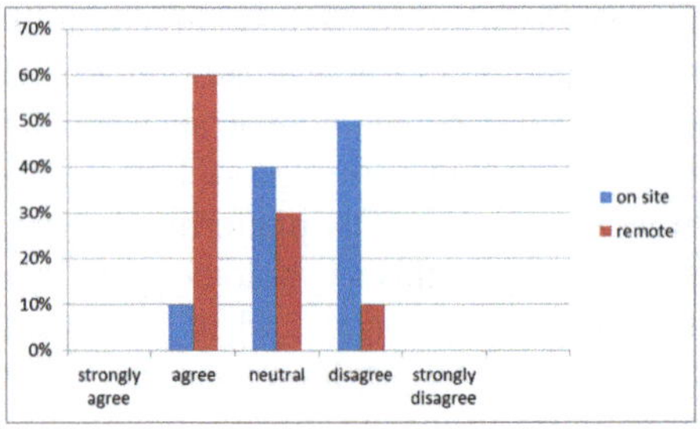

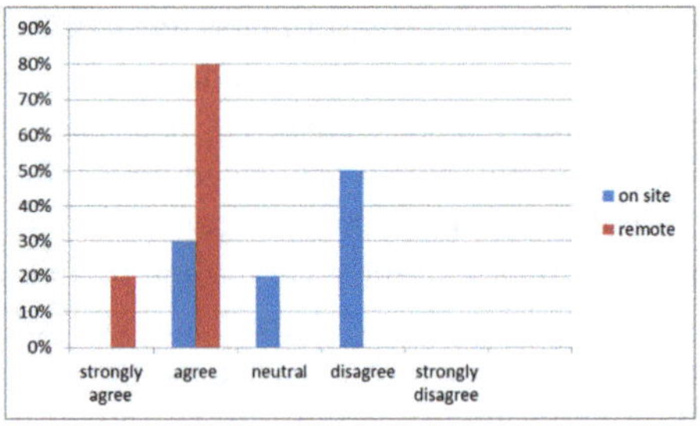

Regarding questions about ease of learning, 70% of students "Agree" on the ease of learning of the NEWTON platform, while just 30% of students "Agree" when referring to Fab Lab machines on-site. Similar percentages were obtained when asking if students could easily remember how to use it. 80% of students "Agree" to this statement when referring to the remote access to Fab Labs and just 20% of students agree when alluding the use of the machines on-site (Table 4).

Table 4. Ease of learning. Answers to the question about learning quickly Fab Lab technologies on-site (top left) or using the NEWTON platform (top right). Answers to the question 'I can easily remember how to use it' Fab Lab technologies on-site (bottom left) or using the NEWTON platform (bottom right)

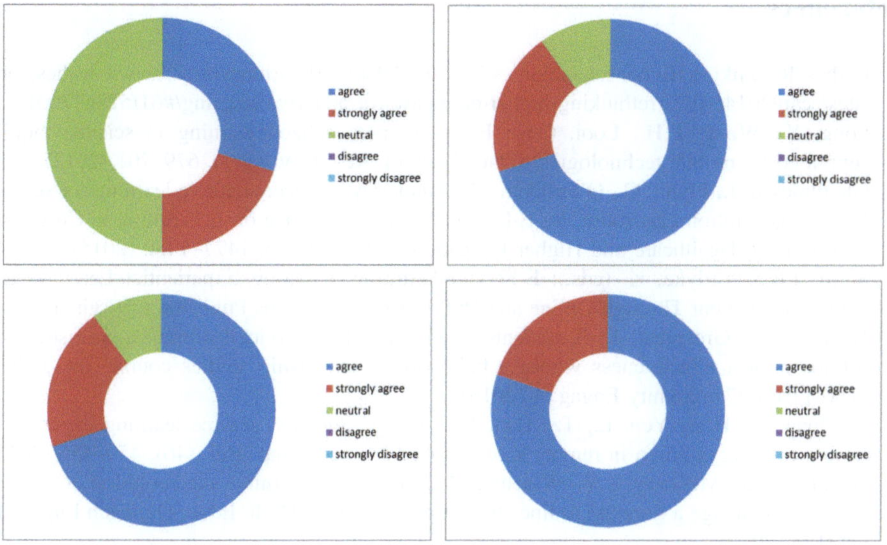

5 Conclusions

Results reveal that most of the participants of service-learning programs that run at the Fab Lab are also interested in e-service-learning activities. The evaluation shows greater interest in e-service-learning in those students who cannot devote all their time to service learning activities, as they have to combine volunteering with university studies. Regarding the comparison between the conventional use of Fab Lab technologies in the field of service-learning and the use of the new NEWTON platform that allow remote access to digital fabrication technologies, students seems to agree on the idea that both methods help them to be more effective and save them time. However, when asked about the ease of use, the need of instructions to use Fab Lab technologies, the safety of their use and the ease of learning, a high percentage of students bet on the use of the NEWTON Project platform.

After the case study, our e-service-learning program will include some activities that allow students to use the NEWTON Project platform to promote active learning and new educational approaches that improve the service-learning programs, allowing students to free service-learning from geographical constraints and provide university students remote access to digital fabrication tools to work with communities in need. Work is ongoing but we expect that this initiative contribute to the development of an e-service-learning program, which will allow students and faculty to experience the benefits of working collaboratively with communities in need.

Acknowledgments. NEWTON project has received funding from the European Union's Horizon 2020 Research and Innovation program under Grant Agreement number 688503.

References

1. Forbes: Rethinking Higher Education: A Case for Adaptive Learning. https://www.forbes.com/sites/ccap/2014/10/22/rethinking-higher-ed-a-case-for-adaptive-learning/#61f528c37001
2. Song, Y., Wong, L.H., Looi, C.K.: Fostering personalized learning in science inquiry supported by mobile technologies. Educ. Technol. Res. Dev. **60**(4), 679–701 (2012)
3. Ghergulescu, I., Flynn, C., O'Sullivan, C.: Adaptemy science: adaptive learning for science for next generation classroom. In: E-Learn: World Conference on E-Learning in Corporate, Government, Healthcare and Higher Education, vol. 2015, pp. 1477–1482 (2015)
4. Strait, J.R., Nordyke, K. (eds.): E-Service-Learning. Creating Experiential Learning and Civic Engagement Through Online and Hybrid Courses. Stylus Publishing, Virginia (2015)
5. Goertzen, B., Greenleaf, J.: A student-led approach to e-service-learning: a case study on service project effectiveness within a fieldwork in leadership studies course. Int. J. Res. Serv.-Learn. Community Engag. **4** (2016)
6. Lih-Juan, C., Hong-Yen, L., Tze-Han, L.: College students' service learning experiences from e-tutoring children in remote areas. Procedia. Soc. Behav. Sci. (46), 450–456 (2012)
7. Waldner, L.S., McGorry, S.Y., Widener, M.C.: E-service-learning: the evolution of service-learning to engage a growing online student population. J. High. Educ. Outreach Engag. **16** (2), 123 (2012)
8. Lorenzo, C., Lorenzo, E.: On active learning and sharing through digital fabrication. In: ICERI2017 Conference Proceedings, vol. 1, pp. 2000–2009 (2017)
9. Lorenzo, C., Lorenzo, E.: When learning happens through a cycle of invention, design and digital fabrication as students bring their ideas to life. In: INTED2018 Conference Proceedings, vol. 1, pp. 7836–7844 (2018)
10. Escario, A: In3 http://fabacademy.org/archives/2015/eu/students/escario_mendezalejandro/
11. Uzunboylu, H., Karagozlu, D.: Flipped classroom: a review of recent literature. World J. Educ. Technol. **7**(2), 142–147 (2015)
12. Bloom, B.S., Engelhart, M.D., Furst, E.J., Hill, W.H., Krathwohl, D.R.: Taxonomy of Educational Objectives: The Classification of Educational Goals. David McKay Company, New York (1956)
13. Stephanie Bell, S.: Project-based learning for the 21st century: skills for the future. Clear. House: J. Educ. Strat., Issues Ideas **83**, 3–43 (2010)
14. Gershenfeld, N.: FAB, The Coming Revolution on Your Desktop. Basic Books, New York (2005)
15. Gershenfeld, N., Gershenfeld, A., Gershenfeld, J.C.: Designing Reality: How to Survive and Thrive in the Third Digital Revolution. Basic Books, New York (2017)
16. Cornetta, G., Touhafi, A., Mateos, F.J., Muntean, G.-M.: A cloud-based architecture for remote access to digital fabrication for education. In: CLOUDTECH Proceedings, Brussels (2018)
17. Montandon, L., Playfoot, J., Ghergulescu, I., Bratu, M., Bogusevschi, D., Mawas, N.E., Rybarova, R.: Multi-dimensional approach for the pedagogical assessment in stem technology enhanced learning. In: EdMedia Proceedings (2018)

Mapping, Measuring and Assessing Internationalization. A Dutch–Russian Perspective

Dmitry Shtykhno[1], Florentin Popescu[2(✉)], Roman Iskandaryan[1], Tijmen Weber[2], and Robert Warmenhoven[2]

[1] Plekhanov Russian University of Economics, Stremyannyy Pereulok, 36, Moscow, Russia
shtykhno.da@rea.ru, iskandaryan.roman@gmail.com
[2] HAN University of Applied Sciences, Ruitenberglaan 31, 6826 CC Arnhem, the Netherlands
{florentin.popescu, tijmen.weber, rob.warmenhoven}@han.nl

Abstract. Delineating the institutional perspective of international programmes, approaches and policies is a very useful method for every higher education institution to measure and assess internationalization as stated by Green [1]. Based on the Hudzik and Stohl [2] Brandenburg and Federkeil [3] and Green [1] models, the international landscape of both Plekhanov Russian University of Economics and HAN University of Applied Sciences (Arnhem Business School), the Netherlands will be measured and assessed around faculty and higher management as institutional strategic policies and plans. This paper discusses indicators those are definitely applied to the array of inputs and outputs of stated internationalization goals of both universities: enhance the global and international dimension of the curriculum, develop dissemination of knowledge and strengthen the value of research, increase international competencies and experience of faculty and stuff, increase students intercultural and international experiences.

Keywords: Mapping internationalization · Measuring and assessing internationalization · International accreditations · Internationalization policies

1 Introduction

At the moment, both Plekhanov Russian University of Economics (PRUE) and Arnhem Business School (ABS) at HAN University of Applied Sciences are busy with assessing their internationalization and research activities for International accreditations: EPAS - EFMD (ABS) and QS reaccreditation (PRUE).

EFMD Program Accreditation System is the principal worldwide system of quality evaluation, development and accreditation of business programs. Consequently, The EPAS procedure depends on a theoretical framework of quality criteria. A global advisory group of EFMD individuals verify and revise these criteria constantly.

W. Karwowski et al. (Eds.): AHFE 2019, AISC 963, pp. 39–51, 2020.
https://doi.org/10.1007/978-3-030-20135-7_4

The EPAS accreditation frame is a model based on inputs and outputs. This model is moving from programme design to programme delivery and finally to programme results. This framework focuses additionally to environmental and institutional contexts. Moreover, the programme and institutional quality assurance procedures are considered. The EPAS evaluation framework is grounded on three transversal pillars, namely internationalization, a corporate outlook, and also ethics, responsibility, sustainability (ERS) [4].

The QS Stars Development Roadmap is based on numerous years of experience gathering institutional information and assembling the QS World University Rankings®. The aim of QS Stars system is to offer a more comprehensive range of criteria for the institutions in order to help them to emerge in their specific domain(s) of excellence. A series of relevant performance parameters are the base of the rating method used by the QS Stars system in order to perform the assessment of the universities. An institution receives Stars awarded by QS Stars system according to different criteria and the result is that the rankings may include also universities that usually are not well-placed.

A QS Stars evaluation can highlight even more the complexity of the ranking outcomes. This rating system stimulates readers and users to understand the ranking outcomes and not just to accept them. The early stage of the application decision making-process at the university can be influenced by the supplementary information ensured by this rating system. Ratings are connected to an individual institution's achievement, but are not conditioned by the achievement of the other institutions. The achievement is based on pre-set thresholds [5].

According to Green [1] universities must be transparent and fair about internationalization. She emphasizes a range of drivers why higher education institutions engage in internationalization:

- To train students to become 'worldwide citizenship' (which can be outlined in multiple ways).
- To train students to get ready for the global workforce.
- To strengthen the quality of research and teaching.
- To enhance the institutional efficiency and competence.
- To increase prestige and visibility.
- To generate income.
- To help in the economic growth at local and regional levels.
- To get solutions to international issues.
- To enhance global understanding and encourage peace all over the world.

Based and adapted from Hudzik and Stohl [2], Brandenburg and Federkeil [3] and Green [1] models, the international landscape of both Plekhanov Russian University of Economics and HAN University of Applied Sciences (Arnhem Business School), the Netherlands is measured and assessed around faculty and higher management as institutional strategic policies and plans. This article discusses those indicators that are definitely applied to the array of inputs and outputs of stated internationalization goals of both universities: *increase the international and intercultural dimension of the curriculum; improve the quality of research and enhance dissemination of knowledge; increase the intercultural and international competencies and experience of faculty and staff; widen students intercultural and international experiences.* These are called the 4 international dimensions on which later on the Inputs, Outputs and Outcomes will be mapped, measured and assessed in detail.

This mapping can be seen as a quality review process tool where the above-mentioned universities undertake a self-study (internal quality control) that map and catalog their international activities and analyses the progress to date and conclude with a comprehensive judgment on recommendation on further directions. A positioning tool regarding internationalization and research activities and efforts.

An assessment of specific university approaches and policies related to the model and the theme utilized is planned by the researchers. Moreover, researchers are planning an assessment of institutional feedback of both universities to a large variety of strategies, problems and policies related to evaluating and measuring internationalization.

This article brings important contributions by trying to establish faculty management by bringing the best practices in the internationalization procedure and at the same time by designing a procedure for evaluating and measuring internationalization in order to enhance the university competitiveness at global scale. The responses to the internationalization procedure of chosen institutions are positioned in the policy background established by these universities. The conceptual framework of this study is analyzing the different categories of internationalization trying to equalize them to the general setting of institutional strategic planning regarding measuring and assessing internationalization.

2 Problem Definition and Research Structure

This article evidences the late advancement of the chosen universities and explores for this in the area of institutional internationalization strategic planning.

The current research investigates how university managers and faculty understand various dimensions or aspects of internationalization process. Moreover, this article points out several of the major problems related to the institutional responses to the effect of internationalization regarding the duties that extend from being local to global in nature. As the literature demonstrates, the strategic planning regarding measuring and evaluating internationalization is a crucial aspect to discover the arguments for institutional returns of universities as complex structures. Consequently, this strategic planning is explored within this paper [2, 3].

Researchers chose a design of the research by combining archives, interviews and observations, weighting more the first two methods. Taking into account the explorative structure of this research, the aim of the interviews is to explain why the interviewee has a certain perspective towards the research topic.

In order to supplement the primary and secondary data obtained on site, documentation and policy for the university was gathered. In the selected university, in order to scale the position on higher education at international, national and local levels different sources of documentary data were utilized for the global policy framework. In order to collect positional and policy information couple of different websites and governments sources were utilized. Moreover, those sources quoted in research articles were used as online searches by accessing different search engines and electronic databases. The documentation obtained from the institution included sources such as annual reports, strategic plans, internationalization policy documents, academic structure charts and websites. This documentation collected from the institution represented the first sources of observation and it was used to understand the vision of the institution policies and strategies.

3 Results

Being able to measure internationalization becomes increasingly important because growing competition has led to a larger emphasis on making the performance of HEIs measurable [6]. Internationalization specifically is becoming increasingly relevant as a performance measure because it is being adopted in various accreditation, rankings, and certifications [6].

QS measures Internationalization criteria by offering in total 150 points accordingly [7]:

- International faculty (25% international faculty individuals) – 20 Points
- International partnerships (50 university research partnerships yielding publications in Scopus™ in the last 5 years or joint-degree programmes with QS top 500 universities in previous 3 years published rankings) – 50 Points
- International students (20% international students) – 20 Points
- Religious facilities (1 place of worship for at least 3 major religious groups (Christianity, Buddhism, Islam, Judaism, Sikhism/Hinduism, or Taoism) OR 1 non-denominational facility for use by any religion) – 10 Points
- Incoming exchange students (2% incoming academic students against total number of students) – 20 Points
- Outgoing exchange students (2% outgoing academic students against total number of students) – 20 Points
- International diversity (50 nationalities scaled down to 5 represented in student body) – 10 Points

For the EFMD/EPAS accreditation systems, internationalization is one of the traversal pillars. "Within the EPAS framework, the teaching materials, delivery methods, including virtual interactions along with the programme content, the qualifications, background and diversity of faculty and students alike, and partnerships, all contribute to the students' international learning experience. The learning experience in itself cultivates other values such as tolerance, cultural sensitivities, between and among faculty and students alike that are equally important in an increasingly changing business and management environment" [8].

The EPAS Internationalization standards reflect the skills, knowledge and intended learning outcomes that need to be reach by the students "to develop as potential international managers; students should be exposed to a diverse mix of students, faculty, teaching materials and international links". Institutions under accreditation need to show how the mix of international partnerships enhances the international learning experience of the programme and what are the links to international organizations (private, public and nonfor-profit) institutions in the context of the international programme [8].

Hudzik and Stohl [2] point out that assessment of international activity focuses on: the institutional aims associated to international activity, how to measure which objectives have been achieved, and how these objectives contribute to the overall values and mission of the institution. They provide a method of assessing internationalization efforts by focusing on three dimensions.

The first is Inputs, which refers to resources available to support internationalization activities. The second dimension is Outputs referring to the number and types of work undertaken in assistance of internationalization. Finally, the third dimension is outcomes which measure the actual impact and end results of the actions undertaken.

Brandenburg and Federkeil [3] offer indicators that are helpful in defining and measuring the degree of internationalization of a HEI. The first group of indicators refer to overall aspects such as the international experiences of staff, resources made available for international activities, and participation in international networks. The second group of indicators refers more specifically to international research experience such as research project with international partners and the number of international publications written. Finally, the third group of indicators covers the international dimension of the curriculum such as number of courses taught in a foreign language, number of international students and lecturers, and the extent of intercultural and international experiences made available to the students.

We use and combine these insights by analyzing the actions undertaken by ABS and PRUE. Specifically, we analyze several of the indicators laid out by Brandenburg and Federkeil [3] using the inputs-outputs-outcomes framework developed by Hudzik and Stohl [2] and Green [1] internationalization drivers. By doing so, we are able to provide a clear overview of activities and consequences of the various actions undertaken to increase the level of internationalization.

For ABS this is measuring 0.0 (EPAS - EFMD accreditation) while for Plekhanov this is measuring 1.0 (QS reaccreditation). As mentioned before, the 4 international dimensions based on the Inputs, Outputs and Outcomes are be mapped, measured and assessed in detail as follows:

Both EPAS and QS have this dimension as part of their assessment. EPAS specifically outlines that the curriculum should demonstrate that it incorporates international aspects [8]. QS is less explicit about it, but does acknowledge the importance of providing a cosmopolitan environment which of course can be strengthened using an international and international curriculum [7]. In Table 1 it can be seen that ABS and PRUE both reach many of the outputs (Table 2).

Table 1. Mapping, measuring and assessing internationalization dimension 1 – increase the intercultural and international dimension of the curriculum

International dimension 1: increase the international and intercultural dimension of the curriculum			
Sample inputs		ABS	PRUE
	Number of courses having international/intercultural/global focus	30	18
	Number and range of foreign language courses	4	12
	Number and proportion of faculty with international and intercultural experiences or expertise	85/90	254/1189
	Number of courses provided in collaboration with an international partner through technology	7	32
Sample outputs	Number and proportion of students enrolled in courses with international and intercultural focus	675/1683	1280/12340
	Number and proportion of students enrolled in language courses at different levels	1683/1683	12340/12340
	Number and proportions of students majoring in programs with an international/global focus	1683/1683	826/12340
Sample outcomes	Demonstrated specific student learning results as proved by portfolios, intercultural competency	✓	✓
	Demonstrated language proficiency	✓	✓
	Career choices or engagements of graduates	✓	✓

Table 2. Mapping, measuring and assessing internationalization dimension 2 – enhance the quality of research and increase dissemination of knowledge

International dimension 2: enhance the quality of research and increase dissemination of knowledge			
Sample inputs		ABS	PRUE
	Number of faculty/researchers with international experiences or expertise	12/90	337/1351
	Amount of funding for international cooperation in research and international projects	180.000 Euro	18 620 000 rub
	Amount of funding from international sponsors	0	8 642 000 rub
	Number of international papers with international partners	13	86
	Number of Cooperation Projects with international professional field	4	98
Sample outputs	Number of publications co-authored with international partners	13	86
	Number of international conferences presentations	25	96
	Number of international projects with international professional field	12	98
Sample outcomes	Recognition, prizes, awards, rankings of institutional or programme international activity	✓	✓
	International accreditations and quality marks internationalization	✓	✓
	Institution's income growth from commercial applications and international funding	×	✓
	Contribution to solving local of global issues for companies in the region or international (Glocal)	✓	✓

EPAS considers the quality of research done by faculty as important, but also stresses that it does not only look at research published in high-ranking journals. Instead, it states that practice-oriented research is equally important [8]. QS more specifically emphasizes the importance of research collaboration with partners as it contributes to academic progress which universally beneficial [7]. Both ABS and PRUE are actively involved in research partnerships with international partners, though only PRUE has been successful at attracting funding from international sponsors.

QS explicitly mentions the importance of having international faculty. Not only because it allows universities to employ best talent, but also because it allows faculty to offer students a different approach [7]. Similar, EPAS stresses the importance of diversity in its faculty because it can help to cultivate values such as tolerance and cultural sensitivities [8]. Both ABS and PRUE have staff with substantial international experience as can be seen in Table 3.

Table 3. Mapping, measuring and assessing internationalization dimension 3 – increase the intercultural and international competencies and experience of faculty and staff

International dimension 3: increase the intercultural and international competencies and experience of faculty and staff			
Sample inputs		ABS	PRUE
	Number and proportion of faculty and staff with international experiences or expertise	90/100	398/2206
	Number and proportion of faculty and staff educated or trained abroad	75/100	123/1189
	Number and proportion of faculty and staff who participated in teaching/projects abroad	50/100	151/1189
	Number and proportion of faculty who are multi-lingual	98/100	246/1189
Sample outputs	Growth in number and proportion of faculty engaged in international cooperation	50/100	151/1189
	Growth in number and proportion of staff engaged with partner universities	35/100	112/1189
	Number of teacher mobility exchange programmes	8	16
Sample outcomes	Enhanced reputation and recognition for the institution international character and work	✓	✓
	Number of international degrees of staff	✓	✓
	Number of staff who followed pedagogical and professionalization courses with international and intercultural dimensions	✓	✓
	Support of HRM policies in place for foreign staff members	✓	✓

Finally, both EPAS and QS recognize that it is not just the faculty that needs international experience, students also benefit from going abroad. QS specifically mentions that universities should foster both inbound and outbound student exchange and adds the importance of attracting international students [7]. EPAS reaches a similar conclusion stating that a good program should have a diverse mix of students with an international perspective who are actively involved in international activities [8]. Once again, both ABS and PRUE have demonstrated that students develop international competencies in several areas.

ABS uses the concept of Preferred Partners (see Table 4) as a strategic focus when working with partner universities where intensive and active collaboration of the following areas is taking place:

- International Joint Students Projects
- Double Degree Programmes
- Curriculum Benchmarking
- Minor Programmes Development
- Transfer Programmes to Master Degrees

Table 4. Mapping, measuring and assessing internationalization dimension 4 – Widen students intercultural and international experiences.

International dimension 4: Widen students intercultural and international experiences			
Sample inputs		ABS	PRUE
	Number and proportion of students with international experiences or expertise via internships	1683/1683	4898/12340
	Number and proportion of students educated or trained abroad (study abroad)	1683/1683	2031/12340
	Number and proportion of students participated in intercultural and international projects through technology	1683/1683	240/12340
	Number and proportion of students who are multi-lingual	1683/1683	5000/12340
	Number of Institutional International Partners (Partner Universities)	123	156
Sample outputs	Growth in number and proportion of students engaged in international projects	1683/1683	391/12340
	Growth in number and proportion of students following courses from/at partner universities	1683/1683	559/12340
	Number of students mobility of exchange programmes (outgoing)	400/1683	290/12340
	Number of double degree programmes	5	30
	Number and proportion of preferred partner universities (intensive collaboration)	8/123	28/156
Sample outcomes	Demonstrated specific student learning outcomes as evidenced by portfolios, intercultural and international competencies	✓	✓
	Demonstrated language proficiency	✓	✓
	Career choices or engagements of graduates	✓	✓

A considerable part of PRUE's faculty members today (around 90%) do not participate in international scientific studies, more than half of faculty members do not carry out neither fundamental nor applied research, publication activity of PRUE's faculty members and academic staff in journals indexed by Scopus and Web of Science scientometric systems is significantly lower than at universities from the reference group. No more than 5% of the faculty members have the level of English proficiency that is enough to deliver lectures and perform international scientific projects. Less than 7% of the faculty members regularly upload materials at scientific internet systems (academia.edu, researchgate.net, Google scholar, RePEc), foreign faculty members account for less than 3%, most of them are CIS citizens.

According to the HAN Institutional Plan 2016–2020, all internationalization activities are geared towards the following objectives:

"In 2020 HAN is internationally recognized as an innovative higher education institute. Our graduates are well prepared for an international and intercultural work environment. The campus is characterized by an international climate, where foreign students are full partners and where multidisciplinary partnerships address social challenges from an international perspective in collaboration with the professional field".

"Internationalization is incorporated into all our goals and activities. We see internationalization not as a separate agenda, but as an integral part of all our education and research activities as well as related support processes (HR, ICT, marketing and communications, etc.)" [9].

The strategic goal of the Plekhanov Russian University of Economics (PRUE) is the development of an effective multi-level system of continuing education in the domain of economics that meets the requirements of the country, society and individual. It requires transformation into university which has a wide national and global recognition which ensure a resource base for its later on sustainable growth. Internationalization is considered as one of the keystones of such transformation. Internationalization strategy was developed in 2013. It has been being implemented aiming the following goals [10]:

- improving the quality, attractiveness and competitiveness of educational services of PRUE in the global and regional educational field;
- enhancing the reputation and value of research activities of the PRUE in the global and regional scientific field;
- ensuring effective participation of the PRUE in the global and major regional processes of development of science and education;
- increasing share of exports of educational and research services in the total volume of services of the PRUE.

PRUE's potential advantages in the improvement of international competitiveness are also measured by the experience of international cooperation in the educational sphere: more than 40-year experience of training of foreign students, more than 110 foreign partner universities from more than 40 countries, more than 20 "double diploma" programs, master's "triple diploma" program, summer schools for foreign students, accreditation from AMBA (Association of MBAs), CIM (Chartered Institute of Marketing, UK), ECBE (European Council of Business Education), CIMA

(Chartered Institute of Management Accountants, UK), EdUniversal Ranking, experience of participation in international cooperation programs Tacis, Tempus, Erasmus+, etc., PRUE educational audit according to EFMD methodology.

For many years internationalization has stopped of being only a theme or central point in economic education. Internationalization it is nowadays the norm in economic education. It is quite right, therefore, that HAN's institutional plan demonstrates strong ambition with regard to internationalization. The HAN Institutional Plan 2016–2020 includes the following strategic priorities:

"Students learn to deal with people from different backgrounds. We achieve this explicit focus on diversity partly by internationalizing education and research. That can be done, for example, by simplifying and strengthening internationalization at home, but also by means of international exchanges and work placements."

"We expressly include internationalization in all our goals and activities. We embed foreign language knowledge and intercultural skills in the curriculum and develop joint education and research programmes with foreign partners, create diversity in the classroom and give students the opportunity to take part in international projects of social relevance" [9].

Both students and professors work on real-life internationalization problems that businesses confront in different work placements, research projects and final assignments. Both knowledge increase and international business activity are stimulated constantly by this "innovation driver". Moreover, it is also the foundation for more defined professional profiles in the individual courses. ABS makes a substantial contribution to the economic and social fabric of the region and aims to gear its education renewal as closely as possible to the professional practice in the region.

There is increasing demand in Arnhem's region for professionals with knowledge of the world outside; customers want customized products and services and their requirements are growing more complex [9]. This demands professionals with a high level of education, the ability to think analytically and creatively and a readiness to work in an international context. Each department and each course fulfills this in its own way, in a manner appropriate for the profession for which it is training its students, but each acts on the basis of the described outlook and the general wish to put ABS firmly on the map both in the region and internationally.

During his or her studies, any ABS student develops knowledge of the international context of his or her professional profile and learns to handle intercultural differences so that he or she is prepared for a professional future in an increasingly globalized world. The ABS mission is to give its students a solid grounding for future professional practice. "Because that will be in a globalized world, we want the future professional to have intercultural skills, knowledge of the international aspects of professional practice and a sense of being a global citizen. Internationalization is no longer merely a theme but is the norm" [11].

Internationalization is an important means of improving the intensity, quality and impact of the connection we make between research, education and the professional field. This will aid both universities' efforts to produce well-qualified professionals with a strong moral compass, who are socially engaged, not afraid to explore new horizons and keen to continue developing. In particular, HAN seeks to train its students in the international and intercultural competencies increasingly demanded by the professional field.

4 Conclusions and Further Research

In the first part of July 2015, an audit focused on methodology QS Stars was started being conducted by a central PRUE. Audit's results were provided after three to the QS Quacquarelli Symonds, which is an international rating agency, in order to verify the obtained results. After the verification and based on those results of audit, in December 2015, 4 starts of QS Stars were awarded to PRUE. This achievement made from PRUE the single one for now and the first university in Russia receiving such a high appreciation.

Still, in order to secure a firmer presence in the QS 4 Stars category, PRUE needs to boost its internationalization and visibility efforts, especially in respect to research. As well as this, it is necessary that more attention is devoted to areas from the Advanced Criteria in order to establish a truly individualized, world-class profile. In 2015, the scores attained in the Research and Internationalization categories further support this conclusion, as the points gained represent 31% and 37% respectively from the maximum and attracted only 2 Stars each. Research in particular is an area that, according to these results, would require more focus. Plekhanov performed well only in respect to the Academic Reputation indicator, where it achieved a maximum score. However, this good academic reputation is not matched by a high number of publications indexed in Scopus, nor by a large number of citations.

Internationalization, the category where Plekhanov attained 56 of the 150 points available, also received 2 Stars and appears as an area where further development is needed. In this category, the indicator with the lowest score is International faculty: 2 out of 20 points, followed by Institutional research collaborations with a fifth of the maximum score. By contrast, within the category, the indicators in which Plekhanov performs better are related to the number of Inbound and Outbound exchange students, as well as International diversity. These results suggest that although the institution is diversifying in terms of student body, these efforts are not yet sufficiently matched by an extensive international collaboration of the academic staff.

Although attracting more international Faculty Staff could be more challenging due to language or the specific national context, increasing the number of international research collaborations could become an area towards which more resources are devoted. In this respect, perhaps the institution could capitalize on the existing collaborations for student exchange programs.

This being said in 2015, the reaccreditation is now in process and this article and the model proposed will help measuring and increasing the importance of research and the international dimension so that higher score (more stars) will be awarded.

The target operating model of PRUE suggests that by 2020 [12]:

- at least 70% of the faculty members will take part in fundamental and applied scientific research;

- at least 50% of the faculty members will regularly post their works in scientific internet systems (academia.edu, researchgate.net, Google scholar, RePEc);
- at least 50% of full-time faculty members will be fluent in at least one foreign language;
- at least 25% of administrators and at least 50% of top executives will be fluent in English;
- a system of international recruitment of faculty members and academic staff will be implemented;
- the proportion of foreign professors and researchers in the total number of the faculty members including Russian citizens with PhD degrees of foreign universities will rise to 14%.

ABS seeks to contribute actively in an ongoing dialogue between business schools and business life at an international level. It aims to share knowledge and best practices in business management and to debate on innovation, best practices and research with the intent to educate knowledgeable and socially-responsible future managers in business. Regarding ABS, they already are a member of the EFMD, which is a promising start towards EPAS accreditation. By becoming involved in the EFMD network and applying for the EPAS accreditation, ABS commit themselves to further and continuously improving the quality of their programmes, services and facilities. Also, EFMD membership would facilitate their ambition to benchmark with other Business Schools and exchange knowledge and experience with companies at an international level while they also strongly believe their students will benefit from the EFMD membership.

According to the Concept Strategic Objectives ABS 2018–2023 [13], ABS is aiming to become recognized for providing high quality education through regular contact with representatives from the professional field, a highly-valued study programme by students, staff, and various accreditation institutions, and finally, having an experienced and diverse staff (e.g. 20% hold a PhD, 25% has a non-Dutch background, and all lecturers have international experience).

There is also an increased emphasis on facilitating and stimulating (applied) research through an extra focus on research based on questions from the professional field, but also ensuring that relevant research findings are adopted within the curriculum. Furthermore, every lecturer that holds a PhD will have to do work on a research project, and each member of the Research Centre will have to publish at least two articles in relevant peer-reviewed journals or conferences. By going into this direction, ABS intends to develop an international environment which attracts global talent. Through such initiatives, it becomes possible to stimulate the development of socially aware global citizens who understand the value of international contacts and evidence-based decisions.

Acknowledgments. We would like to thank the ABS Management Team (Arjan Keunen, Marjon Elshof, Joop Verbrugh) for their involvement, efforts and dedication in providing and sharing internal documents regarding policies and procedures on Internationalization and accreditation in a transparent manner.

References

1. Green, F.M.: Measuring and Assessing Internationalization. NAFSA: Association of International Educators (2012)
2. Hudzik, J.K., Stohl, M.: Modelling assessment of the outcomes and impacts of internationalization. In: Measuring Success in the Internationalisation of Higher Education. EAIE Occasional Paper 22 (2009)
3. Brandenburg, U., Federkeil, G.: How to measure internationality and internationalisation of higher education institutions! Indicators and key figures. Centre for Higher Education Development (2007)
4. EFMD Programme Accreditation System. https://efmdglobal.org/accreditations/business-schools/epas/
5. QS. https://www.topuniversities.com/
6. De Wit, H.: Measuring success in the internationalisation of higher education: An introduction. In: Measuring Success in the Internationalisation of Higher Education. EAIE Occasional Paper 22 (2009)
7. QS.: Rating Universities on Internationalization: QS Stars. https://www.topuniversities.com/qs-stars/qs-stars/rating-universities-internationalization-qs-stars
8. EPAS: 2018 EPAS Standards & Criteria. https://efmdglobal.org/wp-content/uploads/EFMD_Global-Accrediation_EPAS-Standards_and_Criteria.pdf
9. Institutional Plan 2016–2020, HAN University of Applied Sciences
10. The concept of internationalization of the Plekhanov Russian University of Economics, Plekhanov Russian University of Economics (2013)
11. Strategic Internationalization Agenda 2016–2020, HAN University of Applied Sciences
12. Strategy Program 2016–2020, Plekhanov Russian University of Economics
13. Concept Strategic Objectives ABS 2018–2023, HAN University of Applied Sciences

Human Factors in Digital Transformation of Education: Lessons Learned from the Future Gate at Saudi K-12

Yousef Al-Ohali[1,2](✉), Mohammed Alhojailan[1,2], Nikos Palavitsinis[3], Jad Najjar[3], Anastasios Koutoumanos[3], and Ahmed AlSuhaibani[1]

[1] Tatweer Educational Technologies Company, Riyadh 12612, Saudi Arabia
{yohali,asuhaibani}@tetco.sa
[2] King Saud University, Riyadh 11451, Saudi Arabia
malhojailan@ksu.edu.sa
[3] Eummena, Kapeldreef 60, 3001 Leuven, Belgium
{palavitsinis,jnajjar,tk}@eummena.org

Abstract. Future Gate is a country-wide, large-scale initiative that aims to upgrade the quality of school education for all, in the Kingdom of Saudi Arabia, with an optimal combination of educational technologies and change management. Future Gate implements a continuous measurement and monitoring of human behavior towards the delivery of a successful digital transformation in the Saudi K-12 Education. This paper presents an overview of the results of a comprehensive empirical evaluation for the 1st year of Future Gate. The paper focuses on the interpretation of the human behavior in relation to the technologies and equipment introduced to Future Gate beneficiaries. Through this, the aim is to optimise the compatibility amongst all FG actors, technology, and environment. Overall, teachers and students are adopting the FG services, thus embracing the change brought about by the initiative.

Keywords: Evaluation · Education · Human factors · K-12 · Learning Management System

1 Introduction

In the context of the Saudi Arabia Vision 2030[1], the Ministry of Education implements the Future Gate (FG) initiative to catalyse the K-12 digital transformation. Launched in 2017, Future Gate is a country-wide initiative for upgrading the quality of school education for all, with an optimal combination of educational technologies (EdTech) and change management. To this end, the focus from the conception of the initiative has been on Human Factors and the orchestration of processes, relationships, and interactions between all actors within the K-12 community. Future Gate targets more than 6 million students and 500 thousand teachers in 25 thousand schools across all

[1] https://vision2030.gov.sa/en.

W. Karwowski et al. (Eds.): AHFE 2019, AISC 963, pp. 52–64, 2020.
https://doi.org/10.1007/978-3-030-20135-7_5

geographical regions of Saudi Arabia, along with all other involved stakeholders: parents, school leaders, educational supervisors, and policy makers. Its core objectives are: (i) changing the traditional style of education by creating a digitally-enhanced learning environment that eradicates the limitations of a paper-based, teacher-centered education; (ii) creating an enjoyable learning environment that promotes interactions between students and their teachers; (iii) guiding students to the constructive uses of technology by capitalising on their underlying interest in the digital medium; (iv) expanding learning and pedagogical methods to exist beyond the classroom and school environments; and (v) enabling better communication between all those involved in a student's learning process: parents, teachers, school administration and other related stakeholders.

According to the International Ergonomics Association, "Human factors is the scientific discipline concerned with the understanding of interactions among humans and other elements of a system, and the profession that applies theory, principles, data, and other methods to design in order to optimise human well-being and overall system performance" [1]. Human factors are therefore concerned with applying what is known about human behavior, abilities, limitations, and other characteristics to the design of systems, environments, and equipment/technologies. It is also concerned with the design of training programs and instructional materials that support the performance of tasks or the use of technology/equipment [2]. Educational ergonomics is defined as that field of human factors/ergonomic science concerned with the interaction of educational performance and educational design [3].

As discussed in the relevant literature [4–8] human factors in education, involves the application of psychological and physiological principles to the engineering and design of products, processes, and systems. In this context, Future Gate is focusing on human factors and more specifically on how people interact with technologies/equipment and with the environment, in order to understand these interactions, through a comprehensive empirical evaluation. The evaluation results presented in this paper aim to interpret the human behavior behind the technologies/equipment offered through the initiative, to ultimately empower and facilitate all involved actors. This in turn contributes towards the successful delivery of the digital transformation in the K12 education that is supported with continuous measurement and monitoring of human behavior through the transformation life cycle.

This paper, presents and discusses indicative findings of the evaluation that was carried throughout the first year of FG. The paper also aims at discussing data from the evaluation, utilising them, to ultimately optimise the compatibility amongst all FG actors, technology, and environment. The next sections of the paper are structured as follows: Sect. 2 briefly outlines the design and methodology of the evaluation as well as provides details for the data collection. Section 3 presents the main findings and connects them to human factors. Finally, Sect. 4 discusses the conclusions of this work and outlines the work ahead.

2 Evaluation Design and Data Collection

In Future Gate, the evaluation process seeks to: (i) make the Future Gate initiative accountable, towards the Ministry of Education and all other stakeholders that are affected by its outcomes; (ii) serve as a management tool allowing for a continuous improvement of the initiative's implementation, as its results can be used for educated decisions and to fine-tune field actions; (iii) communicate quality criteria and standards, allowing to constitute a message to the education system by stating implicitly or explicitly what is important to achieve; and (iv) minimise human error, increase productivity, and enhance the interaction of users with the services that Future Gate offers.

Towards those goals, a hybrid methodology has been designed for data collection and analysis, utilising both quantitative and qualitative evaluation methods to analyse data collected. Data were collected from a multitude of distribution channels such as learner analytics tools, messages through social apps, SMSs, e-mails, the Future Gate portal, calls, workshops and face-to-face meetings. This approach has allowed for inclusive data collection, reaching out to all the different Future Gate stakeholders, regardless of their familiarisation with technology.

Quantitative data covered fixed response items from either a list of specific options or from a five-point Likert scale. That is in addition to the data on the actual usage that was collected from the learner analytics component of the FG Learning Management System (LMS). Qualitative data collection included open-ended questions and semi-structured interviews with teachers, parents and school leaders. This methodology was based and connected to a set of Key Performance Indicators (KPIs), that were monitored throughout the initiative's implementation [9].

2.1 Key Performance Indicators

Table 1 presents a subset of the indicators that have been monitored throughout the FG first year of implementation. The table also connects each KPI to the aspect of Human Factors that the KPI relates to.

2.2 Data Collection

Data collection was planned throughout the first year of the Future Gate Initiative. At the end of the first year, Future Gate served 104,485 students and 7,275 teachers in 310 schools. Table 2, outlines a subset of the collected datasets, limited to the ones that are relevant to the KPIs discussed herein.

The next paragraphs present the main results of the analysis of the data collected throughout the 1st year, with an emphasis on how the human factors impact different aspects of the targeted digital transformation.

Table 1. Key performance indicators for Future Gate

No	Indicator	Human factor
Actual usage of FG LMS		
1	Authenticated teachers on LMS	Engagement
2	Authenticated students on LMS	Engagement
3	Posts in discussion forums	Engagement
4	Created and submitted assignments	Engagement
5	Number of deposited e-content items	Engagement
6	Attendances recorded	Engagement
7	Lesson preparation	Engagement
Educational technology perception and attitudes		
8	% of teachers that do not think that EdTech can help them	Learning experiences
9	% of teachers who feel EdTech improved their efficiency	Performance
10	% of teachers who feel EdTech enhances teaching experience	Learning experiences
11	% of teachers who feel EdTech improves students' participation	Learning experiences
12	% of teachers who feel EdTech increases student assignments	Performance
13	Teacher use of LMS within the classroom	Engagement
14	Student use of LMS within the classroom	Engagement
15	% of parents that are aware of the FG LMS	Learning experiences
16	% of parents with positive comments on the FG LMS	Learning experiences
User support and training		
17	% of teachers with ICT certification	Learning experiences
18	Teacher confidence in ICT skills	Learning experiences
19	Teacher experience in using ICT	Learning experiences
20	% of teachers requesting further professional training on EdTech	Learning experiences
21	% of FG stakeholders trained on the use of the FG LMS	Learning experiences
22	% of trainees with a training satisfaction higher than 75%	Learning experiences
23	Support tickets related to FG LMS as a % of total tickets	Usability

Table 2. Data sets collected and analysed

No	Dataset	KPIs
1.1	**Usage data** Log files from the FG LMS for 310 schools, divided by region, down to the school level, measuring the sums of specific actions on the FG LMS (such as assignments, posts in forums, lesson preparation, etc.)	1–7
1.2	**Teacher perceptions and attitudes** 300 questionnaires (online) from teachers that are participating in the FG initiative through their schools. Data included demographics, and questions related to teacher experience, and the teachers' work with the LMS	8–14 17–20
1.3	**Parental engagement** Questionnaire distributed at schools and parent forums for Riyadh, Jeddah and Eastern Region (894 responses); Online Survey for parents in the Riyadh, Jeddah and Eastern Regions (7,886 responses) and Online Survey distributed by text message and through the FG LMS (5,205 responses)	15–16
1.4	**User training** Number of trained stakeholders (e.g. school leaders, agents, teachers) versus the total number of targeted trainees	21
1.5	**User satisfaction** 667 Questionnaires (132 school agents, 152 academic supervisors, 211 digitisation coordinator, 105 school leaders, 67 special needs training)	22
1.6	**User support** 30,103 tickets from the call center received via the different channels and classification of the tickets received by the help desk	23

3 Analysis

In this section, we discuss the results related to the actual participation of the K-12 community in FG services and we continue with the analysis of the perception and attitudes of the main stakeholders towards EdTech. This part ends with the measured impact of actions that focused on the human factors, such as user support, motivation,

Table 3. Schools joining the FG initiative during each semester

Region	Schools joining FG on semester 1	Schools joining FG on semester 2	Total
Riyadh	51	6	57
AlQassim	0	40	40
Onaiza	0	10	10
Jeddah	50	32	82
Aseer	0	18	18
Eastern Region	50	30	80
Al-Ahsa	0	23	23
Total	151	159	310

and training. FG services during the 1st year of the initiative were gradually introduced to the educational community over the two semesters, as Table 3 shows.

3.1 Actual Usage of the Future Gate LMS

Tracking and analysing data on the actual usage of the FG LMS allowed us to determine the level of engagement of the FG stakeholders. This has affected the selection of services offered through FG as well as the optimisation of the user experience and user interface design. Moreover, this information allowed for early identification of risks and challenges, as well as educated decisions related to the overall change management involved with the K-12 digital transformation. As shown in Table 4, at the end of the 1st year of the FG implementation, the majority of teachers have activated their accounts (98.2%) across the different regions and the same is valid regarding the activation of their students' accounts (95.7%).

Table 4. Percentage of teachers and students that have activated their FG accounts

Region	Teachers	Students
Riyadh	98.9%	97.4%
AlQassim	98.5%	97%
Onaiza	100%	98.8%
Jeddah	99.2%	91.2%
Aseer	97.2%	88.8%
Eastern Region	96.3%	97.5%
Al-Ahsa	97.1%	98.9%

The most striking result, interestingly, most of these users, actively use the FG services on a regular basis throughout the school year. Table 5, indicates the actions performed on the FG LMS for the three regions that joined the FG Initiative during the 1st semester (Riyadh, Jeddah, Eastern Region), comparing them to the 2nd semester actions for the same regions. In these three regions, the number of users has increased by 62% over the 2nd semester, as more schools connected to FG, but the corresponding actions on the LMS have increased by far more, as the last column indicates.

Table 5. Actions on the LMS for semesters 1 and 2 for initial regions

No	Semester 1	Semester 2	Diff.
Electronic assignments	70,742	285,060	+197%
Interactive content	195,095	578,921	+300%
Discussion forums	36,525	223,480	+510%
Attendance	66,903	335,482	+400%
Lesson preparation	76,293	400,416	+425%

Another interesting insight is how the digital transformation has accelerated throughout the school year, and how the positive signs of first results have attracted the interest of the school community. Table 6 shows the FG LMS actions for the regions that initially joined the Future Gate during the 1st semester, comparing them to the actions on the LMS, coming from the regions that joined the Future Gate during the second semester of 2017 (Qassim, Onaiza, Aseer and Al-Ahsa).

Table 6. Actions on the LMS for initial versus new regions

No	Initial regions	New regions	Diff.
Electronic assignments	70,742	68,735	−3%
Interactive content	195,095	149,668	−23%
Discussion forums	36,525	45,616	+25%
Attendance	66,903	107,642	+61%
Lesson preparation	76,293	97,286	+28%

The regions joining Future Gate in the 1st semester of the 2017–2018 school year, contained 151 schools and 53,108 users, whereas the new regions that joined in the second semester contained 91 schools and 25,720 users (51.5% less users). Although the new regions brought about half the users than the initial regions, the actions these new users generated on the LMS were more, as the last column indicates.

These collected data reveal that the new regions which entered FG during the 2nd semester, directly interacted more with the system than the first regions. We must note here however, that the users from the initial regions continued to increase their participation and interaction with FG services during the 2nd semester. Overall, even greater levels of participation were recorded, compared to that of the schools from the new regions.

This fact shows that the actions that were taken to foster the digital transformation, have worked well and should be continued onto the next period of the initiative. Such actions involved incentives for the most successful teachers, gamification for the student experience within the LMS, continuous support through a dedicated helpdesk. Moreover, a clear activation guide that was communicated to all beneficiaries in terms of what is expected of them weekly.

3.2 Educational Technology Perception and Attitudes

The continuous analysis of the perception and attitudes of the teachers towards the FG services safely steers the implementation of the initiative, to achieve in the digital transformation and improve the overall learning experiences for all throughout K-12. The analysis focuses on the actual use of the FG services within the classroom, taking into consideration the teachers' experience with EdTech as well as their perception of the services offered. The analysis also reflects on the parents' experiences from the FG services.

Related to the reported use of the LMS, 42.6% of the 300 surveyed teachers reported using the system more than 2.5 h per week within their classroom (Table 7). Following up, the teachers reported that 26.4% of their students use the FG LMS within the classroom for more than 2.5 h per week.

Table 7. Reported weekly use of LMS within classroom by teachers for themselves and students

Duration of LMS use	Never	0–30 min	31–90 min	91–150 min	More
Teachers	0%	26.7%	17.1%	13.7%	42.6%
Students	11.2%	41.3%	12.7%	8.4%	26.4%

Despite the fact that surveys have shown that 49% of the teachers have been using EdTech in their teaching practices for more than six (6) years (Table 8), only 48.6% of them had any related professional training [1].

Table 8. Teacher experience in using educational technology

Response	Never	<2 years	2–6 years	6 years
Percentage	11.6%	17.4%	21.9%	49%

Combining the findings from Tables 7 and 8, there is a clear need for specific guidance on how to maximise the benefits that stem from the use of FG services within the classroom. Teachers seem to have enough experience to support FG adoption inside the classroom but still, their reported us of the services is not reaching the expected level. Consequently, this led the authors to look into the attitude of teachers towards technology, to realise how adoption is related to human factors and resistance to change.

When teachers were asked about their opinion on EdTech and teaching as well as its impact on the learners, the following points came up (Table 9). Comprehensively, it seems that the positive effects of EdTech are realised by around 75–80% of the teachers, whereas only 8–13% do not see a positive effect, and around 11–13% are still indecisive about EdTech in general.

When this finding is inserted into the equation of interpreting teachers' behavior, it seems that the teachers are utilising the technology offered by FG in similar levels as they were utilising technology that was at their disposal, pre-FG. This presents a major challenge for the human factors' theory, to address the psychological factor that affects teachers' behavior in regards to this.

Evaluation also tackled parents' awareness on Future Gate, to find that a significant number of survey respondents are aware of the Future Gate initiative (66.48% out of a total of 5,045 parents). As Table 10 indicates, there are similar levels of consistency in knowledge and awareness of the FG initiative that were observed through regional comparative analysis. In the end of year 1 (2nd semester), Riyadh has higher parent

Table 9. Teacher attitudes towards EdTech related to themselves and their students

No	Educational technologies can…	Agree	Neutral	Disagree
1	Help my efficiency when preparing for class	75.9%	13.4%	12.7%
2	Help me with my day-to-day tasks in the class	79.2%	11%	9.8%
3	Help me correct student homework	75.9%	11.1%	13%
4	Engage students more, in class and outside	80.8%	12.3%	6.9%
5	Provide a snapshot of student performance	72.7%	14%	13.3%
6	Enhance my teaching experience	85.9%	7.9%	6.2%
7	Help my teaching in general	75.2%	9.2%	15.6%
8	Increase the participation of students in class	77.5%	14.1%	8.4%
9	Increase the assignments submitted by students	80.8%	11.4%	7.8%
10	Facilitate the assessment process for students	78.2%	12.6%	9.2%
11	Increase the time students work with material	73.6%	11.4%	15%

Table 10. Parent awareness on Future Gate

No	"Are you aware of Future Gate?"	Semester 1	Semester 2
1	Riyadh	59.4%	70.1%
2	Jeddah	66.0%	66.1%
3	Eastern Region	60.6%	64.8%

awareness (70.1%), with Eastern Province and Jeddah having similar rates of awareness (64.8% and 66.1% respectively).

Parents were also asked to assess the Future Gate LMS in terms of its communication effectiveness and the quality of information offered in relation to the students' activities and performance. The data refer to the same initial set of 151 schools that joined Future Gate during the 1st semester and were re-examined in the 2nd semester. The findings presented in Table 11 underline how parents positively assess that the FG LMS has facilitated their communication with the school and made it easier for them to track their children's performance.

Table 11. Parents' assessment of the FG LMS

Survey item	Ranking out of 5 (Jan 2018)	Ranking out of 5 (April 2018)
Has the portal facilitated communication with the school administration and teachers?	3.94	3.88
Has the portal made it easier for you to be updated with your child's performance and behavior?	4.06	3.90
Has the portal enabled you to know the exam schedule and the latest school news?	3.96	4.10

Looking at the parents' findings, it becomes evident that parents are aware of the FG services and are appreciating the manner in which they transform their own experience related to their children's schooling. Although a more in-depth analysis of the actions of the parents on the FG LMS is needed, it seems that parents' appreciation of the FG offering can also be used as a motivating factor for teachers to incorporate the FG in their daily teaching practices.

3.3 User Support and Training

Another significant aspect related to human factors, that has been taken into consideration from the early design of the FG Evaluation methodology, was the user support and training. Due to the fact that teachers are drivers of digital transformation, we have focused on responding to their needs so they feel empowered to maximise the benefits of the introduction of EdTech, both for their daily loaded work-schedules and, mainly, for their students. In addition to that, aiming at streamlining user experience, evaluation monitored the user support that was offered to all the beneficiaries of FG services.

In the context of the teachers' questionnaires, FG teachers were asked about the support they would need to increase their use of EdTech in general, and specifically within the classroom. According to teachers the most important factors that would affect the use of educational technology in the classroom are (i) having a better connection to the internet than what is currently available (24.6%) and (ii) having more computing devices for students (22.8%). Table 12, presents all the factors reported.

Table 12. Factors that would affect the use of educational technology in the classroom

What are the factors that would enhance the use of FG services?	Percentage
Better internet connection	24.6%
More devices for students	22.8%
Climate of support for teacher by school and parents	14.4%
On-the-ground technical support	14.1%
Training for ICT use in teaching	10.3%
More devices for teachers	9.8%
More time to practice	4.1%

Last but not least, related to the support that the teachers would need to incorporate educational technology in their teaching practice (Table 13), the main responses included (i) training from ICT-certified professionals (22.3%) and (ii) having an onsite ICT coordinator for help (21.5%).

Looking at the last two tables, the importance of taking into consideration the human factors is evident. As far as the teachers are concerned, on a personal level, they stated that they would need more training as well as on-site support from an ICT expert. On the topic of training, Table 14, reflects on the number of Future Gate stakeholders that were trained by the Activation Team of the Future Gate initiative, analysed for each semester.

Table 13. Factors that would affect teachers personally to use FG services more

What are the factors that would help you personally to use the FG services more?	Percentage
Training from ICT professionals	22.3%
On-site support from ICT expert	21.5%
More time in my schedule to practice	15.2%
Time to consult with my peers	15.1%
Teacher-led workshops for teachers	15%
Online professional forums	11%

Table 14. FG Beneficiaries trained on the use of FG services during Year 1

Role	Semester 1	Semester 2	Total
Ministry Employees	17	0	17
Digital Transformation Supervisor	157	156	313
Teachers	3,652	3,240	6,892
School Leaders	142	133	275
Student Advisors	131	139	270
Educational Supervisors	1,091	566	1,657
Department of Education	58	147	205
Agents	129	126	255
Total	5,377	4,507	9,884

Training addressed a broad audience from the FG beneficiaries, focusing more on the teachers as the main drivers behind the adoption of the FG services. In total, 6,892 teachers were trained, out of 7,275 teachers that participated in FG during its 1st year. The number of beneficiaries trained, reflects the importance of human factors for the overall success of FG services.

Out of the total users trained, a sample of 667 questionnaires were collected, assessing the training satisfaction level of the trainees across the different regions. The data were collected in the 1st semester, shows that 82.6% of the participants have indicated a satisfaction rate ranging between 76% and 100%, whereas 2.1% indicated a low satisfaction rate ranging between 0%–25% and the remaining 7.5% declared a satisfaction from 26% to 50%. Based on these outcomes, the training offered was redesigned over the 2nd semester, to make sure that trainees needs are met efficiently.

Complementing the training process, FG set up a helpdesk, to support the beneficiaries using the FG services. Table 15 details the tickets activity for the 1st and 2nd semesters in terms of calls received, direct chats, emails and support request forms. The table shows a significant increase in the number of tickets received in the second semester, which could be attributed to (i) the implementation of the second phase where 159 schools were added; (ii) the increase in the number and authentication percentages of FG users; and (iii) the increase in the authenticated accounts as well as the different FG LMS tools used by the different FG users.

Table 15. Call center ticket sources

Ticket source	E-mails and support forums	Direct chat	Calls	Total
Semester 1	4,834	5,005	1,858	11,697
Semester 2	8,032	7,192	3,182	18,406

The next section, reflects on the main conclusions that came out from the data analysis. It also outlines the directions upon which, future work will build upon the work presented in this paper.

4 Conclusions and Future Work

The main conclusions that can be drawn upon from the FG continuous evaluation during its first year of implementation are: (a) the majority of the people that interact directly with the LMS (teachers and students) have activated their accounts; (b) the same users are increasingly adopting EdTech within their teaching and learning practices through the course of the first year of the initiative; and (c) new regions added to the initiative seem to be picking up on all FG services faster than the previous regions, which is a strong indication of the positive impact of the initiative's actions targeting the human factors.

Half of the teachers surveyed have a significant experience in using EdTech in their teaching, and at the same time, almost half of the teachers report not having any professional training on educational technologies. This lack of training indicates a need to support teachers on this aspect, by providing more professional development opportunities. In general, more than 75% of the surveyed teachers are appreciative of the benefits of educational technology for their educational practices. Similarly, more than 75% of the surveyed teachers think that educational technologies can benefit students' overall performance and learning experience. As far as the parents are concerned, 66.48% of the parents surveyed declared that they were aware of the FG initiative and its services, although data show that they are not really engaged within the FG LMS yet.

Related to user support and training, a significant number of support tickets were received during year 1, which were successfully addressed. As the year progressed, tickets increased, which reflects the significant increase in the use of the services offered by Future Gate, by the educational community. 69.7% of the total trainees for year 1, were teachers which is an indicator of the importance of targeting and qualifying the teachers for the success of FG. Overall, 82.6% of total trainees for year 1, believe that the training sessions that were organised were very good (rated with 75% and above).

Overall, Future Gate is progressing towards its goals, aiming to transform K12 education in Saudi Arabia. Teachers and students are adopting the FG services, embracing the change brought about by the initiative. Whilst bigger part of the teaching and learning experience is being moved online, Future Gate aims at examining in greater depth, the actions and interactions of the users.

One limitation of the work presented here, is that it reports on high-level KPIs related to the use of the FG services, which are not analysed in greater depth. Despite that, the authors strongly believe that even these initial findings can be of use for similar initiatives that are trying to inform decisions related to human factors, through day-to-day evaluation and assessment. These findings will allow Future Gate to align the services offered with the human factor of the educational ecosystem for the continuation of the initiative but they will also serve as a starting point as the initiative explores human factors in greater depth.

Building upon this work, Future Gate continues to monitor and assess the use of the services offered to the educational community. Through this monitoring and evaluation, Future Gate aims to examine user interactions with the services, in a greater depth, both in terms of cognitive and organisational ergonomics. To this end, new processes are being set up, allowing for a closer examination of teaching and learning processes, while making sure that these processes do not interfere with the essence of day to day activities of students and teachers.

References

1. International Ergonomics Association: What is ergonomics. https://wwwiea.cc/whats/index.html. Accessed 10 Jan 2019
2. National Research Council. Health Care Comes Home: The Human Factors. National Academies Press (2011)
3. Smith, T.J.: Educational ergonomics: educational design and educational performance. Int. Soc. Occup. Ergon. Saf. (2001)
4. Soong, M.B., Chan, H.C., Chua, B.C., Loh, K.F.: Critical success factors for on-line course resources. Comput. Educ. **36**(2), 101–120 (2001)
5. Campbell, J.O., Bourne, J.R., Mosterman, P.J., Brodersen, A.J.: The effectiveness of learning simulations for electronic laboratories. J. Eng. Educ. **91**(1), 81–87 (2002)
6. Arvanitis, T.N., Petrou, A., Knight, J.F., Savas, S., Sotiriou, S., Gargalakos, M., Gialouri, E.: Human factors and qualitative pedagogical evaluation of a mobile augmented reality system for science education used by learners with physical disabilities. Pers. Ubiquit. Comput. **13** (3), 243–250 (2009)
7. Bhuasiri, W., Xaymoungkhoun, O., Zo, H., Rho, J.J., Ciganek, A.P.: Critical success factors for e-learning in developing countries: a comparative analysis between ICT experts and faculty. Comput. Educ. **58**(2), 843–855 (2012)
8. Smitha, T.J.: Differing systems perspectives on educational quality. In: Proceedings 19th Triennial Congress of the IEA, vol. 9 (2015)
9. Al Ohali, Y., Al Suhaibani, A., Palavitsinis, N., Koutoumanos, A.: Digital transformation of education in the Kingdom of Saudi Arabia: deploying a country-wide Learning Management System for K-12 education. In: 17th European Conference on e-Learning, Athens, Greece, 1–2 November 2018

Digital Skills and New Media and Information Literacy in the Conditions of Digitization

Valentina Milenkova[1(✉)], Dilyana Keranova[2], and Dobrinka Peicheva[1]

[1] Department of Sociology, South-West University "Neofit Rilski", Blagoevgrad, Bulgaria
{vmilenkova,peichevad}@swu.bg
[2] Department of Sociology, South-West University, Blagoevgrad, Bulgaria
dkeranova@swu.bg

Abstract. Objectives of the article are: (1) to explore digital skills and media and information literacy, representing different aspects of the digital person who is responsible for how technology is used. (2) to present that these skills contribute to the development of identity, thinking and decision-making skills, and values. The article is based on several surveys: (1) "Cultural Universals in Academic Environment", (2) "Fundamental transformations in media and communication systems and its reflections in society", DN 05-11 2016/2019, team leader D. Peicheva; (3) "Digital Media Literacy in the context of "Knowledge Society": state and challenges" КП-06-Н25/4, team leader V. Milenkova. Students, who are respondents in surveys indicated that they use the Internet anywhere and feel confident in creating different digital content and on-line products. They are aware of the new dangers of emerging hybrid media wars and see the role of the digital media education and literacy in this direction.

Keywords: Digital skills · University education · New digital media and information literacy · Media ecosystem

1 Introduction

The internet and new technologies create prerequisites for being part of the global network, linked to the upgrading and dynamic flow of information, fast communication and easy access to various institutions. This related both to opportunities and to challenges, deepening inequalities and emerging problems. People of different ages, professions and settlements have different levels of technological literacy, and there are differences between them in terms of access, use and skills they have to work with digital devices and the Internet. Digital skills are an indicator for the development of society, the level of competence of the population, the readiness to accept new challenges in the context of social transformations.

Digital skills form an essential part of key competencies. There is a necessary minimum of digital skills that everyone must possess in order to be able to participate fully in society and in the work process. As the internet is part of our daily life, most people have access to the internet to take advantage of on-line services, training, networking and information opportunities.

W. Karwowski et al. (Eds.): AHFE 2019, AISC 963, pp. 65–72, 2020.
https://doi.org/10.1007/978-3-030-20135-7_6

The widespread use of digital technologies provides people with access to a large amount of information, which creates a need for a set of access, management, integration and evaluation skills.

In recent years, digital competence and skills have been of great interest in discussing what skills people need to have in the knowledge society [1].

The discussion on digital skills is a lesser or greater degree related policies at European and national level relating to expectations qualification of the personnel of the labor market and has its roots in economic competition in which new technology is viewed as an opportunity and approaches for solution [2, 3].

2 Theoretical Consideration

Digital competence is a concept describing the skills associated with knowledge and mastery of technology and application of these skills into professional and social life. In recent years, the use of other terms that have a similar meaning, such as computer skills, information and technology skills, media and information competence [4, 5].

In some cases, terms are narrower, e.g. "internet skills" or "computer skills" [6].

In other cases, the terms are more general, for example '21st Century literacy', where emphasis is placed on the social skills that develop through the knowledge of digital technologies [7].

The wide variety of concepts reflects the rapid development of technology as well as the various fields of interest related to their deployment in all areas of professional, economic and cultural life [8].

Digital technologies also provide impetus for the development of teaching and library research [9]. Approaches to digital skills and competences are dynamic due to emerging new technologies and their use in society [8, 6].

The relationship between competence and skills is defined in the OECD publication: "a competency is more than just knowledge and skills. It involves the ability to meet complex demands, by drawing on and mobilizing psychosocial resources (including skills and attitudes) in a particular context." [10]. In addition, digital competence involves not only digital skills, but also social and emotional skills related to the use and understanding of digital devices.

In this context, the European Commission Punie & Cabrera, defines digital competence as including the confident and critical use of information society technologies for work, leisure and communication. It based on the core ICT skills, that is, the use of computers for the extraction, evaluation, storage, production, presentation and exchange of information, as well as communication and participation in Internet collaboration networks [11].

Digital competences include:

- Competence assessment information and knowledge covering the skills needed to find digital content, make a critical assessment of it and use it for various purposes - in work or entertainment.
- Active Digital Competence - encompasses the skills for producing, validating, editing, enriching and updating digital content [12]

- Fair and legal digital citizenship, this is the legitimate way to use copyrighted content [13]
- Competence to use the right tools for human purposes: these are the technical skills to use different tools, e.g. mobile platforms and devices - to understand their potential and limitations. In addition to "computer skills", "ICT competences" "knowledge" and "digital media literacy" [14, 15]
- Digital competency is described as a confident, critical and creative use of ICT to achieve goals related to work, employability, training, recreation, inclusion and /or participation in society [12]
- Digital competences are seen as a crosscutting key competence enabling people to acquire other key competences such as learning and critical skills, cultural awareness etc. [9, 16]

Digital competence includes the following areas: [12]

- Instrumental use of knowledge
- Additional skills and knowledge for communication and collaboration, information management, training and problem solving and meaningful participation
- Attitudes towards the use of strategic skills in intercultural, critical, creative, responsible and automated ways.
- Culture competence, which covers the understanding of digital culture and to be able to work in a digital environment

Digital competence is the ability to be in line with rapid ICT changes. Digital competence is related to the ability to recognize the necessary information and includes the skills, information to be found, evaluated and used effectively [17]. In summary, the concept of digital competence is an emerging and reinforcing concept related to technology development as well as to the political goals and expectations of citizenship in the knowledge society. It includes a variety of skills in several areas: media and communications, technology and computing, literacy and information science [14].

Digital competence consists of: (1) technical skills for using digital technologies; (2) the ability to use new technologies in a targeted way for work, learning and learning, and for everyday life as a whole in different directions; (3) ability to critically evaluate digital technologies; (4) motivation to participate in digital culture [17].

3 Methodology

The measurement of digital skills shows the connection between people's skills and their different kinds of engagement in online services and activities [18].

The measuring of digital skills in the research included both basic skills needed to use the internet and the skills that are required for understanding and using online content [18].

The research project "*Cultural Universals in Academic Environment*", conducted in 2017 with team leader V. Milenkova, used the self-assessment method to measure digital skills, applied to 120 students from social and humanities specialties in Southwest University (Bulgaria).

The other project aimed at studying digital literacy was the national survey "*Digital Media Literacy in the context of "Knowledge Society": state and challenges"*, № КП-06- H25/4, *2018,* funded by National Science Fund – Bulgaria, with team leader Valentina Milenkova. The sample, which included 250 individuals representing the population, aged 18–65 from settlements across the country. The purpose of the study was to investigate the digital literacy of the population - what are the specific challenges people encounter when working with digital devices, what is the degree of real use of digital devices in work and leisure time.

A questionnaires (in both surveys) were used to collect data on digital skills; they included questions regarding the frequency of use of Internet and the social media as well as questions for self-assessment of knowledge and skills for working in a digital environment.

The questionnaires comprising variety of questions related to digital literacy and skills. The self-assessment method was used to implement a fuller instrumental approach to measuring skills, how students engage in various kinds of online and offline activities, and how these skills can subsequently impact on specific activities. The most frequently used indicators in measuring digital skills include skills for finding information, communicating, creating content, and ensuring safety.

The "Information" variable includes indicators related to surfing, searching and filtering information, storing and retrieving information. The variable refers to the share of students who have used the Internet in the last few months to find information on various topics; read or download online news, receive information from the Internet pages of public and private organizations.

The "Communication" variable involves indicators referring to various degrees of competence, interaction through technology, sharing information and content. It registers the share of people who have used the Internet to receive and send email messages; hold conversations through the Internet; participate in social networks; upload content, which is then shared.

The "Creating content" variable includes indicators pointing to activities such as: creating websites or blogs; preparing presentations, which included images, sound, video, graphs; copying and pasting information from one document to another. The variable indicates the share of people who have engaged in these activities during recent months.

The "Security" variable includes indicators related to familiarity with and application of security equipment and personal data protection, as well as personal security. In this connection, the formulated indicators refer to competencies in solving technical problems and identifying needs and technological solutions for installing new equipment; installing a new operation system or replacing an old one; Internet banking, holding a meeting through a website [18].

4 Results

We present the summarized results of the two studies.

The young adults (18–35 aged) in two surveys shared that they were using the internet everywhere: at home, "in the university", "on work", "in their leisure time".

This is an age group of people aged up to 40 years, which range places them under the category of "digital natives", meaning they have been actively socialized in various digital technologies since childhood. All the respondents registered in at least one social network, such as Facebook, Instagram, Snapchat, or Twitter.

The young adults shared they were almost constantly online on the Internet, more than 30 hours per week, and spent between 25 and 30 hours per week in the social networks.

They use various modifications to access the internet and to work online. These include a computer, laptop, tablet, or smartphone of their own; they also have access to computers in their offices or classrooms (when they study) as well as in the libraries. This is why they can be online regularly and for long periods.

Young adults *all indicated possessing operative skills to use the internet*, including the mobile Internet. These skills include: easily using the Internet, connecting to a Wi-Fi net; downloading applications for a mobile device; keeping track of, and updating, mobile applications; surfing on the Internet, downloading and uploading files; regulating the privacy settings for dealing with computer viruses and problems in the Internet; working with search engines, like Google and Bing.

The surveyed young people indicate they possessed systematic internet skills for quick orientation in a given website, moving from one webpage to another, or easily finding a website they had visited before.

The respondents assessed themselves as having information-related internet skills, such as: easily finding the information they need, easily examining search results and choosing which to check. They also read news and journals online, and were able to use different strategies for finding Information on the Internet.

The young adults were confident about their ability to judge whether a website was trustworthy or not; they could compare websites to decide which ones were truthful; they could evaluate the information found online.

Regarding "*communication internet skills*", the respondents declared they were confidently able to engage in online communication. They freely shared comments in blogs and social networks and they were convinced sharing online was risk-free. They were more confident about, and better able to work with Facebook and Messenger than with Twitter and YouTube – perhaps because Twitter is less popular in Bulgaria among online users, while YouTube mainly used to watch films and listen to music.

Creating content: regarding this skill, the surveyed students declared they could create and upload online content and could make changes in already created content. They stated they had little confidence they could personally create websites and preferred to turn to specialists for this. On the other hand, they knew how to share online video content, write comments on various contents, or create new products out of existing images, music, and video.

Can be said, that the most important differentiating criterion for digital literacy was the independent variable "age". From the point of view of age, it can be said that young people in Bulgarian conditions confirm the thesis of belonging to "digital natives", i.e. those born after 1980 have a predisposition to digitization, they have a higher digital literacy due to the greater density of their digital media environment. After 40 years, and especially after 50+, we have witnessed a progressive increase in the difficulties and challenges of using digital devices. Older generations most easily master computer skills, especially if they are part of their work environment. The less people use computers in their professional activities, the lower their digital literacy. On the other hand, knowing the possibilities of the internet is becoming even more challenging for older generations. In this sense, we need to highlight the selectivity of the Internet and the betting of another discriminatory indicator of social inequality: Internet access.

5 Discussion and Conclusions

The students at Southwest University and generally young people (18–38) spend much time online in the internet and the social networks. They pointed out they used internet everywhere. They made a realistic assessment of their skills in creating content and their informational and formal skills. They know how to create and upload online content and make changes in already created content. They feel confident they can create different kinds of content and online products, including confidence in their skills regarding privacy and knowledge about security equipment and personal data protection; however, they find they have yet more to learn in this field. Overall, their self-assessment is high, which indicates their good knowledge of the digital environment.

The participation of population in the modern digital environment includes access to computers, electronic resources, and other information products and services. It should be taken into account that the acquiring of digital skills for work in an interactive environment improves their efficiency with respect to creatively and innovatively pursuing education activity. Achieving greater effectiveness in training and acquiring knowledge involves the successful use of information and communication technologies, based on acquired skills for seeking and finding useful information and resources, as well as the capacity to analyze and combine the obtained information, to share and discuss different ideas and viewpoints, which they may comment on with their fellow students and teachers.

In this connection, based on the results of the two survey presented, it may be said enabling adaptation to the challenges of digital society requires finding more effect pathways to engaging, encouraging, and motivating people to assimilate good theoretical and practical knowledge and skills for working with information and communication technologies. Hence, the efforts of people should generally be guided towards optimizing various ways of using ICT and interactive communication in their work; this may improve their capacity for critical thinking, effective communication and joint problem solving.

The results of the two surveys conducted indicate which specific skills should be the focus of greater effort in order to make the work of the teachers and the role of education more effective.

The results of the conducted content analysis of traditional and online media in frame of the research project "Fundamental changes in the media and communication ecosystem and their reflections in society" to the NSF support the placement of these skills as well at the center of the efforts of teachers to cope with the problems in the media ecosystem [19].

Engaging young people in short-term or long-term courses could also contribute to their acquirement of important knowledge and qualities related to digital skills. One must not underestimate the conducting of seminars and lecture courses, the possibility of access to online tutorials, electronic textbooks and other means of enhancing digital competence; through these means, it is possible to develop extensive skills for seeking, identifying and critically assessing and using information, and for young people's more independent and creative behavior in a digital environment. For the formation of digital culture, it is of great importance to saturate the work environment with computers, and to include the Internet in the specific responsibilities of those performing different activities. The results showed that the working environment as well as age are identified as factors of digital literacy of people. In this sense, complicating the responsibilities, the work and the nature of the obligations becomes a prerequisite for the formation of digital literacy of the population.

Acknowledgements. The article has been developed in the framework of the research project " Digital Media Literacy in the context of "Knowledge Society": state and challenges" № КП–06–Н25/4, funded by National Science Fund – Bulgaria and research project DN-05/11-14.12.2016 "Fundamental changes in the media and communication ecosystem and their reflections in society" to the NSF.

References

1. Sefton-Green, J., Nixon, H., Erstad, O.: Reviewing approaches and perspectives on "digital literacy". Pedagogies **4**(2), 107–125 (2009)
2. Punie, Y.: Learning spaces: an ICT-enabled model of future learning in the knowledge-based society. Eur. J. Educ. **42**, 185–199 (2007)
3. OECD: Are the New Millennium Learners Making the Grade? Technology use and educational performance in PISA. Centre for Educational Research and Innovation (2010)
4. Adeyemon, E.: Integrating digital literacies into outreach services for underserved youth populations. Ref. Libr. **50**(1), 85–98 (2009)
5. Krumsvik, R.: Situated learning and teachers' digital competence. Educ. Inf. Technol. **13**(4), 279–290 (2008)
6. Jones-Kavalier, B., Flannigan, S.L.: Connecting the digital dots: literacy of the 21st century. Teach. Libr. **35**(3), 13–16 (2008)
7. Jenkins, H., Clinton, K., Purushotma, P., Robinson, A.J., Weigel, M.: Confronting the Challenges of Participatory Culture: Media Education for the 21st Century, the John D and Catherine T MacArthur Foundation (2006). http://www.digitallearning.macfound.org/atf/cf/%7B7E45C7E0-A3E0-4B89-AC9C-E807E1B0AE4E%7D/JENKINS_WHITE_PAPER.PDF
8. Aviram, R., Eshet-Alkalai, Y.: Towards a theory of digital literacy: three scenarios for the next steps. Eur. J. Open Distance E-Learn (2006). http://www.eurodl.org/materials/contrib/2006/Aharon_Aviram.htm
9. Ala-Mutka, K., Punie, Y., Redecker, C.: Digital Competence for Lifelong Learning. Luxemburg, Office for Official Publications of the European Communities (2008). http://ftp.jrc.es/EURdoc/JRC48708.TN.pdf

10. OECD: The OECD Program Definition and Selection of Competencies. The definition and selection of key competencies. Executive summary. 30 June 2005. http://www.oecd.org/dataoecd/47/61/35070367.pdf
11. Punie, Y., Cabrera, M. (eds.): The future of ICT and learning in the knowledge society. European Commission, Luxembourg (2006)
12. Ilomäki, L., Kantosalo, A., Lakkala, M.: What is digital competence? Linked portal. European Schoolnet, Brussels (2011). http://linked.eun.org/web/guest/in-depth3
13. Eshet-Alkalai, Y., Chajut, E.: Changes over time in digital literacy. Cyber Psychol. Behav. **12**(6), 713–1715 (2009)
14. Peicheva, D., Milenkova, V.: Knowledge society and digital media literacy: foundations for social inclusion and realization in Bulgarian context. In: Quality of Life, vol. 1, pp. 50–74 (2017)
15. Bawden, D.: Origin and concepts of digital literacy. In: Lankshear, C., Knobel, M. (eds.) Digital Literacies: Concepts, Policies, and Practices. Peter Lang Publishing, Inc., New York (2008)
16. Milenkova, V., Peicheva, D., Marinov, M.: Towards defining media socialization as a basis for digital society. Int. J. Cogn. Res. Sci. Eng. Educ. **6**(2), 21–32 (2018)
17. Correia, A.M.R., Teixeira, J.C.: Information literacy: an integrated concept for a safer internet. Online Inf. Rev. **27**(5), 311–320 (2003)
18. Measuring Digital Skills across the EU: EU wide indicators of Digital Competence (2014). https://ec.europa.eu/digital-single-market/en/news/measuring-digital-skills-across-eu-eu-wide-indicators-digital-competence
19. Peicheva D., Milenkova, V., Manov, B., Raycheva, L.: The ICT, E-reading and media ecosystem. In: Callaos, N., Welsch, F. (eds.) The 11th International Multi Conference on Society, Cybernetics and Informatics, Proceedings III S, pp. 83–87 (2017)

Behavioral and Cognitive Science

The Modality Effect of Cognitive Load Theory

Juan C. Castro-Alonso[1(✉)] and John Sweller[2]

[1] Center for Advanced Research in Education,
Universidad de Chile, Santiago, Chile
jccastro@ciae.uchile.cl
[2] School of Education, University of New South Wales, Sydney, Australia
j.sweller@unsw.edu.au

Abstract. The modality effect, which has been investigated by cognitive load theory, predicts that learning from visualizations supplemented with written text should be less effective than learning from the same visualizations supplemented with comparable spoken text. An explanation of the effect assumes a degree of separation between the processing of visuospatial and auditory information. Due to this separability, learning only from visuospatial information (visualizations and visual text) is more likely to overload visuospatial processing, as compared to learning from visuospatial and auditory information (visualizations and auditory text), in which both the visuospatial and the auditory processors share the load of the learning material. The aims of this review chapter are to: (a) describe the modality effect, (b) provide supporting evidence using computer multimedia about STEM topics, and (c) describe studies indicating the separability of visuospatial and auditory processing. We finish by suggesting future directions for research on the modality effect.

Keywords: Modality effect or modality principle · Cognitive load theory · STEM education · Multicomponent working memory · Multimedia learning

1 Introduction

For over 30 years, cognitive load theory [1] and the related cognitive theory of multimedia learning [2] have published several educational principles based on randomized experiments with control of variables. Under these strict methodological conditions, many instructional principles or effects have been investigated. We focus on one of these principles, the *modality effect* or modality principle [e.g., 3], which guides the design of multimedia instructional resources that combine visualizations (e.g. animations, videos, photos, diagrams) and texts (e.g., on-screen, printed, narrated). The modality effect occurs when multimedia that depict visualizations associated with written text is less effective for learning than multimedia that depict the same visualizations associated with comparable narrated text [4].

Cognitive load theory explains the modality effect by building on Baddeley and Hitch's multicomponent model of working memory [see 5]. In this model, working memory includes two systems with limited capacity: (a) the *phonological loop*, managing the processing of auditory information, and (b) the *visuospatial sketch pad*, dealing with visual and spatial information. The multicomponent model indicates that

W. Karwowski et al. (Eds.): AHFE 2019, AISC 963, pp. 75–84, 2020.
https://doi.org/10.1007/978-3-030-20135-7_7

auditory and visuospatial information to a substantial degree tend to be processed separately in these limited systems [cf. 6]. Due to this separability, learning only from visuospatial information (images supplemented with written texts) is more likely to overload the limited visuospatial sketch pad, as compared to learning from visuospatial and auditory information (images supplemented with spoken text), in which both the visuospatial sketch pad and the phonological loop share the cognitive load of the learning material.

Hence, the modality effect is produced when an overloaded system does not have enough capacity to deal with learning, as compared to two less overloaded systems. For example, Fig. 1 shows two multimedia formats to teach the shape of bacteria. The format on the left (Fig. 1a) supplements the images with spoken text, whereas the design on the right (Fig. 1b) supplements the images with written text, so it is more likely to overload the visuospatial processor. (Note that cognitive load theory usually investigates more complex educational materials, likely to overload the visuospatial or auditory systems [see 7], rather than the simple example we are providing).

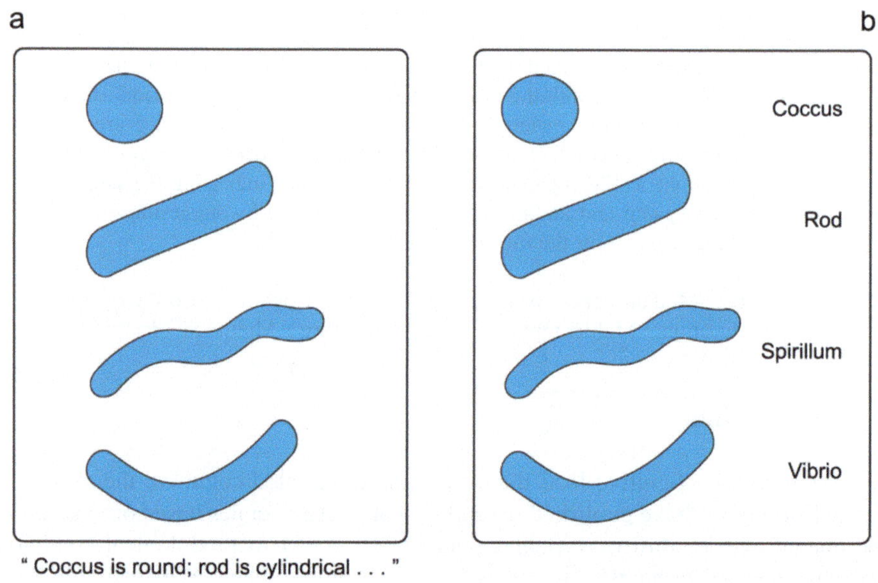

Fig. 1. Example of formats that **a** use both the visuospatial sketch pad and the phonological loop (with the shapes spoken rather than written), and **b** use solely the visuospatial sketch pad (at the risk of overloading it).

From the diverse materials and topics where the modality effect has been investigated, in the next section we provide examples of supporting results on computer multimedia dealing with STEM (science, technology, engineering, and mathematics) topics.

2 STEM Multimedia Evidence for the Modality Effect

In a meta-analysis of the modality effect [8], in which 43 effect sizes and more than 1,900 participants were investigated, there was an overall effect size of $d = 0.72$. According to benchmarks for behavioral sciences [9], this size corresponds to a medium to large effect. Notably, when the meta-analysis [8] compared different instructional disciplines, a larger effect size of $d = 1.20$ was reported for the science domains.

Additional evidence for the modality effect using STEM multimedia has included a variety of participants and learning content. For example, Experiment 1 reported in [10] compared the instructional effectiveness of visual vs. auditory explanations added to meteorology multimedia. In the study, where university students had to learn lightning formation, it was observed that those given auditory texts outperformed those shown the texts on-screen. Also, in [11], Experiment 1 assessed first-year apprentices learning to read a fusion diagram for soldering. Results showed that a group receiving the animated visuals supplemented with auditory explanations showed higher test scores and lower self-ratings of cognitive load than a group given animated visuals supplemented with written text explanations. In an experiment where university participants studied lightning formation from different animations [12, Experiment 1], results showed that animations with narrations obtained significantly higher retention and transfer scores than animations with on-screen texts.

Concerning biological sciences, in [13] the authors investigated university students learning about fish movements. For both static and animated displays, it was observed that narrated multimedia outperformed written multimedia in tests of retention and transfer. Also, Experiment 2 in [14] reported university participants who studied through computer static pictures and texts describing the structure and function of an enzyme. Randomly, half of the sample studied from the pictures supplemented with narrations and the other half received supplementary written texts. Results showed higher recall scores for the group studying with images and narrations. A similar direction of effects for the modality effect, although non-significant, was observed for comprehension and transfer tests. Lastly, [15] investigated the capacity of university students to learn a health science first-aid procedure using two animation formats. Half of the participants were randomly assigned to animations and written texts (subtitles below the depiction), and the other half watched the animations with narrated texts. Results of the behavioral performance test showed that the narrated versions outperformed the written text formats.

Thus, there are diverse studies using STEM multimedia learning supporting the modality effect. As described above, the effect assumes that working memory can be at least partially separated into different processors. Evidence for this separability can help predict the effectiveness of the modality effect under different learning conditions. Next, we describe two areas that have supported the separation of processing between visuospatial and auditory information.

3 Working Memory Separability that Allows the Modality Effect

As reviewed in [16], there are at least two research areas that show evidence for the separability between the visuospatial and auditory subcomponents of working memory, which will be termed here: (a) *selective interference*, and (b) *modality organization*. Concerning selective interference, this research shows selective impairment in processing visuospatial information when receiving additional visuospatial information (but not when receiving extra auditory information), and selective impairment in processing auditory information when receiving additional auditory information (but not when receiving extra visuospatial information).

In a classic example of selective interference [17], male university students spoke out letters while memorizing other letters. The information to be memorized was presented either in an auditory or visual modality. It was observed that visually shown letters were remembered for a longer period, because, in contrast to auditorily shown letters, they were not interfered by the auditory speaking aloud process. In an example from multimedia research [18], Experiment 1 tested female education undergraduates learning about the cardiovascular system from a multimedia module. While learning the contents, students were required to respond as rapidly as possible to a change in color of an on-screen element. Showing selective interference, students were slower in responding to the visual change when the multimedia included on-screen text, as compared to narrated multimedia. In other words, students were more impaired in the color visuospatial tasks by the visual text than by the auditory text of the multimedia. Similar multimedia are shown in Fig. 2, which depicts information based on the design in [18]. In these replicas, the on-screen text is expected to produce a slower student's response when the star at the top-right changes color from purple to green (Fig. 2a), compared to the response when there is a narration (Fig. 2b).

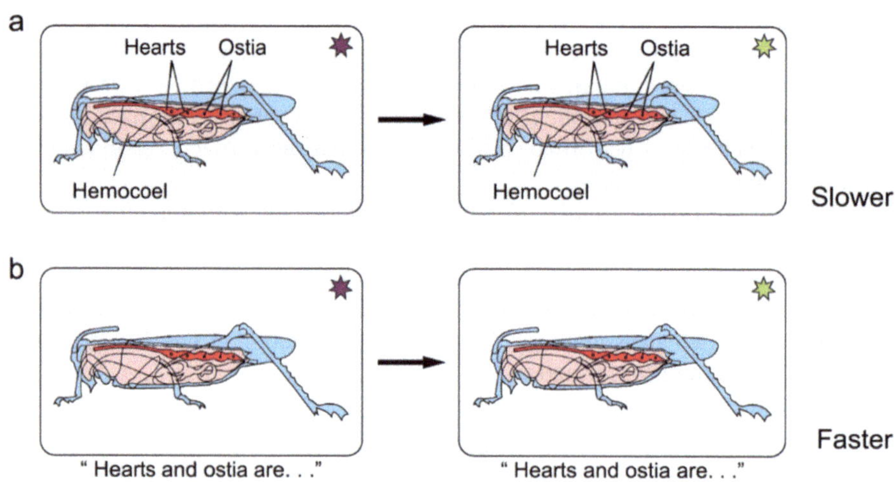

Fig. 2. With on-screen text **a** the response to the star changing color should be slower than with auditory text **b**.

As another example, in two dual-task experiments investigating visuospatial processing [19], undergraduates were asked to generate and rotate mental images while executing other simultaneous tasks. Results showed that the simultaneous task of speaking a word (auditory task) did not interfere with the visuospatial tasks. In contrast, the simultaneous task of localizing the source of a sound (spatial task) was deleterious to processing the mental images.

The investigations by Robinson and colleagues are also part of the selective interference research. The authors observed more interference between visuospatial and verbal processing when the verbal information was presented in visuospatial configurations [cf. 6]. For example, in [20] they reported four experiments with university students learning zoology categories from different verbal displays, including less visuospatial (written paragraphs) and more visuospatial configurations (graphic organizers and concept maps). To produce processing interference, the students were also given verbal and visual working memory tasks. As predicted, test scores on zoology information memorized from graphic organizers or concept maps (high in visuospatial information) were lower when attempting a visual working memory task, as compared to a verbal working memory task. In contrast, these visuospatial interferences were not observed when the information was memorized from paragraphs, which relied less on visuospatial organization than graphic organizers and concept maps. In a follow-up with stricter controls [21], the findings were replicated.

To exemplify these experiments, Fig. 3 provides different visuospatial configurations for texts about two penguin species. The facts comparing the species are given either as paragraphs (Fig. 3a), as a graphic organizer (Fig. 3b), or as a concept map (Fig. 3c). It can be predicted that paragraphs will be less affected by visuospatial interference than the graphic organizer or the concept map. In short, these investigations of selective interference support the hypothesis that visuospatial information (including texts presented in visuospatial configurations, such as concept maps), tends to be processed separately from verbal information (including texts presented as written paragraphs).

a

The King Penguin has a height of 90 cm, a lifespan of 30 years, and a weight of 14 kg.

The Magellanic Penguin has a height of 65 cm, a lifespan of 18 years, and a weight of 5 kg.

b

Species	King	Magellanic
Height	90 cm	65 cm
Lifespan	30 years	18 years
Weight	14 kg	5 kg

c

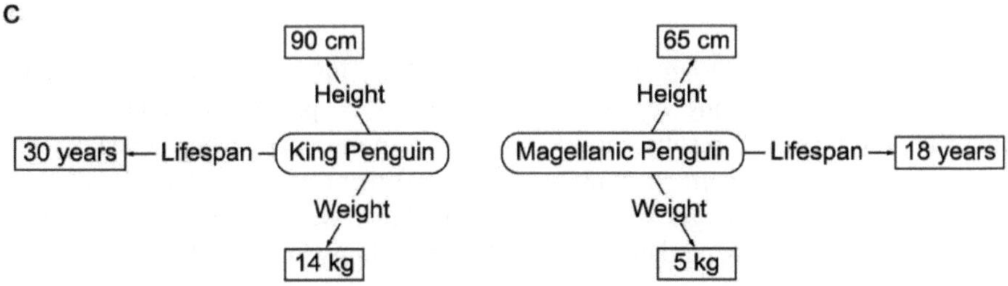

Fig. 3. Comparable texts configured as **a** paragraphs, **b** graphic organizer, and **c** concept map.

In addition to selective interference investigations, another area that supports the separability of processing visuospatial and auditory information is modality organization. This research shows that, when retrieving elements from memory, the modality of the memorized items is pervasive and is even stronger than semantic or other types of grouping categories. As a classic example, in [22] the author employed an original strategy to present sets of four simultaneous words to university students. Two words were shown visually, and the two remaining were presented auditorily (one word per ear). Students had to report the four words from memory. The results showed that participants reported the words in blocks of the same modality (i.e., auditory & auditory, visual & visual), even though the four stimuli were simultaneous. In a follow-up study, it was observed that this modality order was stronger than an associative or semantic order between pairs of words of different modalities. For example, when a pair such as *girl* (auditory) & *boy* (visual) was presented as stimuli, the association girl–boy was weaker than the modality of each word. The words were always reported in modality-determined blocks, even if this order broke the associations between word pairs. Figure 4 shows examples of stimuli and responses when the modality of the stimuli was congruent with the semantic association of the word pairs (Fig. 4a, girl and boy in auditory modality) or when it was non-congruent (Fig. 4b, boy and day in visual modality).

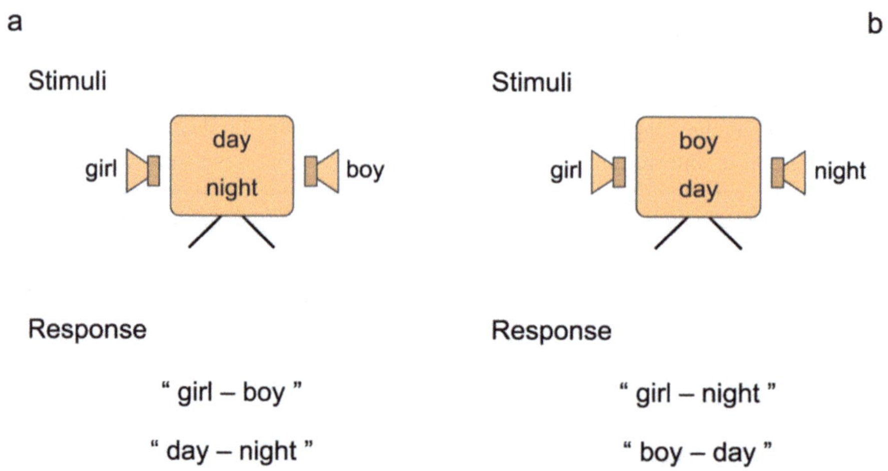

Fig. 4. The modality organization shown in stimuli and response time, when the modality of the stimuli is either congruent **a**, or non-congruent **b** with the semantic association.

Another study related to modality organization [23], investigated the response times of undergraduates receiving repeated and novel words, presented either visually or auditorily. As expected, results showed a faster response time when repeated words were presented (same word in studying and test times), compared to novel words. Notably, the effect was larger when the same modality was used in studying and test times, compared to cross modality results. These findings indicate that processing the modality of a word occurs earlier than processing its meaning. Analogous findings with numerical stimuli were reported in [24], regarding an experiment with university students. The students received a series of interspersed auditory and visual one-digit numbers. It was observed that participants who reported the digits grouped according to modality outperformed those grouping the digits according to their presentation order and mixing the modalities. In other words, it was more efficient to memorize the two modalities in parallel rather than intermixing them.

In conclusion, modality organization research supports that items presented visuospatially or auditorily tend to be remembered attached to the modality in which they were presented. Thus, both modality organization and selective interference investigations support the suggestion that visuospatial processing in working memory is at least to some extent independent from auditory processing. As described next, future directions for research in these areas of investigation may inform cognitive load theory and the modality effect.

4 Future Directions for Research

A possible direction for further study of the modality effect concerns different designs of visualizations and texts. For example, under visualization conditions that demand more working memory, such as transient animations [25], the modality effect should be larger than under less demanding conditions. The design of the auditory text can also be important, as voice narrations can be more effective than machine narrations [26].

In addition, selective interference and modality organization research may inform more fine-grained analyses to investigate the modality effect. For example, selective interference can include the *time* factor, as giving a space or lapse of time can be effective to allow working memory to replenish resources and avoid interference [cf. 27].

Also, there are current investigations relating the modality effect to other effects of cognitive load theory or the cognitive theory of multimedia learning [see 4]. For example, links have been established with the *redundancy effect* [see 28], the *expertise reversal effect* [see 29], the *transient information effect* [see 30], and the *signaling principle* [see 31].

For example, if the narration merely reiterates information in the visualization, this will likely produce a negative redundancy effect [28] instead of a positive modality effect. Also, related to the redundancy [28] and expertise reversal effects [7, 29], if an on-screen text contains information already known by the students, then it is more appropriate to discard it rather than to present it as narration. Similarly, boundary conditions for the modality effect entail the integration of findings from the modality effect (supporting narrated texts) and the transient information effect (discouraging transient narrated texts) [e.g., 32].

Regarding the signaling principle [31], sometimes a short on-screen text that signals important visual information can be an effective learning asset. For example, in [15], short labels placed at relevant areas of the screen had positive effects on learning. Future research should investigate the most appropriate length, format, and placement to produce these positive signaling effects of texts, while avoiding a cognitive overload of the visuospatial processor.

Lastly, there are known factors that influence multimedia learning, which may also affect the modality effect, such as gender [33] and visuospatial processing or spatial ability [34].

5 Conclusion

The modality effect has been investigated by cognitive load theory and the cognitive theory of multimedia learning. The effect has shown that learning from visualizations supplemented with written text is less effective than learning from the same visualizations supplemented with comparable narrations. These findings assume that there is a degree of separation in the working memory processing of visuospatial and auditory information, and that written text will tend to overload the visuospatial processor more than auditory text, because written text will be processed simultaneously with the visual learning elements. As auditory text removes the load from the visuospatial processor and onto the auditory processor, it leaves more capacity for the visuospatial processor to deal with the visual learning elements. The modality effect has been supported by diverse educational multimedia depicting STEM concepts. Similarly, the foundation of the effect, that is the separability between visuospatial and auditory processing, has also been systematically investigated. We described two research areas showing this separability, namely, selective interference and modality organization. Future directions for research, including boundary conditions, will provide additional support for the modality effect and its relationship with other findings of cognitive load theory and the derived cognitive theory of multimedia learning.

Acknowledgments. Support from PIA–CONICYT Basal Funds for Centers of Excellence Project FB0003, and CONICYT Fondecyt 11180255, is gratefully acknowledged. We thank Ignacio Jarabran for helping with the illustrations.

References

1. Sweller, J., Ayres, P., Kalyuga, S.: Cognitive Load Theory. Springer, New York (2011)
2. Mayer, R.E. (ed.): The Cambridge Handbook of Multimedia Learning, 2nd edn. Cambridge University Press, New York (2014)
3. Mousavi, S.Y., Low, R., Sweller, J.: Reducing cognitive load by mixing auditory and visual presentation modes. J. Educ. Psychol. **87**(2), 319–334 (1995)
4. Low, R., Sweller, J.: The modality principle in multimedia learning. In: Mayer, R.E. (ed.) The Cambridge Handbook of Multimedia Learning, 2nd edn., pp. 227–246. Cambridge University Press, New York (2014)
5. Baddeley, A.: Working memory. Science **255**(5044), 556–559 (1992)
6. Clark, J., Paivio, A.: Dual coding theory and education. Educ. Psychol. Rev. **3**(3), 149–210 (1991)
7. Chen, O., Kalyuga, S., Sweller, J.: The expertise reversal effect is a variant of the more general element interactivity effect. Educ. Psychol. Rev. **29**(2), 393–405 (2017)
8. Ginns, P.: Meta-analysis of the modality effect. Learn. Instr. **15**(4), 313–331 (2005)
9. Cohen, J.: Statistical Power Analysis for the Behavioral Sciences. Erlbaum, Hillsdale (1988)

10. Schmidt-Weigand, F., Kohnert, A., Glowalla, U.: A closer look at split visual attention in system- and self-paced instruction in multimedia learning. Learn. Instr. **20**(2), 100–110 (2010)
11. Kalyuga, S., Chandler, P., Sweller, J.: Managing split-attention and redundancy in multimedia instruction. Appl. Cogn. Psychol. **13**(4), 351–371 (1999)
12. Moreno, R., Mayer, R.E.: Cognitive principles of multimedia learning: the role of modality and contiguity. J. Educ. Psychol. **91**(2), 358–368 (1999)
13. Kühl, T., Scheiter, K., Gerjets, P., Edelmann, J.: The influence of text modality on learning with static and dynamic visualizations. Comput. Hum. Behav. **27**(1), 29–35 (2011)
14. Seufert, T., Schütze, M., Brünken, R.: Memory characteristics and modality in multimedia learning: an aptitude-treatment-interaction study. Learn. Instr. **19**(1), 28–42 (2009)
15. de Koning, B.B., van Hooijdonk, C.M.J., Lagerwerf, L.: Verbal redundancy in a procedural animation: on-screen labels improve retention but not behavioral performance. Comput. Educ. **107**, 45–53 (2017)
16. Penney, C.G.: Modality effects and the structure of short-term verbal memory. Mem. Cogn. **17**(4), 398–422 (1989)
17. Kroll, N.E.A., Parks, T., Parkinson, S.R., Bieber, S.L., Johnson, A.L.: Short-term memory while shadowing: recall of visually and of aurally presented letters. J. Exp. Psychol. **85**(2), 220–224 (1970)
18. Brünken, R., Steinbacher, S., Plass, J.L., Leutner, D.: Assessment of cognitive load in multimedia learning using dual-task methodology. Exp. Psychol. **49**(2), 109–119 (2002)
19. Bruyer, R., Scailquin, J.-C.: The visuospatial sketchpad for mental images: testing the multicomponent model of working memory. Acta Psychol. **98**(1), 17–36 (1998)
20. Robinson, D.H., Katayama, A.D., Fan, A.-C.: Evidence for conjoint retention of information encoded from spatial adjunct displays. Contemp. Educ. Psychol. **21**(3), 221–239 (1996)
21. Robinson, D.H., Robinson, S.L., Katayama, A.D.: When words are represented in memory like pictures: evidence for spatial encoding of study materials. Contemp. Educ. Psychol. **24** (1), 38–54 (1999)
22. Murdock Jr., B.B.: Four-channel effects in short-term memory. Psychon. Sci. **24**(4), 197–198 (1971)
23. Kirsner, K., Smith, M.C.: Modality effects in word identification. Mem. Cogn. **2**(4), 637–640 (1974)
24. Penney, C.G.: Order of report in bisensory verbal short-term memory. Can. J. Psychol. **34** (2), 190–195 (1980)
25. Castro-Alonso, J.C., Ayres, P., Wong, M., Paas, F.: Learning symbols from permanent and transient visual presentations: don't overplay the hand. Comput. Educ. **116**, 1–13 (2018)
26. Mayer, R.E., DaPra, C.S.: An embodiment effect in computer-based learning with animated pedagogical agents. J. Exp. Psychol. Appl. **18**(3), 239–252 (2012)
27. Chen, O., Castro-Alonso, J.C., Paas, F., Sweller, J.: Extending cognitive load theory to incorporate working memory resource depletion: evidence from the spacing effect. Educ. Psychol. Rev. **30**(2), 483–501 (2018)
28. Kalyuga, S., Sweller, J.: The redundancy principle in multimedia learning. In: Mayer, R.E. (ed.) The Cambridge Handbook of Multimedia Learning, 2nd edn., pp. 247–262. Cambridge University Press, New York (2014)
29. Kalyuga, S., Ayres, P., Chandler, P., Sweller, J.: The expertise reversal effect. Educ. Psychol. **38**(1), 23–31 (2003)
30. Ayres, P., Paas, F.: Making instructional animations more effective: a cognitive load approach. Appl. Cogn. Psychol. **21**(6), 695–700 (2007)

31. van Gog, T.: The signaling (or cueing) principle in multimedia learning. In: Mayer, R.E. (ed.) The Cambridge Handbook of Multimedia Learning, 2nd edn., pp. 263–278. Cambridge University Press, New York (2014)
32. Leahy, W., Sweller, J.: Cognitive load theory, modality of presentation and the transient information effect. Appl. Cogn. Psychol. **25**(6), 943–951 (2011)
33. Castro-Alonso, J.C., Wong, M., Adesope, O.O., Ayres, P., Paas, F.: Gender imbalance in instructional dynamic versus static visualizations: a meta-analysis. Educ. Psychol. Rev. (2019). Advance Online Publication
34. Castro-Alonso, J.C., Uttal, D.H.: Spatial ability for university biology education. In: Nazir, S., Teperi, A.-M., Polak-Sopińska, A. (eds.) Advances in Human Factors in Training, Education, and Learning Sciences: Proceedings of the AHFE 2018 International Conference on Human Factors in Training, Education, and Learning Sciences, pp. 283–291. Springer (2019)

How Can Robotics Be Integrated into the Field of Care and Acceptance by the Population for the Use of Artificial Intelligence Be Strengthened?

Ann Kathrin Stinder(✉), Alexander-Maximilian Frische, and Daniel Schilberg

Bochum University of Applied Sciences,
Lennershofstraße 140, 44801 Bochum, Germany
{Ann.Stinder,Alexander-Maximilian.Frische,
Daniel.Schilberg}@hs-bochum.de

Abstract. The interviews with experts from the field of nursing science and robotics, which were conducted as part of a project study, were intended to provide an overview of the current status of the use of AI, especially with focus on acceptance research. It was found that there is a change towards a service society. During this change, acceptance would have to be created partly trough participation and education, in addition to the focus on transparency in action. The focus here is on the degree of autonomy of the robots and the overcoming of technical knowledge barriers.

Keywords: Artificial intelligence · Health care · Demographic change · Acceptance research

1 Introduction

In the social context, a development can be measured that is based on the knowledge and information of many - the turning to a world in which artificial intelligence (AI) is used to simplify and support life in various areas. On the one hand, this progress, especially on the technical-research side, is linked to high expectations and a positive approach, and, on the other hand, it is also linked to people's fear to interact and/or be replaced by machines. There are various sectors that already have points of contact with AI. One of these sectors is the area of care, which is the subject of this work. Here it is necessary, also due to demographic change, to find solutions which will be satisfactory for all parties involved, i.e. nursing staff, nursing cases, relatives, doctors, etc.

As part of a project study at Bochum University of Applied Sciences, we conducted expert interviews for this purpose, both with users from the nursing sciences and with researchers and developing professors from universities. The key question behind the project was "How can robotics be integrated into the field of care and acceptance by the population for the use of artificial intelligence be strengthened?". Questions were evaluated after Mayring [1] and compared with current literature.

W. Karwowski et al. (Eds.): AHFE 2019, AISC 963, pp. 85–94, 2020.
https://doi.org/10.1007/978-3-030-20135-7_8

2 Acceptance Among the Population and Application Areas of Artificial Intelligence

Looking at certain areas of AI research, such as autonomous driving, the rapid progress of self-thinking or acting systems can be recognized, and, simultaneously, the acceptance of the population, to be accompanied by such systems in life, increases. According to a study by McKinsey [2], the majority of those surveyed see the future viability of autonomous systems as the technology of the future, with only 25% also seeing a greater risk. The general orientation towards these systems can be justified from the available data mainly by the fact that increased comfort is expected and the risk of accidents is not considered significant [2].

The mobility sector is only one of the possible areas of influence for the use of AI, which is of those sectors that will have a great influence on the development of society as a whole when a change takes place. An equally large research area with an equal range of participants is also the care sector, in which artificial intelligence can be and has already been used in different ways. The following question is whether the simplification of life through self-learning and autonomous systems in the field of nursing would be accepted. In particular, the differences in the extent of using AI or machine support in general need to be considered.

Nursing Profession. The care sector poses a challenge as it is a social sector that brings together different social, technical and practical components. The interviewees describe this field as one in which, in addition to the technical challenge of washing or caring for patients' illnesses, cognitive and physical demands are also at stake. These often also require a certain sensitivity in dealing with people, sometimes also intuitive and tactful decisions. One of our interviewees told us: "the carers have a certain professional identity, namely, strange and peculiar as that may sound, but that identity is called "I am trained to help other people", and that identity should in any case be preserved." Since this is a human-oriented profession, it is difficult to act according to certain schemes, as the problems are characterized by their complexity. Here individual problem solutions are often needed, whereby the caregiver himself with all his tasks cannot simply be replaced by a robot and generally only partial areas should be considered. We were told that "you can have someone to optimize the plan, optimize the food distribution, you can even let me print the lunch. That's now also possible. But only very far in the future you will have someone, so you won't see a nurse who's… not a robotic nurse who gives medication." The consideration of establishing the new technology in this area and investing in new systems would arise from the general concern that there would not be enough specialists available to reliably guarantee good care in the coming years. This is partly due to weak wages, which are not rising despite the high demand for skilled workers. This shortage of skilled workers is also due to the lack of attractiveness of the profession, as it is a question of heavy work. One of our interview partners commented: "of course, this is attractive for the employers, because they think you can save employees, but in Germany, when it comes to the number of employees, we are at the lowest minimum. So, if you imagine that during the night in an old people's home there is often one caregiver responsible for 80 residents, internationally one is responsible for twelve."

A general social picture also shows that there is a need to educate new people in the nursing profession. In 2015, only 55 out of 100,000 inhabitants in Germany were graduates in this field [3]. Here the low salary could be a crucial factor, since frequently, this stands in no relation to the actual work and responsibility. In addition, the conditions in hospitals and old people's homes are not acceptable to many carers, so that they are more inclined to retrain and embark on a new profession. Thus e.g. the circumstances of the body-referred measures are criticized, as well as the lack of personnel or the lacking qualification of the coworkers [4]. These factors and demographic change, i.e. an ageing society, may mean that even more elderly people will not be able to receive adequate care in the future. Independent regulation according to market economy principles can be ruled out. The mechanisms to counteract a prevailing shortage of skilled workers would have had taken effect before. Furthermore, salaries would have had to be risen earlier. However, due to the mostly predominant profit maximization, savings can also be seen on the part of the operators of such companies, which lead to prices being depressed and no additional people being hired. This also does not promote the acceptance of the occupational field, as people have a hard job and additional pressure because of tightly calculated personal. Although there is too much responsibility transferred to individual persons. The employees of the university hospital in Tübingen, for example, complained about this: "Increasingly, however, we have to observe that nursing staff are reaching their limits due to the increasing number of patients with ever shorter periods of rest and many simultaneous processes. Many employees* working in the clinics are reaching their mental and physical limits due to the workload compression and the increasing qualitative demands" [5].

Chance of Using Artificial Intelligence. The interviewed people see the chances of a successful use of artificial intelligence particularly in technical and physical tasks. In the process, the nursing staff could be relieved of their workload by means of pick-up and drop-off services and autonomous warning systems. Moreover, in order to relieve the burden on nursing homes, AI support can guarantee a longer life in one's own home. Thus, the responsibility and care need not be primarily borne by the relatives. This could have the consequence that the use of this technology on the part of the affected persons enjoys a higher acceptance, because these do not give up their autonomy to overtaxed care forces. Therefore, the interviewees see the chance to guarantee the social life of the older generations by giving the nursing staff more satisfaction and more time for the cognitive confrontation with the patients. In addition, acceptance by carers can be promoted by the fact that these systems do not serve to fill their vacancies, but only to assist them in a supportive way. The question of an ethical basis for action would be difficult to answer. One of the reasons for this would be that the current state of development had its origins in Japan, where ethical questions would only be considered much later, which is why a realistic use prognosis could only be found in accompanying care.

This is consistent with the results of Broadbent et al, who in turn have found that humans would rather accept robots if they did not feel restricted in their own autonomy [6]. In fact, a high level of comfort in old age is considered desirable by the population here as well [7]. This suggests that support from robots could be helpful, especially if care at home was possible [7].

In the ideal case, the use of artificial intelligence could on the one hand increase happiness in old age, since the degree of autonomy would be higher if care was provided in one's own home than in nursing homes. On the other hand this could reduce the costs for care places and make them available for qualified staff. These qualified workers would then be the main actors in the use and handling of the technology. A common path with human careers and their technical assistants could therefore be a good goal to cover the sector sufficiently. This objective also introduces a further factor for successful acceptance, since the autonomy of the caregivers would not be given up, but only additionally supported by autonomous systems. Above all, this would be due to the use in the own house, since thereby the age-conditioned removal into a nursing home with unknown environment would be omitted.

2.1 Requirements for Robot Integration

There are already robots in use when it comes to surgical treatments [7]. Certainly the systems operating here are not autonomous and still controlled by humans. Here it is worth analyzing to what extent autonomous action on the part of technology would be possible and which forms of autonomation and automation are accepted by the population. Particularly regarding to demographic change, it is important to know whether and how older people can learn to accept newer technologies. A first step into the joint work of man and machine is the "so-called human-robot collaboration (HRC)", in which the robots work together with humans and thus combine the "skills [of man] with the precise functions of the powerful robot" [8]. The reduction of fears in general is an important topic, which could be served e.g. by robots like Paro.[1]

But the inclusion of society in the opportunities and risks of the new technology could also reduce fears. Above all, the potential of this new technology can be seen in places, where it is perceived that the possibilities of care could be more individual and geared more strongly towards one's own needs, than it is the case today. Such a change represents an intrusion into the private sphere which should not be construed as an attack. The researchers should therefore explain to the general public how the systems work and how the data protections will be implemented. Also, in which way other possible technological consequences may appear. The active participation of the population can have a positive effect on development, since acceptance would already arise in research and society would not merely be confronted with the results.

2.2 Approaches to Acceptance Research

The acceptance of new technologies depends on various natural factors such as age, gender or nationality [6]. In addition, however, the previous field of experience also plays a decisive role. The above-mentioned studies were carried out primarily with older people (30+) who will no longer be as affected by the development as younger people (<25). In these networks, algorithms are partly self-evident due to their ubiquity in networks such as Facebook or Google. Although these algorithms cannot be equated

[1] Further information: http://www.parorobots.com/.

with artificial intelligence, they represent a part of social change in the direction of independently acting machines.

Due to the higher influence of technology in the world of experience, it can be assumed that even a later influence of self-learning and self-dealing systems would not be evaluated as negatively as it has been in the current studies. This is also reflected in the use of language-based assistance systems, which seem to be so integrated into current life that some people are no longer even aware that they use artificial intelligence in their everyday lives [9].

2.3 Humanoid Robots and Omnipresence of Autonomous Systems

The experts interviewed only consider that the humanoid robots would be meaningful if there is an environment with complex interactions and tasks. The experts agree that differentiation is more advantageous in comparison with human beings, so that artificial intelligence can continue to be seen as an instrument and not as a personality in its own right. On the other hand, humanoid systems could be helpful when it comes to touching and communication processes. It is advantageous, but not necessarily needed for acceptance.

The development towards a self-evident attitude relating to intelligent machines seems to be influenced by the fact that the systems are not human beings and do not act like humans. Broadbent et al. show that the more human robots become, the less they are accepted by humans, since only 19% of 2000 test persons in a representative study would prefer a human-like robot [6]. According to Eberl [10], this happens because the machines, if they look like humans, should behave like humans. However, this is not possible down to the smallest detail, which is why a machine system as such can be better accepted [10]. Certainly, you should still pay attention to the equipment, since the optical design influences the expectations towards the device [6]. One example of such a robot would be the Care-O-bot 3 from Fraunhofer Research, which is already capable of transporting laundry and carrying heavy crates on its own, "even using the elevator" [10]. This robot also strikes when something conspicuous happens and a caregiver is needed. This could therefore be seen as one of the first steps in which technology will find its way into everyday life. This would have been achieved if the AI had arrived where language assistants already are today, i.e. in people's everyday lives. The normality in dealing with these systems could only be influenced by successful acceptance promotion processes. It can already be seen in previous industrial and technical developments that time is an important criterion for acceptance and that such changes must slowly be introduced into society in order to avoid encountering too much resistance.

It is not surprising anyway, that mankind is defending itself against development, which it has also done before with the most varied of developments. For example, there has already been "unrest among workers [...] with the introduction of mechanical looms at the beginning of the 19th century" [10]. These fears are therefore part of human history and have become visible since the beginning of the industrial revolution. This will often also be based on the fear of change and the consequences which cannot be seen and are partly uninfluenceable. In the field of artificial intelligence research, it will be the fear of losing one's job mainly, guiding people to contrary thoughts towards the development. The current change, also known as the digital revolution or industry

4.0, seems to be even greater than it was in the days of factory automation. It now seems to be much more about cross-sectoral processes that could affect far more people than it did at the time of the introduction of industrial robots. Eberl even says that almost half of all activities can be automated and that particularly the service sector would be affected. This is a development that hits the "core of our economies" [10]. However, sectors, in which a switch to the smart machine would be justified logically, do not always suffer from a certain lack of specialized workforce, as it just seems to be the case in the caring sector. This result could give rise to the idea that fundamental new social models are needed, which would be associated with a far greater acceptance barrier than if it was only a question of abolishing jobs that are stressful and monotonous in terms of health.

2.4 Data Protection and Access to Big Data

According to Frey and Osborne [11], access to Big Data is the main reason why not only routine tasks such as assembly line work are replaced, but also more complex tasks soon may be performed better by computers. The authors emphasize specifically the ability of the systems to process large amounts of data and to be always ready to be used by humans, since the basic satisfaction of needs is no longer necessary for machines [11]. These data represent a reservoir that can strongly influence acceptance research, because the use of Big Data could also lead to (power) abuse of the information. George Orwell [12], for example, wrote a dystopia about the misuse of data for power purposes, in which the entire population was monitored and controlled by data systems. In fact, a study by the opinion research institute TNS Infratest shows that many people do not trust companies when it comes to handling their data. Data handling in the care sector is regarded as trustworthy by only 43% [13].

In a further development of automation, data protection should be expanded in order to find solutions for the handling of data in the sense of the satisfaction of every citizen. This is particularly important in the area of care, as a lot of sensitive data about the patient is collected, such as patient files or personal information, the confidentiality of which should be guaranteed. This digitalization and archiving are mentioned as difficult in the interviews. The data volumes that first have to be processed by the robot in order to guarantee a fixed operational reliability are problematic. Although the observation sensors, such as camera and microphone, cannot be programmed to store data on them, relevant disease data could also be recorded, which would have to be reviewed first by physicians etc. Here there should be a right to ignorance, which prevents the data streams and data transfer. One of our interviewees clarified: "When I bring a robot home, I want it not to tell me where I have my dirty corners immediately. That means I have to be able to trust my robot like I trust a good, loyal member of staff or whatever." Indeed, the experts also speak here of an unresearched factual situation, which will initially be based on grey zone decisions. This could mean that the legal situation has yet to be settled in order to establish a global security network. Similar to the question of state surveillance by such systems, one of our interviewees also asked, "To whom is the machine I have paid loyal right now?" In addition to the form of a planned monitoring, the topic of the user is also considered, since IT systems do not always run error-free and a takeover by third parties in extreme cases could not always be excluded.

The agreement of all participants to the logging of their actions would also have to be secured beforehand, whereby decisions regarding consent would again come to the fore at this point, since people in need of long-term care sometimes no longer have the perception of being able to make such a decision. An example of a meaningful possibility here would be the use of a model that functions similarly to an organ donation card (status 2018). This would mean that in the time in which care by AI is not yet taken for granted, prior consent would have to have taken place in order to be maintained automatically and also to agree to the necessary data processing.

2.5 Policy and Financing of These Systems

Since many jobs could be replaced by this digitization, the financial situation of the population must also be viewed in a different light. Discussions that focus on a change in society as a whole should and must take place. One of these discussions is, for example, that about an unconditional basic income. The joint development of AI systems with the population is important for the recognition of social interrelationships. Also, the financing of external systems will not be insignificant, after all the appropriate infrastructure is important for full functionality.

The interviewees do not see the need for infrastructural measures, since artificial intelligence should be designed to overcome obstacles of a practical nature, "either the robots can climb stairs or, if the robots cannot climb stairs, robots that do not have to climb stairs will be used." Also, additional barriers through robot-compatible infrastructure, especially in places where people are "on the move with wheelchairs or rollators", would not be effective, unless they were, for example, virtual rails. In addition, the existing buildings are difficult to change due to building regulations, and from a financial point of view such changes are often not profitable enough. Also, the "investments in such facilities are partly very strongly regulated by law and also limited." This would mean that planned changes would only take effect if they were responsible for a high increase in efficiency. Integration into existing supply systems is therefore only possible with a high level of user participation. This could, however, be an important aspect for the promotion of acceptance, since technology needs to keep in mind the personal living conditions and creates new, joint fields of action. Also, new fields of work could be created through planned investments, which could counteract the fears of the population regarding the shortage of jobs. On the other hand, politicians also have a great responsibility to identify the lack of personnel in the nursing sector and to no longer look only for specialists in the MINT sectors, but also in the nursing and social sectors, where many employees are needed. Necessary structures should be created and reacted to the needs of the future, an affinity to "technical companions" could be present and, like WiFi, even desired by patients. This would then perhaps also happen out of a willingness to pay for such systems, which would ensure self-financing.

Policies also need to respond to the changes, "create the appropriate legal framework" and "develop a strategy" to meet the challenges of a new digital age [14]. A good political approach to industry 4.0 should prevent mass unemployment and encourage a restructuring of the system. In particular, the increased demand in the MINT occupations sector should be met by a good educational offer [15]. The legal framework, on the other hand, should clarify who is responsible for faulty behavior, be it that an autonomously driving car has injured a human being, be it that a robot has reacted too late in the case of a patient's nursing. Here, too, politicians need clear guidelines and instructions so that the population can adapt to them. A functioning legal system should provide acceptance. The consideration of further investing in possible ISO standard certifications also remains, which could have an acceptance-creating effect above all because they are applied internationally and therefore make uniform standards visible to the user with this certification for products from imports. Similar to TÜV certification, this is a characteristic that can provide sufficient transparency and inspire confidence. E.g., there is a generalized standard roadmap about industry 4.0 [16].

This assumption is also supported by the scientists, even if the current research on the legal situation is not sufficiently clear. Nevertheless, they speak of an initially important producer liability, which under certain circumstances would be transferable to the users. The users should be informed by the manufacturer about possible risks in order to guarantee a duty of supervision. However, this type of legal protection is not to be equated with the implementation responsibility of the nursing staff, which should be properly trained so that a flawless application can take place. Should the financial and political means be lacking at this point, the degree of autonomy of the devices would have to be reconsidered. Overall, however, the interviewees see a positive development, as they consider the errors of the robots to be less significant than those of humans: "If there are now individual aspects in which a machine produces considerably fewer errors, shouldn't we use it?"

3 Possible Consequences of the Use of Artificial Intelligence

In summary the results show, that the use of artificial intelligence will have a strong impact on the way society has been interacting up to now. It should be noted here that there may be a distinction between the acting systems, which act partly autonomously, partly sub-autonomously, and are not initially designed to make the occupational field of nursing appear completely redundant.

The survey showed that the occupational activities that require physical work are likely to disappear within the next few decades. The use of algorithms could also ensure that occupational branches, the monotonous, simple procedures such as writing invoices or documentation, are no longer carried out by people. This could lead to mass unemployment, or the new occupational fields created by this form could become the focus of vocational training. These effects would particularly affect the low-wage sector, which would no longer have to struggle with a shortage of skilled workers. This, in turn, means that society as a community must look at the modern labor market and ensure a fair (re-)distribution of labor resources if there will be as much change as had been predicted. For the care sector in particular, this means that the care profession would need a new image, which promotes more interest in society, so that demographic change does not lead to an isolation of the older generations. This interest could be strengthened by the use of AI. The sciences must be a controlling authority that researches both information and possible applications and technical consequences and risks, so that a social existence is ensured for the majority of the population, but especially for those affected in the nursing profession and in the nursing facilities. This also includes that the use of the new systems should not favor the partial effect in which the use of technology would further displace human beings.

3.1 The Acceptance and Resulting Recommendations for Action

It can also be concluded that the changes in society caused by the use of artificial intelligence are to be regarded as so fundamental that there is a need for a high level of acceptance of this technology and its consequences, so that there will be no instability in society as a whole. It would therefore be important to note that it is not solely in the hands of individuals to find and strengthen a meaningful approach to AI. Politicians, entrepreneurs, users, etc. are all equally involved in this new technical revolution. If the barriers created by the new technology are tackled in time, the scandals on the part of society are likely to be eliminated. For this purpose, however, the individual users and affected parties should be questioned, and the focus should not be on maximizing the profits of individual companies, but on the interest of society as a whole. Furthermore, the integration of personalities into organized participation processes should take place. After all, this would not be a change that would only affect individuals, but, since it was mainly care for the elderly that was addressed here, anyone who grows old enough to have to be cared for.

For a high level of acceptance, education in society seems to be necessary. First of all the redistribution of the labor force in the individual occupations should be addressed in more detail. It is not enough to say that new occupations are also emerging, as these are sectoral changes. This means that not every occupational field that could be eliminated by this development would also be replaced by a similar occupational field. A retraining for those affected from the low-wage sector could be associated with great difficulties. This also requires a discussion in society that encourages people to rethink new concepts of a basic pension that catches those who have been left behind. The individual boundaries of single personalities should be questioned in studies and the legal framework should be formulated in a way that is comprehensible to the individual. Transparency in research is therefore a decisive factor. Especially the handling of the resource "personal data" should be logically explainable for the public and not openly formulated for wrong actions, so that there would be no further dissatisfaction in the population due to the bad handling of nursing patients. In the best-case scenario, there would then be an unenforced change in society that is initiated and implemented in joint processes. Mainly the aspect of society as a community should become clear in a positive future scenario: Focusing on individual groups that increase their personal profit margins would promote AI, but not acceptance, so that too large a part of the population could only watch how they are no longer included in the changes. The resulting restlessness should be avoided by prior acceptance and a common technical-human future should be created. The change there can be regarded as indispensable – The only question is in which way change is arranged.

References

1. Mayring, P.: Qualitative content analysis: Theoretical foundation, basic procedures and software solution, Klagenfurt (2014). http://nbn-resolving.de/urn:nbn:de:0168-ssoar-3951732. Accessed 3 June 2018
2. McKinsey & Company: Smart Moves Required – Der Weg künstlicher Intelligenz im Mobilitätssektor (2014). https://www.mckinsey.de/files/_smart_moves_required_de.pdf. Accessed 26 Apr 2018

3. BMWi. (n.d.): Anzahl der Pflegeabsolventen in ausgewählten Ländern im Jahr 2015 (je 100.000 Einwohner). In Statista - Das Statistik-Portal (2015). https://de.statista.com/statistik/daten/studie/553420/umfrage/anzahl-der-pflegeabsolventen-in-ausgewaehlten-laendern/. Accessed 17 June 2018
4. Medizinischer Dienst des Spitzenverbandes Bund der Krankenkassen e.V. (MDS): 5. Pflege-Qualitätsbericht des MDS nach § 114A ABS. 6 SGB XI. Qualität in der ambulanten und stationären Pfege. Inpuncto:asmuth druck + medien gmbH, Köln (2017)
5. Bräuchle, F.; Bresch, F.; Eckmann, D., et al.: Offener Brief zum Pflegenotstand am Universitätsklinikum Tübingen (2017). https://bawue.verdi.de/++file++5a795745e58deb525ac0ac76/download/Offener%20Brief%20Pflegenotstand%20TUe%201.2018.cleaned.pdf. Accessed 17 June 2018
6. Broadbent, E., Stafford, R., MacDonald, B.: Acceptance of healthcare robots for the older population: Review and future directions. Int. J. Soc. Robot. **1**, 319–330 (2009). https://doi.org/10.1007/s12369-009-0030-6
7. Kirchberg, J., Mees, T., Weitz, J.: Robotik im Operationssaal. Aus der Nische in die breite Anwendung. In: Der Chirurg 2016, vol. 87, pp. 1025–1032 (2016). https://doi.org/10.1007/s00104-016-0313-z
8. Breuer, J.: Industrie 4.0 – kollaborierende Roboter. In: DGUV Forum: Fachzeitschrift für Prävention, Rehabilitation und Entschädigung 2015, 3, p. 10 (2015)
9. An, M.: Artificial Intelligence Is Here – People Just Don't Realize It. https://cdn2.hubspot.net/hubfs/53/assets/hubspot.com/research/reports/AI%20Is%20Here_HubSpot.pdf?t=1524732684673. Accessed 26 Apr 2018
10. Eberl, U.: Smarte Maschinen. Wie künstliche Intelligenz unser Leben verändert. Carl Hanser Verlag, München (2016)
11. Frey, C.B., Osborne, M.A.: The future of employment: how susceptible are jobs to computerisation? (2013). https://www.oxfordmartin.ox.ac.uk/downloads/academic/The_Future_of_Employment.pdf. Accessed 3 May 2018
12. Blair, E. (Hrsg.) (Orwell, G.) (1949). Nineteen Eighty-Four. Penguin Student Editions, England (2000)
13. Vodafone Institute for Society and Communications: Big Data. A European Survey on the Opportunities and Risks of Data Analytics (2016). http://www.vodafone-institut.de/wp-content/uploads/2016/01/VodafoneInstitute-Survey-BigData-en.pdf. Accessed 3 May 2018
14. Bundesministerium für Bildung und Forschung (BMBF): Deutschlands Leitlinien für ein zukunftsfähiges Innovationssystem (2018). https://www.bmbf.de/de/deutschlands-leitlinien-fuer-ein-zukunftsfaehiges-innovationssystem-6020.html. Accessed 3 May 2018
15. Bundesagentur für Arbeit (BFA): So viele offene MINT-Stellen wie nie. Der Bedarf an Fachkräften ist im MINT-Bereich (Mathematik, Informatik, Naturwissenschaften, Technik) auf einem Höchststand. Das meldet jetzt das IW Köln (2017). https://www.arbeitsagentur.de/news-mint-fachkraeftebedarf. Accessed 3 May 2018
16. DIN e.V.: Deutsche Komission Elektrotechnik (DKE): Deutsche Normungsroadmap 4.0. Version 3 (2018). https://www.dke.de/resource/blob/7b1a6b3764d64b8669d97cba2852bc73/deutsche-normungs-roadmap-industrie-4-0-version-3-0-data.pdf. Accessed 3 June 2018

A Study of Metacognitive Problem Solving in Undergraduate Engineering Students

Lisa Jo Elliott(✉), Heather C. Lum, Faisal Aqlan, Richard Zhao, and Catherine D. Lasher

The Behrend College, The Pennsylvania State University, 4951 College Drive, Erie, PA, USA
{lje12,hcl11,fua11,richardzhao,cdl5208}@psu.edu

Abstract. One of the key challenges in engineering education is the problem of teaching future engineers' professional skills. Engineering students need to know what they do and do not know. This is termed metacognition. There is still quite a bit that we do not know about how metacognition develops in classroom settings. In this study, we discuss an exploration of these issues using both physical and virtual reality (VR) simulations of manufacturing systems; which are performed by student teams. We discuss the incorporation of measures of metacognition into a model of conflict and error to predict what types of experiences may be most helpful to produce improved metacognition in engineering students.

Keywords: Metacognition · Engineering education · Virtual reality

1 Introduction

Two million manufacturing jobs may go unfilled in the next decade because of a skills gap and a decline in education in areas such as engineering [1, 2]. One such skill, problem solving, is most critical. Engineers are required to have both technical skills, which are part of the traditional engineering curriculum, and non-technical skills which are often not taught or fostered. These non-technical skills can determine a future engineer's success when working in teams, deriving novel solutions to ill-defined problems, and determining the viability of any solution. In most cases, engineering students are taught structured strategies for solving problems. These steps include: (1) defining the problem, (2) understanding the process, (3) identifying root causes, (4) developing solutions and sustaining the improvement. In addition, several methodologies help problem solving such as Lean, Six Sigma, Design for Six Sigma, and Business Process Re-Engineering [3]. However, when the problem is ill-defined, the process muddled, or solution is unclear, problem solving strategies need to be redeveloped and reassessed. Regardless of the methodology or strategy, engineers must know what questions to ask, when to ask them, and how to convey the proper information to other engineers on their teams as well as other stakeholders.

When working in teams, engineers face challenges. Within the team, there must be a shared understanding within and between each team member. Knowing what is known within themselves and each other is termed metacognition [4]. Developing this

W. Karwowski et al. (Eds.): AHFE 2019, AISC 963, pp. 95–102, 2020.
https://doi.org/10.1007/978-3-030-20135-7_9

skill in new engineers is a key component of the emerging engineering curriculum as demanded by employers. Training engineering students to understand this concept is key to deriving the best solution for a problem [5].

Physical drawings and models that engineers are accustomed to working with are explicit. They can be viewed, manipulated, and easily discussed. Metacognition is implicit and involves the construction of a mental representation by the problem solver [6]. This makes metacognition particularly difficult to teach and examine. In the engineering curriculum, many solutions are possible for a given problem. However, with given affordances and constraints, only one optimal solution exists. It is up to the curriculum to instruct students in ways of finding that one optimal solution. While there are explicit methods, if an engineer does not use implicit methods and think critically, a less than optimal solution is easily chosen. In psychology, research has suggested that one of the reasons for choosing a less than optimal solution and believing it to be the optimal solution is the lack of a proper mental representation constructed through a good understanding of what is known and what needs to be known. To date, there is little research on how design and manufacturing solutions are determined when students solve problems in class and how their teamwork and questions contribute to their mental representations are made during the classroom exercise [7].

Knowing what is known and what needs to be known guides the students' decision of when they are ready to solve a problem. Flavell defined this idea in a study of school children under the age of 12 years old who were asked to study a set of items until they could recall them without error (i.e., the problem) [8, 9]. In this study, older students were able to accurately assess when they knew the items perfectly. When tested, they were able to recall the items without any errors. Younger children thought that they were able to accurately assess when they knew the items perfectly. When tested, the younger children's performance was less than perfect.

In a subsequent study by Markman [10] in children under the age of 12, they were asked to evaluate instructions and detect any errors in the simple instructions. The researchers incorporated obvious errors and omissions. The younger students were surprisingly poor at detecting the errors and thought that they understood the instructions until they started the procedure. In both cases, students believed that they had memorized and understood, but they had not. The monitoring of their own memory and comprehension was flawed as they built mental representations of the problem space.

Understanding how knowledge is integrated and monitored is termed metacognition. According to Flavell, metacognition plays an important role in communication, comprehension, writing, attention, memory, problem solving, and social cognition as well as self-control, self-instruction. social learning, personality development, and education (p. 906) [5]. In education, we see that educators need additional tools developed to help engineering students improve this process as they solve complex and multi-layered problems. The parts of metacognition are complex. In Flavell's model, there are four subcategories of metacognition: knowledge, experiences, goals, and actions. Metacognitive knowledge is the idea that other people are those with separate thoughts, tasks, and experiences. For example, you may believe that another person is better or worse at a particular task (i.e. Jeff is better at algebra than I am). Metacognitive goals are the objectives to be attained in the problem space. Metacognitive actions are the behaviors to attain the goals.

Within this framework exists a person's thoughts about themselves and other persons, their opinions, understanding, and beliefs about the task, and their approach in using these resources to attain the goal with strategies that have worked in the past, what they have learned currently, and what they can adapt from similar situations. Metacognition is an interaction of all of these moving parts: understanding of self and other persons, tasks, strategies, metacognitive knowledge, metacognitive experiences, goals, and actions of the self and others [5]. Metacognition can be acquired, it can fail, and it can be inaccurate. The lack of metacognition can lead a person to believe that they have all the knowledge that they need to solve a problem, when they do not. This would lead someone to select one course of action over another better course of action. It can influence your communication style, critical thinking, decision making, and problem solving [11].

Garrison and Akyol [12] discuss the role of metacognition in collaborative environments such as engineering in which self-regulated learning plays a role. Metacognition mediates knowledge construction and collaboration. Students must be aware of each other's metacognition in order to construct meaning. In this case, students as co-learners constantly assess internal and external conditions. They may ask for help or provide help to realize the learning goal.

Zohar and Lustov [13] recognize that teaching metacognition leads to teaching higher order thinking which establishes ideas about causation in problem solving. However, teaching strategies require a learners' self-knowledge of their own judgment of learning and feeling of knowing. Back to the two studies with the children by Flavell [8, 9] and the one by Markman [10]. It was clear that the younger children had not yet developed an accurate judgment of learning which caused them to make errors in both the memory recall task [8, 9] and the detecting errors in instructions task [10].

Judgment of learning and feeling of knowing are impacted by the limited resources in working memory [14]. Working memory integrates the current state with past, regulates attention, and allocates cognitive resources during learning. When a student is learning something for the first time, their working memory may allocate full attention to the task as they integrate the instructions with strategies, tasks, goals, and previous experience. As they begin to learn the task, students with good metacognitive strategies will monitor for uncertainty in an uncertainty monitoring state. This type of monitoring requires intensity in attention and good self-regulation. As this is an internal and implicit process, it is difficult to verbalize what mental representations are being constructed, how, why, and when. Often, it is only when an important component is lacking, that individuals know their representation is lacking. Because of the internal implicit nature of metacognition, observation and some self-report has been the primary way to measure it [15]. For this study, we will have students engage in a manufacturing exercise individually and in teams in such a way that metacognition increases within and between students. We will integrate an improved measurement paradigm to detect changes during the exercise.

2 Simulations

We expect that up to ten teams of four students in each team will participate in this study. We are using physical and virtual reality simulations of two different manufacturing systems, craft production and mass production, as a framework. In both manufacturing processes, the students manufacture a car made of Legos building blocks from a larger Lego kit. Students will start individually in the craft production exercise in either the virtual reality or physical simulation. In this portion of the task, they will learn their role, expectations, and overall goals of the task. Following this, students will then come together to take part in the mass production exercise. During each exercise, we will be observing the participants' interactions including the questions that they ask and answer and how their knowledge base and overall metacognition changes from the beginning to the end of the simulation. Then, they will report their own understanding of the exercise and their perception of others'. During the virtual reality part of the exercise, we will use eye-tracking to model their knowledge change. The use of eye-tracking in this type of study is novel.

Within the virtual reality simulation, an eye-tracker will record fixation points, latencies, and saccades. The latencies and fixations points will add to our ability to model attention and tie it to their answers on the metacognitive measures as shown in Van Gog and Jarodzka [16]. As participants use the virtual reality game, we expect that their performance will begin to approach an expert's performance in the same game as their metacognition improves as measured by the Engineering Design Metacognitive Questionnaire (EDMQ) developed by Lawanto [17] as well as the Metacognitive Awareness Inventory (MAI) [12] and Group Style Inventory (GSI) [18]. Other questionnaires such as Flow State Scale (FSS) will also be used [19] (Fig. 1).

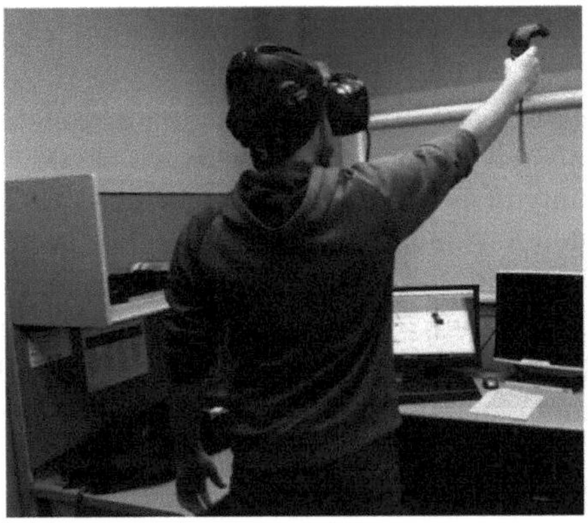

Fig. 1. Student in the virtual reality engineering simulation.

3 Analysis

We are incorporating the raw eye tracking data as a measure of metacognition into a model of conflict and error to predict what types of experiences are most beneficial when training metacognitive skills. The raw eye tracking data analysis using signal detection theory (SDT) [20] as an approach to differentiating stimuli and quantifying a student's performance as it approaches expert performance over the course of the simulation. For example, initially we expect that the student will survey the Lego car parts that are available and focus on one or two options to use to manufacture their car. As the student considers the affordances and constraints of each choice, he will vacillate between the choices eventually settling on one and choosing that one. Expert performance is similar, but the vacillation time is less as the expert knows which choice is optimal.

Through comparing the student data to the expert data, we obtain a more accurate estimate of what the student is considering and how the metacognition is developing in terms of sensitivity (an observer's ability to discriminate stimuli) and response bias (an observer's standards for producing different behavioral responses) [20]. As the virtual reality simulation progresses, they view and choose items to construct the car. It is during this viewing of items that they attend to, the amount of time that they attend to them, the order in which they attend to them and their choice of attention to each of the items. Let us discuss the details of how that score will be derived.

Eye-tracking and gaze-tracking tools have made tremendous strides in providing information on a number of perceptual and cognitive processes, including focus and attention, information processing, and cognitive workload [21]. An eye-tracker is a device that measures eye movements, pupil size, focus, and other characteristics of one or both eyes while engaged in a given task. Through the measurement of these characteristics, an eye tracker enables researchers to track an individual's eye movement patterns across an entire task. In this study, participants were equipped with a virtual reality headset outfitted with an eye-tracker within the mark. This eye-tracker model utilizes a scene camera to record the direction in which the participant is looking, and a second camera pointed at the participant's eye is used to examine fixations. The information from the eye-tracker collects two basic measures: gaze fixation and saccade. Previous literature has established that these measures are sufficient measures of attention and information processing as it relates to learning in knowledge change and metacognition [22]. The Fixation will measure the amount of attention in terms of location (area of interest- AOI) and in terms of time [23] and the Saccade will measure the length of time for which items on the screen are attended to [24]. Within these two measures, there is spatial and temporal information as seen in Table 1.

Calculating these measures alone for each participant would be helpful but not as useful as when the measures are compared to the performance of an expert with knowledge of the optimal solution and a complete and accurate mental representation of the problem space (i.e. the subject matter expert or SME). We want the participant performance to approach expert performance as their information processing behavior starts to match the behavior of the SME in the same virtual reality exercise. The SME sets the optimum areas of interest (AOI) and benchmarks the gaze points and optimum loiter. The eye tracking data will be segregated into the following categories and processed as described in A–D.

Table 1. Measures of eye-tracking.

	Spatial measure	Temporal measure
Fixation (attention- correct item)	Gaze Point- as an X,Y coordinate as the center of the optimum Area of Interest/AOI (attention to correct item)	Time stamp of gaze point – gathered each 16.7 ms
Saccade (order of processing, correct length of time)	Order of gaze points- what the participant looks at first, second, third, and so forth	Loiter of fixation- the amount of time the participant gazes before the next saccade (attention-correct length of time)

A. Attention Location Difference Score-ALDS. The root mean square or RMS value from the center of the SME's AOI for each salient item will be calculated. The RMS value from the center of the participant's AOI for each salient item will be calculated. The difference between the two scores will be calculated to derive the Item Location difference score or ALDS.

B. Attention Time Difference Score-ATDS. The amount of time that the SME loiters on a particular item before moving onto the next item indicates the amount of information processing time needed to incorporate the item and decide. This loiter time between large saccades will be calculated and then each participant's loiter time will be subtracted from the SME's ideal time to find the attention time difference score or ATDS.

C. Number of Fixations Difference-Fxd. As outlined in the original document, the number of fixations also plays a factor and can moderate duration time by distraction. The number of fixations difference before a major saccade will be Fxd.

These measures will comprise the individual's eye tracking performance.

$$Participant\ Individual\ Eye\ Tracking\ Performance = ALDS + ATDS + Fxd$$

We expect that this performance equation will change as the participant engages in continued game play and will begin to approach the SME model of performance. For example, a very good model of performance in the game may be an ALDS of 13 (the center of the AOI was within thirteen pixels of the center of the expert AOI), ATDS = 53 (fixated about 53 ms longer than the expert), Fxd = 2 (two additional fixations than the expert). The score for this participant would be 68 (13 + 53 + 2). This would be a fairly good score as earlier in the exercise the same participant had a farther AOI (139 pixels) and a longer difference in fixation ATDS = 106, with more 18 more fixations Fxd = 18. This score of 163 (39 + 106 + 18) essentially described performance more distant from the ideal of 68 and could be described as a 95-point improvement. With this model, participant's micro-improvement in the tasks can be observed over individual items and over sections of the game temporally.

When we incorporate this into a Conflict (C) and Error (E) model we can predict the team performance by examining the divergence of acceptance of a specific task, the task time to solve the problem, the disparity between the expected state and the actual state. Conflicts and errors are very common and can occur in any problem solving process. We will focus on conflicts occurring during the problem solving process among collaborative undergraduate engineering student teams. Such C & Es can impact team performance and lead to ineffective solutions. Suppose that there is a conflict between two collaborative engineering teams working to solve a given problem related to a manufacturing process in product design. Error is the deviation between intentions and actions. We define the conflict as the competition between two or more simultaneously activated response tendencies represented by the difficulty level of the problem. If a student considers a given task easy whereas another student considers it difficult, a conflict, will occur. An error, will occur if a student did a task wrong, reflected by his/her performance certainty. If an error is not caught during the problem solving process, the error can lead to wrong or ineffective solutions. We will be measuring the flow of information between the collaborating team members, the probabilities of the perception of difficulty and the conflicts. We expect that this model will predict collaborative problem solving and an improvement in metacognition between and within team members.

Acknowledgments. This material is based upon work supported by the National Science Foundation under Grant No. (1830741). Awarded 8/1/18.

References

1. Deloitte. https://www2.deloitte.com/
2. Levesque, K., Laird, J., Hensley, E., Choy, S.P., Cataldi, E.F., Hudson, L.: Career and Technical Education in the United States: 1990 to 2005. Statistical Analysis Report. NCES 2008-035. National Center for Education Statistics (2008)
3. Aqlan, F., Al-Fandi, L.: Prioritizing process improvement initiatives in manufacturing environments. Int. J. Prod. Econ. **196**, 261–268 (2018)
4. Sengul, S., Katranci, Y.: Meta-cognitive aspects of solving indefinite integral problems. Procedia - Soc. Behav. Sci **197**, 622–629 (2015)
5. García Peñalvo, F.J.: Entrepreneurial and problem solving skills in software engineers. J. Inf. Technol. Res. **8**, iv–vi (2015)
6. Bickhard, M.H.: Why children don't have to solve the frame problems: cognitive representations are not encodings. Dev. Rev. **21**, 224–262 (2001)
7. Visser, W.: The Cognitive Artifacts of Designing. CRC Press, Boca Raton (2006)
8. Flavell, J.H.: Metacognition and cognitive monitoring: a new area of cognitive-developmental inquiry. Am. Psychol. **34**, 906 (1979)
9. Flavell, J.H., Friedrichs, A.G., Hoyt, J.D.: Developmental changes in memorization processes. Cogn. Psychol. **1**, 324–340 (1970)
10. Markman, E.M.: Realizing that you don't understand: a preliminary investigation. Child Dev. **48**(3), 986–992 (1977)
11. Schraw, G., Dennison, R.: Assessing meta-cognitive awareness. Contemp. Educ. Psychol. **19**, 460–475 (1994)
12. Garrison, D.R., Akyol, Z.: Toward the development of a metacognition construct for communities of inquiry. Internet High. Educ. **24**, 66–71 (2015)

13. Zohar, A., Lustov, E.: Challenges in Addressing Metacognition in Professional Development Programs in the Context of Instruction of Higher-Order Thinking. Contemporary Pedagogies in Teacher Education and Development. Intech Open (2018)
14. Coutinho, M.V., Redford, J.S., Church, B.A., Zakrzewski, A.C., Couchman, J.J., Smith, J.D.: The interplay between uncertainty monitoring and working memory: can metacognition become automatic? Mem. Cogn. **43**, 990–1006 (2015)
15. Akturk, A.O., Sahin, I.: Literature review on metacognition and its measurement. Procedia Soc. Behav. Sci. **15**, 3731–3736 (2011)
16. van Gog, T., Jarodzka, H.: Eye tracking as a tool to study and enhance cognitive and metacognitive processes in computer-based learning environments. In: International Handbook of Metacognition and Learning Technologies, pp. 143–156. Springer (2013)
17. Lawanto, O., Santoso, H.B.: Development and validation of the engineering design metacognitive questionnaire. In: American Society of Engineering Education (ASEE) Annual Conference (2014)
18. Cooke, R.A., Szumal, J.L.: Measuring normative beliefs and shared behavioral expectations in organizations: the reliability and validity of the organizational culture inventory. Psychol. Rep. **72**, 1299–1330 (1993)
19. Jackson, S.A., Marsh, H.W.: Development and validation of a scale to measure optimal experience: the flow state scale. J. Sport. Exerc. Psychol. **18**, 17–35 (1996)
20. Green, D.M., Swets, J.A.: Signal Detection Theory and Psychophysics. Wiley, New York (1966)
21. Salvucci, D.D., Goldberg, J.H.: Identifying fixations and saccades in eye-tracking protocols. In: Proceedings of the Eye Tracking Research and Applications Symposium, pp. 71–78. ACM Press, New York (2000)
22. Rayner, K.: The 35th Sir Frederick Bartlett lecture: eye movements and attention in reading, scene perception, and visual search. Q. J. Exp. Psychol. **62**, 1457–1506 (2009)
23. Costa-Gomes, M., Crawford, V.P., Broseta, B.: Cognition and behavior in normal-form games: an experimental study. Econ. **69**, 1193–1235 (2001)
24. Orquin, J.L., Ashby, N.J.S., Clarke, A.D.F.: Areas of interest as a signal detection problem in behavioral eye-tracking research. J. Behav. Decis. Mak. **29**, 103–115 (2016)

The Effect of Screen Size on Reading Speed: A Comparison of Three Screens to Print

Lisa Jo Elliott(✉), Medina Ljubijanac, and Danielle Wieczorek

The Behrend College, The Pennsylvania State University-Erie,
4951 College Dr. Erie, Erie, PA 16563, USA
{lje12,mil5474,dnw5128}@psu.edu

Abstract. Many students are starting to rely on screens to consume their textbooks rather than paper. Does the screen influence reading speed according to its size or when compared to traditional paper? In this study, participants started the experiment with a different device and read four short stories for 10 min each. Participants read on a 17″ computer screen, a paperback book, a 10″ tablet and a 3″ cell phone screen. The number of words they read were counted and the actual words read per minute were calculated. No significant effect of reading speed was found across the screen sizes or the book. Individual participant's reading speed was remarkably consistent across all the devices suggesting that reading speed does not vary with presentation mode for a brief period.

Keywords: Reading speed · Screen size · Book · Tablet · Cell phone · Font size · e-Textbooks · Textbooks

1 Introduction

Significant improvements have been made in electronic displays which have improved the clarity. Display technology has moved away from CRT (Cathode Ray Tube) screens toward LCD (liquid crystal display) screens. LCD technology allows for the better reproduction of images and fonts. This has reduced the amount of hardware needed and allowed engineers to develop smaller screens with sharper resolution. Screens are typically measured diagonally from top right to bottom left.

Presently, several devices are marketed as reading devices (i.e. Kindle, iPad) for pleasure and for use as a replacement to textbooks. We know from previous studies that some of these devices will enhance the perception of letter forms by supporting various design factors [1]. Comparing the devices during reading will suggest attributes that contribute positively and negatively to usability and to reading.

The processes involved in the perception of letters is a topic of recent concern because of the proliferation of electronic reading devices and the suggestion that these devices would be suitable replacements for textbooks [2]. Previous research indicates that perception of objects is a natural process that humans can do fairly easily (e.g., [3, 4]). Historically, some studies in this area have found that people have difficulty reading on electronic screens.

W. Karwowski et al. (Eds.): AHFE 2019, AISC 963, pp. 103–109, 2020.
https://doi.org/10.1007/978-3-030-20135-7_10

One of the studies done by [5] Gould, Alfaro, Finn, Haupt and Minuto (1987) suggested that the difference between reading on a CRT (Cathode Ray Tube) screen and paper was primarily due to characteristics of the font reproduction on the screen. As the font increased a resemblance to the paper-based font, reading speed improved. Subsequent studies have introduced other variables and consensus has been elusive. One group argues that paper will never be replaced because of its inherent qualities of portability, long battery life, and the physical sensation of holding a book [6]. Others suggest that it is simply a matter of time before all reading material will be electronic and we should adapt [6]. Up until recently, the consensus that reading speed is severely hampered by a display with poor quality of font reproduction has been a subject of debate. This argument was valid when the CRT screen was the only screen type available. As screen technology has improved with liquid crystal displays (LCDs), the ability reproduce fonts to a near perfect match to paper has improved as well.

Some of the more recent studies found no difference between screen sizes and paper. For example, Georgiev and colleagues [7] compared the speed of reading between paper, a computer, and a mobile device to see if there was a difference between the three. They tested for an interaction between font size and the type of screen size. In this study, 54 women and 70 men who were about 23 years old participated. First, they read on paper in three different font sizes; 10 point, 12 point and 14 point font. As the font size increased on paper so did the reading speed for both men and women. The next test was done on the computer with the same fonts in the same sizes. In the computer results, the speed was faster when the font increased as well. The results from the computer showed that reading off of the computer was faster than reading off of the paper.

The next study, [8] Dündar and Akçayır, tested elementary school students in the United States on their comprehension and reading speed on both printed and electronic devices. Twenty students were divided into two groups. One group read off paper and the other read off a tablet. Qualitative and quantitative data was collected through reading speed and reading understanding. Speed was calculated by the number of words that the student read correctly out loud. Comprehension levels were calculated by a level of understanding chart. The students opinions were collected from interviews afterwards. They stated that they liked the novelty of the tablet and the ability to flip the screen. Even though there was a preference for the tablet, there was no significant difference in the reading speed and reading comprehension between the tablet and paper.

Foasberg [9] wanted to find out which college students preferred; electronic or paper when reading and why they preferred one over the other at different times. She had students keep a diary and record information about their reading habits for 12 days. They recorded (1) Why they were reading, (2) What was the format, (3) The location where they were reading, and (4) Length of the reading. Students preferred paper-based reading when it was for a class and they preferred electronic based reading when it was a short article or recreational reading. The genre of the readings was also recorded; researchers found religious genres took the longest time to read while non-academic article took the shortest amount of time.

Siegenthaler and colleagues [10] wanted to test to see if the printed format and an e-book format created the same reading response. In this experiment, eye movement was recorded while reading on either an e-book screen or on paper. Because of the eye tracking system, participants were 60 cm away from the reading material. The participants were given two hours for the reading. In between both reading formats, they were asked to talk about their experiences and give their opinions. They found that the eye movements for the e-book is very similar to the eye movement on print. Participants reported no difficulty with either of the two formats.

Tveit and Mangen [11] wanted to see if there was a gender difference in reading on a screen or book in high school students. There were 143 students, 71 boys and 72 girls with an average age of 15 years old. The experimenters wanted to see if the students would lose interest or gain interest when switching devices mid-story. The experimenters also collected demographics, native language, and reading habits. Students were given a story to read and were given 15 min. Then, they were told to stop and switch as they continued to read the same story. Thirteen percent of the participants did not notice a difference when reading on the two different devices. The students who enjoyed reading in their spare time reported that they preferred the paperbound print version over the e-book version.

People read on screens of many different sizes throughout the day. Individuals read long emails, browse webpages, documents, and view forms on small format screens. Throughout a typical day, a single individual may read text on five or more screens of different sizes. They may read on the display screen on their phones (3-in. LCD screens), their laptops (17-in. screens), their car's navigation system (5-in. screens), their tablet (10-in. screens) and their television (25-in. screen). The amount of text read on each different type of screen may vary. However, a person is likely to be reading several paragraphs of text on a small phone screen, a tablet screen, or on a laptop screen for about ten minutes at a time. This type of reading may be to get instructions, receive a message, learn the latest news, or read a short poem.

We wondered if the size of the screens had an effect on reading speed. We also wondered if the reading material had an effect on reading speed on the different screens. Would a person read faster or slower with the same book on different screens? Would a person read faster or slower if they were reading a novel from the 1900s compared to a text written in the 2000s?

2 Method

Design. We constructed a mixed design experiment with screen size as the factor. The screen size consisted of four levels, mobile phone (3″), tablet (10″), laptop (17″), and paper material (approximately 5″ × 8″ book). The dependent measure was the number of words that a participant could read in ten minutes. Participants read four stories in the same book. They read stories in either the book *The Dubliners* [12] or the book *Blink* [13]. *The Dubliners* was written in 1914 and *Blink* was written in 2006.

Participants. Sixty-four participants between the ages of 18–37 years old volunteered for the study. These participants were undergraduate college students who received course credit for their participation in the study. They had a mean age of 20 years with a standard deviation of 4.76 years. Overall, 13 men and 42 women (9 participants did not indicate gender) participated in the experiment. Participants did not deviate from the norm on the Nelson Denny reading comprehension test with a mean score of 50% (*SD* = 20%). There was a medium correlation found between reading speed and the score on the Nelson Denny reading test r (53) = 0.28 indicating that the participants were average adult readers.

Materials. We used the two books. In the book, *The Dubliners* [12], participants read the chapters; "The Sisters", "Two Gallants", "A Little Cloud", and "An Encounter". An electronic version of the book is available through Project Gutenberg (http://www.gutenberg.org/ebooks/2814?msg=welcome_stranger). In the book, *Blink* [13], participants read the chapters; "The Statue that Didn't Look Right", "The Theory of Thin-Slicing", "The Locked Door", and "The Warren Harding Error".

Procedure. The participant was greeted and consented, then they filled out a sheet of demographic information and then took a standardized reading test, the Nelson Denny [14]. Then, they were seated at a computer desk with the first device. The device that they started with was determined by the pre-coded condition to which they were assigned. Participants were unaware of the different presentation order or which book they would read.

The participant read the chapter for 10 min. The research assistant kept time using the stop watch function on a cell phone. At the end of ten minutes, the research assistant asked the participant to stop and keep their finger on the place where they stopped reading. The research assistant recorded the last four words that the participant read. Then, the participant rated their satisfaction with the device. Afterwards, they were given the next device and read the next chapter for ten minutes. Once all four devices had been used, the participant was debriefed. Participants read either the book *Blink* or the book *The Dubliners* on all three screens and the book.

3 Results

Reading speed was calculated by determining where the last four words that the participant read were located within the story using Microsoft Word's word count function. If the four words that were recorded occurred more than twice in a given story, that data was marked as undetermined and excluded from the data set. There were seven instances of this in the data set.

For the first hypothesis, we wanted to know if participants read faster on any of the three devices or the book? We found participants did not read significantly faster on any screen or the paperback F (3, 138) = 0.04, p = 0.989. Overall, participants read at an average speed of 179 wpm (*SD* = 63 wpm).

For the second hypothesis, we wanted to know if participants read the content of one novel faster than the other? We found that participants read the book *Blink* faster than the book *The Dubliners*, $t\,(221) = 1.77\,p = 0.03$. The participants read the *Blink* book faster by an average of 75 words per minute. This made sense given that there are approximately 90 years between the books and writing styles have changed dramatically.

After the study, we wanted to know if dyslexia or another reading disability would influence the data. Seven participants reported a diagnosis of a reading disability. We conducted a separate analysis and found that their reading speed was not significantly different by device $F(3,18) = 1.03, p = 0.40$. Their overall mean reading speed was 164 wpm ($SD = 69$ wpm). Table 1 shows the means and standard deviations for each device.

Table 1. Means and (standard deviations) for each device.

	Reported a reading disability	No disability
Book	156 (69)	174 (58)
Tablet	146 (71)	174 (62)
Computer	146 (62)	173 (51)
Phone	157 (65)	172 (43)

The range of reading speeds between participants is shown in Table 2. All participants were enrolled as undergraduate students in a state college either in Florida or in Missouri. Participants were remarkably consistent in their speed across all devices, hence the non-significant findings.

Table 2. Minimum and maximum reading speeds by device.

	Minimum	Maximum
Book	84	325
Tablet	70	453
Computer	75	391
Phone	85	396

In the satisfaction questionnaire, participants preferred to read a paperback book with nearly half of the participants putting this as their top choice. Thirteen participants chose the tablet as their top choice, three participants chose the computer while two participants chose the mobile phone. Fifty-three percent of the participants reported that they had rented an e-textbook in the past. Of these participants, 18 of them had rented an e-textbook more than once.

4 Discussion

We found that screen size does *not* have an effect on reading speed when the participants read for a short amount of time. This is supported by previous research. For example, in [8] Dündar and Akçayır they asked children to read on a computer screen and on a printed piece of paper. There was no significant difference in speed between the two devices ($t = 0.811$, $p > .05$) or in reading comprehension ($t = 0.67$, $p > .05$). Georgiev [7] found that the computer was faster than paper. Seigenthaler [10] found similar eye tracking patterns in reading on a screen and reading on paper. Foasberg [9] has demonstrated that people are adapting to reading on screens.

Even in Gould and colleagues' work in 1987 [5], 40% of the participants showed no difference between the computer and printed reading material on a Cathode Ray Tube (CRT) screen. We believe that as technology improved and the frequency of reading on screens increased, the population has become accustomed to reading on screens.

In this study, reading for short durations of time reflected no difference in processing speed across different screen sizes or between the screen sizes and a similarly sized book. In this study, participants' screens had the same font size as was the default setting of 14 point which is slightly larger than *The Dubliners* paperback book font size of 11 point and is the same font size as the *Blink*'s paperback book font size of 14 point. In this instance, the participants did not mention that *The Dubliners* font size was small, this difference did not impact their ability to read the book. We also noticed that none of the participants changed the font size on the screens even though they were instructed that they could do so if they wished.

This study contributes a final determination that reading on small or large screens does not impact reading speed in comparison to paper. However, this applies with the current population of adult readers and applies to short reading times. We did not investigate longer reading times. Readers in general prefer paper books or a book sized device such as a tablet. We hope that this work will assist instructors and technologists when deciding on the best way to present text.

Acknowledgments. Thanks to the students who worked in the lab on this project: Jenny Stanhope, Caresse Owens, Andrea Sias, Marvin Valarezo and mentor, Dr. Tom Sanocki.

References

1. Sanocki, T., Dyson, M.C.: Letter processing and font information during reading: Beyond distinctiveness, where vision meets design. Atten. Percept. Psychophys. **74**, 132–145 (2012)
2. Young, J.R.: How Kindle could change the textbook market. Chron. High. Educ. **55**, A4 (2009)
3. Greene, M.R., Oliva, A.: The briefest of glances: The time course of natural scene understanding. Psychol. Sci. **20**, 464–472 (2009)
4. Kirchner, H., Thorpe, S.J.: Ultra-rapid object detection with saccadic eye movements: Visual processing speed revisited. Vision. Res. **46**, 1762–1776 (2006)
5. Gould, J.D., Alfaro, L., Finn, R., Haupt, B., Minuto, A.: Reading from CRT displays can be as fast as reading from paper. Hum. Factors **29**, 497–517 (1987)
6. Dillon, A.: Reading from paper versus screens: A critical review of the empirical literature. Ergon. **35**, 1297–1326 (1992)
7. Georgiev, T.: Investigation of the user's text reading speed on mobile devices. In: Proceedings of the 13th International Conference on Computer Systems and Technologies, pp. 329–336. ACM (2012)
8. Dündar, H., Akçayır, M.: Tablet vs. paper: The effect on learners' reading performance. Int. Electron. J. Elem. Educ. **4**, 441–450 (2017)
9. Foasberg, N.M.: Student reading practices in print and electronic media. College & Research Libraries **75**, 705–723 (2014)

10. Siegenthaler, E., Wurtz, P., Bergamin, P., Groner, R.: Comparing reading processes on E-ink displays and print. Disp. **32**, 268–273 (2011)
11. Tveit, Å.K., Mangen, A.: A joker in the class: Teenage readers' attitudes and preferences to reading on different devices. Libr. & Inf. Sci. Res. **36**, 179–184 (2014)
12. Joyce, J.: Dubliners. 1914. Penguin, Harmondworth (1992)
13. Gladwell, M.: Blink: The Power of Thinking Without Thinking. Back Bay Books, Boston, MA (2006)
14. Brown, J.I.: The Nelson-Denny Reading Test. Riverside Publishing, Itasca, IL (1960)

Future of Training, Education and Learning Sciences

Engaging Programming Students Through Simpler User Interfaces

Blessing Leonard and Giovanni Vincenti(✉)

University of Baltimore, Baltimore, MD, USA
{blessing.leonard,gvincenti}@ubalt.edu

Abstract. We used line_explorer, a novel programming instructional tool to administer tests, which involved assigning a math problem based on a programming concept (loops) and the accompanying instructions for solving it to students. There were two display options for the instructions: the first provided all instructional information at once, while the other displayed the instructional information in phases. Findings from the study confirmed that an interface can affect the comfort level of the user, and thus can influence the effectiveness of the tool. Results from the research will aid in identifying parameters that need to be improved or adapted to line_explorer and other programming instructional tools, to encourage better comprehension of programming concepts, as well as improve the willingness of students to use these instructional tools.

Keywords: Human-centered design · Cognitive load · CS education

1 Introduction

Background. Human-centered design can be defined as the process of involving basic human functionalities in the design process, rather than waiting until the product is completed to force users to act according to its will. This approach ensures the users have a pleasant first experience with the program and thus are willing to keep using it. line_explorer aims to provide a comfortable user interface that will help reduce the discomfort associated with programming by allowing "the user to act as the code compiler and step through each line of code seeing the logic flow and what values get assigned to variables in the program." [1].

Findings from a line_explorer pre-design survey [2] suggests that although the average student is not excited about programming, they are able to get the work done; and while they were fine with accessing instructional videos on their mobile devices, students preferred computer-based platforms to mobile-based ones for interactive instructional tools. line_explorer is expected to function as a support tool for instructors of beginner to intermediate courses, and for students to review the material on their own before coming to class [1], therefore effectively creating an interactive support system that can be used to implement the flipped classroom model [3] in computer programming courses.

Project Goals. The goal of this research was to identify the impact (if any), of different amounts of information displayed to students as they work through a programming concept. Given that this is the second phase of testing line_explorer, some features were

W. Karwowski et al. (Eds.): AHFE 2019, AISC 963, pp. 113–121, 2020.
https://doi.org/10.1007/978-3-030-20135-7_11

changed and tested against the initial version to look for any differences (positive or negative) in the reception of the programming tool. The hypothesis was that a reduction in the amount of information displayed to the student at the launch of the program would positively affect their reception of its usefulness. It should be noted that this research focused solely on line_explorer's evaluation mode. To help decide the kind of changes necessary to test the hypothesis, some research was done on the effect of cognitive load on interfaces and instructional tools. A demonstration mode—used for instructional purposes—also exists. This mode is discussed towards the end of the manuscript.

2 Cognitive Load

Oviatt defines cognitive load as the "mental resources a person has available for solving problems or completing tasks at a given time." [4]. Including cognitive load theory in the designing phase ensures the creation of interfaces that will decrease the amount of resources used on peripheral processing, so that the bulk of the mental resources used will be directed towards the main task at hand.

The first concept that was at the core of the redesign process was the need to provide only the number of features required to get the work done, as unnecessary features often act as a form of distraction [4]. For instance, the presence of a multi-colored pen on an interface where the color of the pen does not affect the result of the task at hand, can lead to the mind being occupied with figuring out: how many colors there are, which colors it likes best, what each color means, if and how the functionality of the cursor changes depending on its color, etc. before even trying to focus on the task itself. In this case, there is a high cognitive load as the working memory is fully engaged, but not because of the work at hand.

The second element that drove this project's redesign is the need to keep the interface of the new tool as close as possible to that which the students (users) are already familiar with [4]. The importance of not jarring the students with a completely different interface is to prevent the distraction of processing the difference(s) between what they are used to, and the new design before them. Hence, making it harder to concentrate on the main task, and resulting in a negative toll on the time and quality of the task completion.

The article also suggested that users' ability to communicate multimodally be considered [4]. For instance, an audio-visual or verbal-visual demanding tool will be easier to use compared to an audio-verbal demanding tool. This is because the same processor manages the audio and verbal mental resources, while a different processor in the working memory manages the visual demands. Thus, cross modality was advocated for over intra-modality.

3 Methodology

To test the hypothesis that less information leads to more productivity, we created a modified version of line_explorer, and tested it by comparing the users' performance and reactions with the preexisting version. Figure 1 shows the original interface

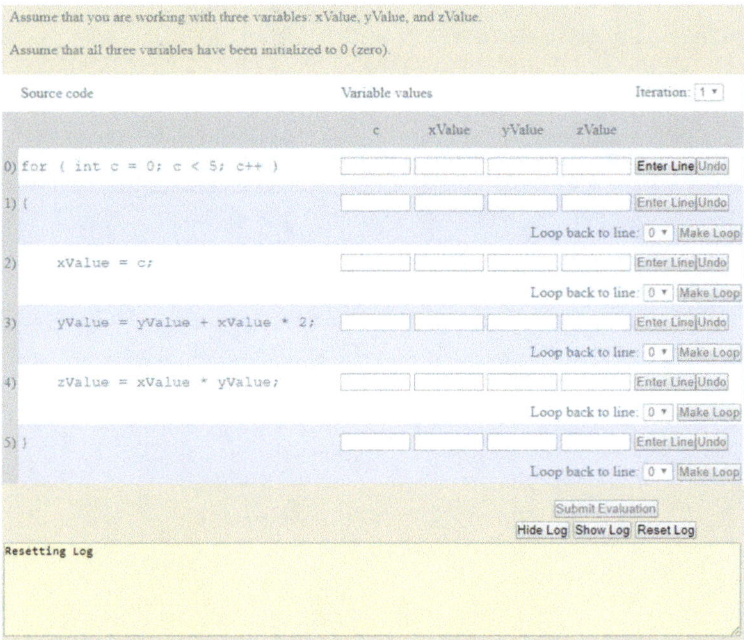

Fig. 1. Original interface used for the control group (Interface A).

(Interface A) while Fig. 2 shows the modified version (Interface B). The new interface focused on increasing the amount of information displayed based on the user's progress in solving the given problem. Both interfaces were preloaded with the same concept: a for loop requiring five iterations and whose statements were a set of instructions for solving the math problem contained in the loop. The test required basic programming and math skills—such as addition and multiplication—to work through the example.

The revised model removed some features completely, starting from the progress log and its controls (at the bottom of the screen). This feature's original purpose was to give students a way to trace their work, but users did not utilize it at all in previous tests [5]. Then, we moved the Make Loop function from every line to the last line of each iteration. The original purpose of this feature was to test whether the students would know when a new iteration of the loop would begin. In previous tests users did not use this feature as well, other than when appropriate [5].

The 59 test subjects were students enrolled in three courses within the Applied Information Technology program at the University of Baltimore: COSC 151 (Computer Programming I), COSC 251 (Computer Programming II), and COSC 351 (Object-Oriented Programming). The first two courses are modeled after typical CS1/CS2 courses, and the last course covers up to linear data structures. All classes include a lecture and in-class coding practice, and students participated in this experiment after loops were introduced. Participants were enrolled in the following academic programs: Applied Information Technology (AIT): 24 students, Digital Communication (DCOM): one student, Psychology (PSYC): two students, Forensic Science (FSCS): two students,

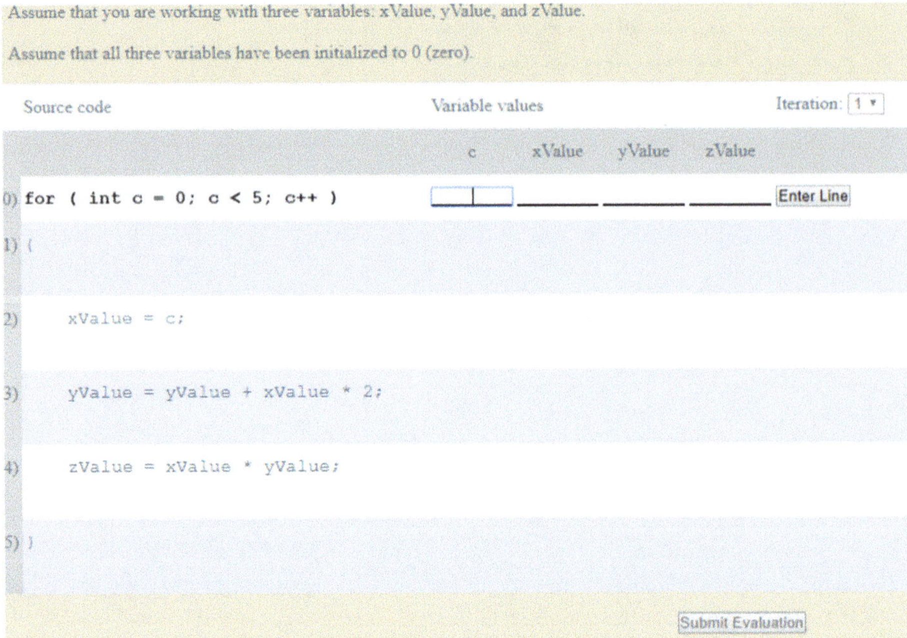

Fig. 2. Modified interface used for the experimental group (Interface B).

Information Science/Systems (INSS): six students, Business Administration (BUAD): one student, Simulation and Game Design (SGD): 19 students, Information and Interaction Design (IID): one student, and three unknown majors.

Prior to the commencement of the test, each participant received an access code after a verbal explanation of the purpose of line_explorer, the reason for the test, and the need to go through five iterations of the loop. The test experience included a questionnaire, an instructional video that went through the first iteration of the tested interface (based on the access code entered), the test itself, an instructional video of the untested interface, and finally a poll to vote on the preferred interface and provide feedback on the experience and line_explorer. The tests looked to measure the amount of time taken to complete the test, the accuracy of the final answer, the interface with the highest votes, and the impact of factors such as major, experience, and comfort level on these measurements.

4 Results and Discussion

For COSC151, three voted for Interface A, and out of the three, two rated themselves as having mid-level programming experience (rated at 3 out of 5) and low comfort level (rated at 2 out of 5) while the third had low experience (rated at 1 out of 5) with a slightly higher comfort level (rated at 2 out of 5). Major did not influence decisions, as the majors were diverse. However, two INSS majors preferred Interface A—one of them was the participant with the lower experience. There were 20 total entries, five of them did not report a preference, and the remaining 12 were for Interface B, as reported in Table 1.

Table 1. UI preference by major for the course COSC 151.

Major	Interface A	Interface B	No answer
AIT	1	5	1
DCOM	0	1	0
PSYC	0	1	1
FSCS	0	1	1
Unknown	0	1	0
INSS	2	1	1
BUAD	0	1	0
SGD	0	1	1

The comfort and experience rating range for those who voted for Interface B was between low and mid-level, but for the two exceptions of higher comfort ratings (averaging around 4). Table 2 provides a visual representation of the average experience and comfort level ratings per major for COSC 151.

Table 2. Average experience and comfort level ratings by major for the course COSC 151 on a scale of 1 (low) to 5 (high).

Major	Experience	Comfort
AIT	2.29	2.86
DCOM	3	2
PSYC	1.5	1.5
FSCS	1	1
Unknown	1	3
INSS	2.2	2.2
BUAD	1	3
SGD	1	1

For COSC251, 6 out of 17 voted for Interface A, one did not vote, and the remaining 10 voted for Interface B (Table 3). This group consisted mainly of Applied Information Technology (AIT) majors (12 students), three Simulation and Game Design (SGD), one Information and Interaction Design, and one unspecified major.

Those that voted for Interface A had mid to high experience and comfort levels (3-5 for experience, and up to 4 for comfort level) except one with a rating of 2 for both. All SGD majors voted for Interface B. The highest experience and comfort level (5 for both) was an Information and Interaction Design major and the participant voted for

Table 3. UI preference by major for the course COSC 251.

Major	Interface A	Interface B	No answer
AIT	5	6	1
Unknown	1	0	0
IID	0	1	0
SGD	0	3	0

Table 4. Average experience and comfort level ratings by major for the course COSC 251 on a scale of 1 (low) to 5 (high).

Major	Experience	Comfort
AIT	2.75	2.75
Unknown	4	4
IID	5	5
SGD	2.33	3.33

Interface B. All but the unknown major that voted for interface A were AIT majors. Table 4 provides a visual representation of the average experience and comfort level ratings per major for COSC 251.

Out of the 22 participants from COSC 351, four voted for Interface A, 15 for Interface B, and three did not vote, as reported in Table 5. All four that voted for Interface A tested Interface B, two of the nonvoters tested Interface B, and 11 of those who voted for Interface B tested Interface A. SGD made up most of the class and had the highest acceptance rate (11 votes) for Interface B (save for INSS' 100% acceptance rate from their two votes). Three participants successfully completed the task and all tested Interface A but voted for Interface B. The recurring reasons for selecting Interface B over A were the fact that it was "clean", less intimidating, and less distracting.

Table 5. UI preference by major for course COSC 351.

Major	Interface A	Interface B	No answer
AIT	2	2	1
INSS	0	2	0
SGD	2	11	1
Unknown	0	0	1

Only one of the eight (overall) successful candidates voted for Interface A, thereby supporting the finding that students tend to deliver expected products regardless of circumstances but are open to new and more relatable processes of learning programming concepts. Five out of these eight participants gave themselves higher ratings for comfort while their experience ratings fluctuated between low and high, and participants from COSC 251 and 351 seemed to spend substantially more time trying to solve the problem than those from COSC 151. The average, minimum, and maximum time spent—in seconds—on problem solving for each of the three courses are reported in Table 6.

Table 6. Time spent (in seconds) on problem-solving by course.

Course	Average	Minimum	Maximum
COSC 151	141.19	41	389
COSC 251	379.82	103	737
COSC 351	256.58	29	622

Although the results slightly derailed from the initial hypothesis (the impact of interface layout on the effectiveness of line_explorer due to cognitive load), it was able to confirm that the interface affects the comfort level of the user, and thus can influence the effectiveness of the tool. These findings, as shown in Table 7 also suggest the provision of more information (as seen in Interface A) for intermediate users, and less information (as seen in Interface B) for novices and users that are more proficient with basic programming knowledge.

Table 7. UI preference by course.

Class	Interface A	Interface B	No answer
COSC 151	3	12	5
COSC 251	6	10	1
COSC 351	4	15	3
Total	13	37	9

In the suggested improvements section of the post-test questionnaire, students provided the following suggestions for improving line-explorer:

- Ability to shuffle between iterations;
- Removing interactivity components from lines that do not contain instructions, such as lines with opening or closing braces;
- Restricting the effects of the undo button to a line—so that correcting an error on a previous line does not erase all progress made after that line;
- Clearer instructions; and
- Creating a function to allow the review of instructional videos while working on the problem.

5 Demonstration Mode

The Demonstration mode, reported in Fig. 3, allows students to practice interpreting code by giving them the option to enter expected values contained in variables for each line. The overall look is very much in line with what we already described in the Evaluation mode, with the instructions panel above the work area, lines of code clearly labeled and easily readable, and a series of input boxes where students can enter expected values for each variable.

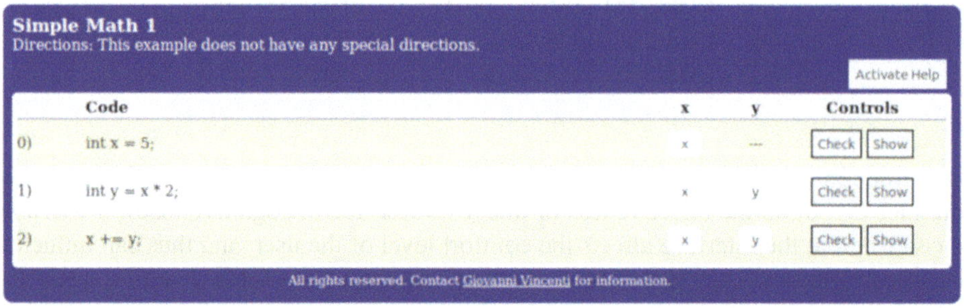

Fig. 3. The updated interface of line_explorer in Demonstration mode.

An important element focuses on teaching students about the scoping of variables. On the right side of the panel, users will be able to enter an input only for variables that actually exist. This should allow users to visualize how scoping works, and when variables exist or do not exist. Since the main goal of Demonstration mode is to allow students to explore lines, and people have different learning preferences, we have added two "Help" features, shown in Fig. 4, which the user can toggle at any point.

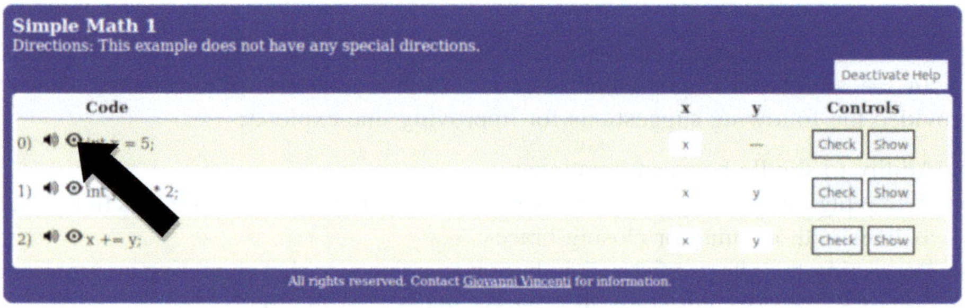

Fig. 4. Help features available in Demonstration mode.

The icons that appear are of a sound speaker and a stylized eye. By pressing the sound speaker, the system will utilize the voice synthesizer available in the browser to vocalize the verbal description of each line. By selecting the eye icon, the system will instead display the same verbal description on a screen overlay, shown in Fig. 5.

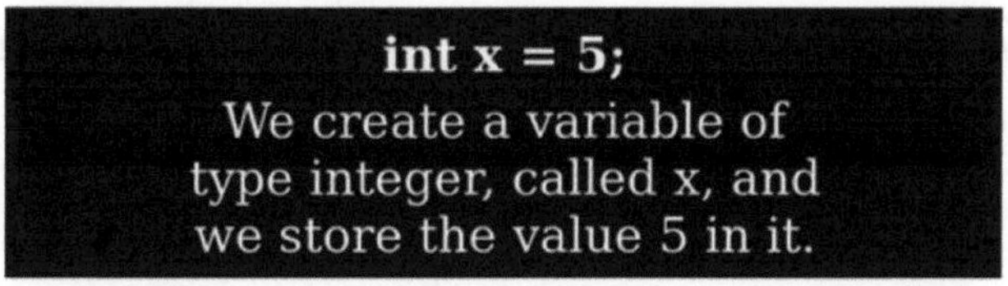

Fig. 5. Visual description of a line of code that the students can access through the Help feature available in Demonstration mode.

As we intend to keep both visual and functional consistencies between the two modes, we will apply the changes reported to the Demonstration mode to the Evaluation one.

6 Conclusions

Since all participants that provided the correct final answer were from the COSC 251 and 351 pools, we can infer that problem-solving progress and final answer accuracy depend more on experience and familiarity than on cognitive load. A sense of comfort on the part of the user is also projected to play a major role in the effectiveness of this tool, perhaps even more than experience; and based on time spent on the assigned task, it can be inferred that experience and comfort level also play a role in the user's desire to work with line_explorer. As testing and data analysis continues, we expect to build future work on findings and feedback from this and previous testing results.

References

1. Vincenti, G., Braman, J., Hilberg, J.S.: Teaching introductory programming through reusable learning objects: a pilot study. J. Comput. Sci. Coll. **28**(3), 38–45 (2013)
2. Vincenti, G., Hilberg, J.S., Braman, J.: Student preferences and concerns about supplemental instructional material in CS0/CS1/CS2 courses. Int. J. E-Learn. **16**(4), 417–441 (2017)
3. Tucker, B.: The flipped classroom. Educ. Next **12**(1), 82–83 (2012)
4. Oviatt, S.: Human-Centered Design Meets Cognitive Load Theory: Designing Interfaces that Help People Think, Journal of the Impossible (2006). https://dl.acm.org/citation.cfm?id=1180831. Accessed 21 Feb 2018
5. Vincenti, G., Hilberg, J.S., Braman, J., Satzinger, M., Cao, L.: Assessing the Usability of a Novel System for Programming Education. https://arxiv.org/abs/1711.05649. Accessed 13 Feb 2018

Automated Design of Competence-Oriented Student Models with Application of Tutoring Integrated Expert Systems

Galina V. Rybina(✉) and Elena S. Fontalina

National Research Nuclear University MEPhI, 115409 Moscow, Russia
galina@ailab.mephi.ru

Abstract. This paper provides a methodological and technological experience description in automated construction competence-oriented student models, young specialists in the field of knowledge engineering, in particular, specialists of the profession "system analyst" with use of tutoring integrated expert systems.

Keywords: Tutoring integrated expert systems · Competence-oriented model of a specialist · Engineer knowledge · Student model · AT-TECHNOLOGY workbench

1 Introduction

Implementation of advanced conceptions of conversion to new professional teaching technologies using computers for professionals in IT field implies a versatile common application of high professional educational standards and the automatical designing technology for the intelligent tutoring systems (ITS), in particular, tutoring integrated expert systems (IES) of various architectural types [1, 2]. ITS architectures procure the progressive tools of the students' intelligent training, controlling and testing, they also save the competence-oriented models in use for future professionals as the result.

The main problem is the automated computing of students' competencies, indicating that they will know (that is, the trainee's theoretical readiness) and be able to (that is, the ability to use theory when solving practical problems) the learner has to complete the learning process for a specific educational program (direction), as well as how he is able to apply the knowledge, skills and personal qualities obtained at the university for successful professional career.

A similar experience has already been accumulated in the Cybernetics Department of MEPhI with training professionals in Applied Mathematics and Informatics. This technique is currently used for the Software Engineering students. This was able to be possible as part of a research project, which plan is to create an intelligent technology for constructing of a wide class of IES, including web-IES [1–3].

The development of IES and web-IES with advanced intelligent tutoring, controlling and testing tools for students, through designing and software support to the individual competence-oriented models of students (with the view to the personality's psychological profile), adaptive tutoring models and explanations, models of

W. Karwowski et al. (Eds.): AHFE 2019, AISC 963, pp. 122–130, 2020.
https://doi.org/10.1007/978-3-030-20135-7_12

course/discipline ontologies. etc. was enabled by the problem-oriented methodology of IES designing [1, 2] and its supporting unique AT-TECHNOLOGY workbench. The basic provisions of the problem-oriented methodology and the specification of functional options of various versions of AT-TECHNOLOGY workbench are available in several monographs [1, 2, 4] and numerous papers, e.g. [3–6, 8–10] etc.

For the implementation of intelligent tutoring based on the development and use in the educational process of tutoring IES and web-IES, the professional standards for the Information Technology industry [6] were used quite effectively as a basic information and methodological resource for building models of professional competencies, in particular, for such professions as "software engineer", "system analyst", "IT system specialist", "software architect", etc.

Among the listed professions, such professions as "system analyst" should be highlighted, the demand for which in the modern market of high information technologies approaches the demand for programmers. Today, these specialists in the field of knowledge engineering and technologies for building intelligent systems are called differently - knowledge engineers, knowledge analytics, cognitive engineers, task designers, and much attention is paid to their training in various universities. For example, a systematic view of the training of knowledge engineers is contained in [11, 12], which summarizes the experience of authors in the training of knowledge engineers and business analysts over the past 15 years. Many interesting results on this problem are given in [13–22] and other works.

In general, as the experience of the development and use of tutoring IES and web-IES in the educational process showed, the main problems in the formation of professional and universal competences are:

- Selective selection at each stage of training (bachelor, master) of the knowledge, skills and abilities that students should acquire (applied ontologies of courses/disciplines, generalized ontologies of individual areas of training are used);
- Improvement of methods of control and testing, conducted both with the purpose of forming current competence-oriented models of students, and upon completion of training (using the web-testing of students with the generation of variants based on a genetic algorithm);
- Effectively taking into account the personal characteristics of trainees when selecting and shaping learning strategies and influences, including the development of special corrective learning influences aimed at developing individual trainee personal characteristics (the results of psychological tests of trainees are used together with various types of learning interactions);
- The use of additional (repeated) training on the basis of the identified gaps in knowledge and skills, etc. (the sets of learning interactions are used for different clusters of students).

As for the information necessary for the formation of social and personal competences (from the group of universal competences), taking into account the personal characteristics of the trainees, here you can partially use the information presented in the professional standards in the job description "self-development" for each specialty. In addition, to identify personal characteristics, there are a large number of psychological tests, surveys, there are special sites, etc. For example, in the context of tutoring

IES and web-IES [1–5] for building models of students, the possibility of identifying about 20 personal characteristics and their correlation with an individual learning model. The main problem here is the search and selective selection of expert information, signaling the degree of manifestation of specific competence for each of the personal characteristics.

It should be noted that at present, there is no universal classification of competencies, however, the generally accepted point of view is the allocation of professional and universal competencies. Further specification depends on the specifics of the profession, the traditions of the university that trains specialists in this field, and other features.

The purpose of this work is to analyze the methodological and technological experience of computer-aided construction of using competence-oriented models of students, future specialists in the field of knowledge engineering (the profession "systems analyst") using the tutoring IES and web-IES.

2 The Dynamic Designing of Competence-Oriented Models for Future Professionals Based on the Review of the Monitoring of Tutoring IES Students' Functioning Processes

Since 2008, tutoring IES and web-IES, developed in the Intellectual Systems and Technologies laboratory of the Cybernetics Department of the National Research Nuclear University MEPhI, have been actively used for automated support of basic courses/disciplines in the areas of Applied Mathematics and Computer Science and Software Engineering. including: Introduction to Intelligent Systems, Intelligent Dialog Systems, Dynamic Intelligent Systems, Designing the Knowledge-Based Cybernetic Systems, Modern Intelligent System Architectures, Intelligent Information Systems.

For all these courses and disciplines, using the basic tools of the AT-TECHNOLOGY complex, the corresponding applied ontologies are implemented and dynamically supported, which together form the "Intelligent Systems and Technologies" generalized ontology as an educational and methodological basis for the training of knowledge engineers. Significant methodological and technological experience has been accumulated in the automated management of a significant number of individual models of students in the above disciplines (more than 2,700 models) and corresponding training models, the joint analysis of which allows us to predict the so-called "ideal" model of a young specialist, in particular, a system analyst (knowledge engineer).

The possibility of implementing such a forecast is largely determined by the characteristics of the development and use of tutoring IES and web-IES associated with automation of almost all the processes that occur during the training and monitoring of knowledge/skills of students. At the same time, all the information about the students, the topics of the courses/disciplines, the results of the training, the results of the students' control, individual recommendations based on the results of the training, etc. is in a single environment and at any time available to the student and/or supervising the learning process, which is ensured by special means of monitoring the process of functioning of the tutoring IES.

In accordance with [2, 4] and other works, the monitoring of the functioning of tutoring IES is considered from the standpoint of two aspects. One of them is connected with the place and role of the IES trainers from the point of view of the actual organization of studies in higher education, i.e. use of tutoring IES to support typical stages of training: conducting classes (lectures, seminars, laboratory work), conducting periodic monitoring activities, both during training and in the control points of training provided by the curriculum of a particular course/discipline, as well as control efforts as part of credits and exams.

Another aspect is the consideration of a set of functional tasks characteristic of intellectual learning. The implementation of typical tasks of intellectual education has been repeatedly considered in various papers, in particular [2, 4–6] and others, therefore, we note that in the context of the generalized ontology "Intellectual systems and technologies" and the creation of a single ontological space of knowledge and skills at present it was possible to implement, practically, the full set of functional tasks characteristic of the intellectual technology of learning, namely [2, 4] and others:

- *Individual planning* of the methodology for studying a specific training course (specifying on the basis of ontologies of courses/disciplines of a personal trajectory/learning strategy, individual control and identification of "problem areas" in students' knowledge and skills, optimization of individual learning, taking into account the psychological portrait of the learner, etc.);
- *Intelligent analysis* of learning tasks (modeling the reasoning of students solving learning tasks of various types, including using non-formalized methods, identifying types of errors and causes of their manifestation in knowledge and skills instead of their finding, feedback through dynamic updating of students' knowledge and skills, forecast grades on exams, etc.);
- *Intelligent decision support* (using technology of traditional expert systems (EC) and PEC for intellectual assistance at each stage of solving educational problems, including extended explanations like "how?" and "why?", the choice of solutions, a hint of the next stage of the decision, etc.).

Thus, the monitoring of the functioning of tutoring IES and web-IES in this case is associated with "tracking" and analyzing all the processes of building for each student an personalized model of the student in the relevant discipline by identifying the current level of knowledge/skills using web testing and other methods, as well as the formation of a psychological portrait of the student's personality as an important component of the student model.

It should be noted that, in accordance with the problem-oriented methodology, the basis of the approach for constructing the current competence-based model of the student is a dynamic comparison of the results of web testing with the corresponding fragment of the applied ontology of the course/discipline. The result is the so-called "problematic areas" [1, 2] in the knowledge of students in individual sections/subsections and the construction of current competencies, jointly reflecting the state of the student model not only in terms of knowledge level, but also providing a conceptual and technological connection with the processes of identifying skills solve some types of educational non-formalized tasks recommended in [1, 9] or training in knowledge engineering, for example [1, 2]. It is also necessary to constantly form lists of students (contingents) with high and/or low indicators of knowledge/skills, conduct systematic statistical data processing, as well as ensure the generation of current and final reports (statements) for departments and deans.

The final term logs that reflect the students the competence-oriented models of students contain complete information about the students - assessments obtained during the control measures related to the identification of knowledge and skills, the current level of professional competence, information about passing psychological testing, information about independent work, the final forecast grades, as well as a real grade obtained in the exam (statements are formed for all students enrolled in a particular course/discipline).

An important place in the formation of the future specialist's model is given to the analytical and statistical processing of the results of the use of training PECs. By introducing special parameters that characterize both the individual learner and a specific contingent (cluster) of learners. These parameters were formulated by an expert on the basis of an analysis of a fairly representative amount of data (about 2000 student models) and focused primarily on the basic structure of the student model [1, 2], the components of which are: the student knowledge model, the student skills model, psychological portrait, model competences and other components).

Experience has shown that the parameters (indicators) formed as a result of [2] were the most popular from the point of view of building competence-oriented models of future specialists:

- Analysis of the "problematic areas" of each student in specific courses/disciplines and their clustering;
- Individual training planning (typology and sequence of learning influences, the influence of learning influences to increase the level of knowledge, the search for the most effective training impacts);
- Calculating the correlation between current levels of knowledge and skills on relevant topics of the course/discipline;
- Taking into account the psychological portrait of the student (personal degree of achievement of the target competencies for specific courses/disciplines, etc.);
- Forecast assessment on the exam according to the results of the semester (analysis of reasoning in solving specific training problems).

In addition, a number of parameters are used to process information for the entire contingent of students (group, stream, etc.), namely: the cumulative analysis of "problem areas" for specific courses/disciplines and their clustering; assessment and clustering of individual training plans for specific courses/disciplines; forecast of the results of the examination session (connection of levels of knowledge and skills and assessment for the exam in the context of the course, analysis and clustering of psychotypes of students, etc.).

Now we will take a closer look at practical examples of building an "ideal" model of a specialist graduate in the profession "System Analyst" (knowledge engineer) in accordance with the basic competencies of the current Federal State Educational Standard for Higher Professional Education.

3 Particular Features of Training Knowledge Engineers Based on Use of Tutoring IES

The success of knowledge engineering associated with the development of models and methods for transferring knowledge from various sources of knowledge into a computer program, known as the expert system (ES), and in a broader sense - knowledge-based systems (KBS) [2, 10, 12, 19], led to the emergence of a new profession, in which not only professional competencies, but also individual personal characteristics are actively demanded. Therefore, the training of knowledge engineers along with the inculcation of professional knowledge, skills and abilities in the field of the development of (KBS) is impossible today without taking into account his psychological portrait.

Let's give some examples from the experience of practical use of tutoring IES and web IES in the educational process of MEPhI, focusing on the technology of automated formation of a competence-oriented model of a specialist in the field of knowledge engineering, taking into account the total set of curricula and generalized ontology of Intelligent Systems and Technologies in Software Engineering training domain.

Shaping the Specialists' Professional Competencies

According to the Federal State Educational Standard 3+, the two following competencies are used as the base ones for training knowledge engineers:

- PK-1 is the formalization capacity in his or her subject area, with the view to the limitations on study methods in use;
- PK-2 is the ability to use methods and instrumental means of study of professional business items.

The achievement of these target competencies is facilitated by the common ontological space of knowledge and skills, which is formed by the applied ontologies of courses/disciplines of several tutoring IES and web IES [2–7, 20]. It is important to note that the general competency model, which is a component of the ontology base model in the form of a semantic network [2], is applied in applied ontologies of courses/disciplines as a hierarchy of subject/problem-oriented private competences (with weights) reflecting the methodology of teaching specific courses.

Let's briefly review some of the approaches used to identify individual professional competencies. Section 1 already noted the peculiarities of students' knowledge revealing processes carried out during control events by dynamically building the current competence-based model of the student, which is based on the analysis of answers to questions from special web tests and subsequent comparisons with the fragment of course/discipline ontology [2].

Now we focus on the problems of identifying the ability to apply theoretical knowledge in practical sessions. Experience has shown that the training of knowledge engineers today is impossible without imparting the skills and abilities to solve practical tasks related to the ability to build on the principle of "self-expert" models of the simplest situations of a problem area based on products, frames and semantic networks [12].

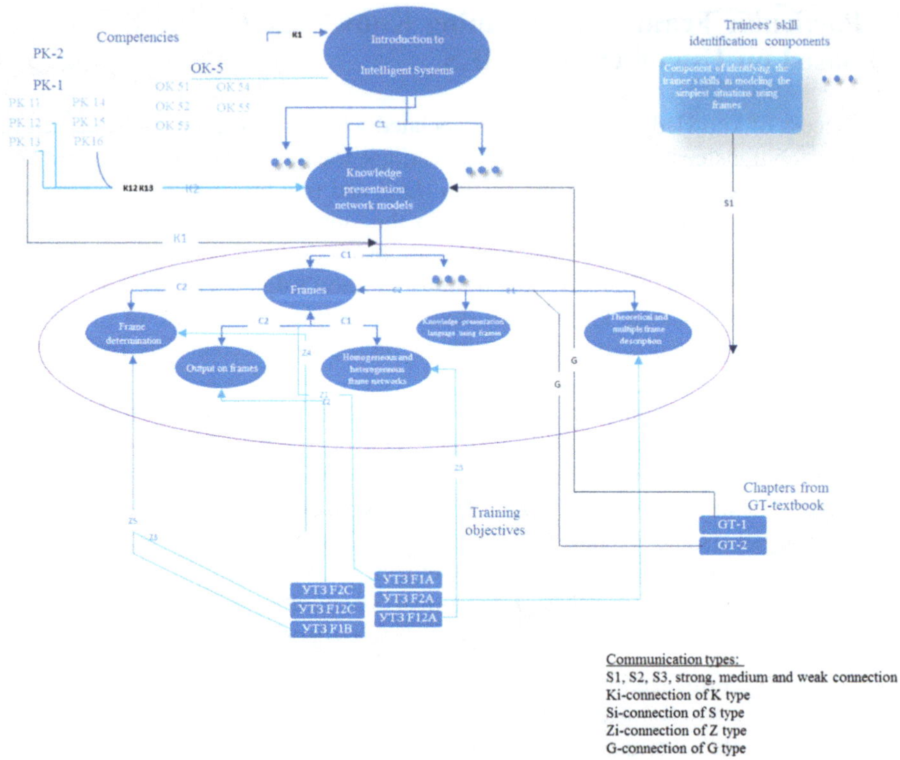

Fig. 1. «Introduction to Intelligent Systems» ontology fragment

Therefore, in the process of developing the tutoring IES, special software was developed and tested in practice in the educational process of MEPhI and other universities. These tools implement "manual" methods for solving non-formalized tasks (NF problems), in particular, presented in [19].

In Fig. 1. the fragment of the applied ontology, Introduction to intelligent systems, is presented, which includes the elements: Network Models of Knowledge Presentation, Frames, Frame Definition, Theoretical and Multiple Frame Description, Notion of Frame Prototype and Frame Copy, Homogeneous and Heterogeneous Frame Networks, LKP (language of knowledge presentation) based on frames, Release on Frames Such competencies as PK12 and PK13, which make part of the target competence PK1, are referred to this course/discipline ontology fragment.

To identify the competencies of type PK2, laboratory workshops and practical exercises related to the instrumental and technological aspects of training knowledge engineers, and focused on studying the technology of designing (KBS) (ES) using the modern arsenal of instrumental tools [1, 2, 10, 19].

Thus, with the help of the "Introduction to intelligent systems" ontology and special tools related to the construction of the current models of the student, a certain set of professional requirements (criteria) imposed on the competences of the future knowledge engineer is dynamically formed.

Now let's consider the formation of common cultural competencies and psychological portrait of a future specialist in the field of knowledge engineering.

3.1 Shaping the Psychological Profile of a Future Knowledge Engineer (in the Context of General Cultural Competencies)

As shown in [19], in knowledge engineering, an important place is given to the psychological aspect associated with the processes of knowledge extraction, since it is he who determines the success and effectiveness of the interaction of a knowledge engineer with the main source of knowledge — an expert. In modern works, for example [2, 11, 12, 19–22] and others, it is proposed to take into account a number of personal characteristics or their combination in the form of a psychological portrait when determining the so-called "ideal" pair of knowledge engineer - an expert for organizing collective work in creation of the problem area model.

For these purposes, about two dozen different author texts are currently used in tutoring IES and web-IES, and the optimal test configuration for psychological tests of trainees, depending on the type of competencies to be detected, is performed using a special psychological test generator.

4 Conclusion

Thus, the methodical and technological experience gained in training specialists in "Applied Mathematics and Informatics" and "Software Engineering" in the domain of automated designing using tutoring IES and web-IES of competence-oriented models of knowledge engineers enable to promptly and efficiently review, adjust (focusing on the most modern innovations in the professional sphere) and predict the level and quality of the graduate professionals' cohort. This approach lays the foundation for not only in relations with employers and potential customers, but also allows to plan targeted training of specialists in various areas, starting with junior courses.

The work was performed with the Russian Foundation for Basic Research support (Project No. 18-01-00457).

References

1. Rybina, G.V.: The Theory and Technology of Integrated Expert Systems Construction: Monograph. Nauchtekhlitizdat, Moscow (2008)
2. Rybina, G.V.: Intellectual Systems: From A to Z: A Series of Monographs in Three Books. b.1: Knowledge-Based Systems. Integrated Expert Systems. Nauchtekhlitizdat, Moscow (2014)
3. Rybina, G.V., Fontalina, E.S.: Certain approaches to automated designing of competence-oriented models for knowledge engineers using the tutoring integrated expert systems. In: Advances in Intelligent and Soft Computing, vol. 785, pp. 161–173 (2019)
4. Rybina, G.V.: Intellectual Systems: From A to Z: A Series of Monographs in Three Books. b. 2: Intelligent Dialogue Systems. Dynamic Intelligent Systems, 164 p. Nauchtekhlitizdat, Moscow (2015)

5. Rybina, G.V.: Modern approaches to intelligent computer learning based on the development and use of tutoring integrated expert systems. Instrum. Syst. Manage. Control Diagn. **5**, 10–15 (2010)
6. Rybina, G.V.: Intelligent tutoring systems based on integrated expert systems: the development and use. Inf. Measuring Control Syst. **10**, 4–16 (2011)
7. Rybina, G.V., Sergienko, E.S.: Intellectual training based on integrated expert systems: the modeling skills of trainees to solve complex educational problems. Inf. Measuring Control Syst. **1**, 31–39 (2015)
8. The portal of the Federal state educational standards of higher education. www.fgosvo.ru
9. Rybina, G.V., Blokhin Yu., M.: Methods and tools for the intelligent planning: application to manage build processes integrated expert systems. Artif. Intell. Decis. Making, **1**, 75–93 (2015)
10. Rybina, G.V., Blokhin Yu. M., Danyakin, I.D.: Intelligent technology of integrated expert systems construction. Inf. Meas. Oper. Syst. **1**, 3–19 (2015)
11. Gavrilova, T.A., Leshcheva, I.A.: Systematic view on training of knowledge engineers and business analysts. In: Proceedings of Fifteenth National Conference on Artificial Intelligence with International Participation of KII – 2016, 3 volumes, vol. 1, pp. 16–23. Universum, Smolensk (2016)
12. Gavrilova, T.A.: Knowledge Engineering. Models and Methods, 324 p. LAN Publishing House, SPb. (2016)
13. Fedyaev, O.I.: Model of the system of training and employment of specialists on the basis of software agents with neural network architecture. In: Proceedings of Fifteenth National Conference on Artificial Intelligence with the International Participation of KII-2016, 3 volumes, vol. 2, pp. 372–381. Universum, Smolensk (2016)
14. Kirichek, A.V., Morozova, A.V.: The Axiomatic basis of qualitative multidimensional model of social-professional competence of the young specialist. Proc. Samara Sci. Cent., Russ. Acad. Sci. **13**(4(4)), 1232–1235 (2011)
15. Alekseeva, A.V., Mamyrbekov, A.K., Moshegov, A.T., Pashkov, P.M.: Competence approach to the training of analysts, engineers. In: Engineering Enterprises and Knowledge Management (SP&UZ-2017): Collection of Scientific Proceedings of XX All-Russian Scientific Conference, vol. 1, pp. 82–86. FGBOU VO "REU them. G.V. Plekhanova", Moscow (2017)
16. Koblova, N.Oh., Barbara A.D.: The implementation of a comprehensive approach in the process of additional training of University students. Mod. Res. Soc. Probl. **6**(38) (2014)
17. Khokhlova, M.: The competence of the competitive young specialist: opinions of employers and graduates. A Telesc. **2**(104), 26–31 (2014)
18. Bobrov, L.K., Medyankina I.: Competence approach to training business information managers. In: Engineering of Enterprises and Knowledge Management (IP&UZ-2017): Collection of Scientific Papers of the XX Anniversary All-Russian Scientific Conference, vol. 1, pp. 93–99. FGBOU VO "REU them. G. V. Plekhanova", Moscow (2014)
19. Rybina, G.V.: Fundamentals of Intelligent Systems, 432 p. Finance and Statistics, Moscow (2014)
20. Rybina, G.V., Fontalina, E.S.: Automated construction of young specialists models with the use of tutoring integrated expert systems. In: Proceedings of IV International Conference on Information Technologies in Engineering Education, pp. 41–44 (2018)
21. Rybina, G.V.: Intelligent System: from A to Z. Monograph Series in 3 Books: Book 3. Problem-Specialized Intellectual Systems. Tools for Building Intelligent Systems, 180 p. Nauchtekhlitizdat, Moscow (2015)
22. Rybina, G.V., Sergienko, E.S., Sorokin, I.A., Belova, A.S., Bochkova, T.V.: Ontological approach to the educational process through the development and use of tutoring integrated expert systems. Instrum. Systems Manag. Control. Diagn. **2**, 13–24 (2017)

A Teaching Experience of the Human-Computer Interaction Course in a Master Program

Freddy Paz[1(✉)], Freddy A. Paz[2], Arturo Moquillaza[1,3], and Fiorella Falconi[1]

[1] Pontificia Universidad Católica del Perú, San Miguel, Lima 32, Lima, Peru
{fpaz, amoquillaza}@pucp.pe, ffalconit@pucp.edu.pe
[2] Universidad Nacional Pedro Ruiz Gallo, Lambayeque, Peru
freddypazsifuentes@yahoo.es
[3] Universidad San Ignacio de Loyola, Lima 12, Lima, Peru
miguel.moquillaza@usil.pe

Abstract. Learning-by-doing is an educational theory which establishes that the students learn in a more effective way if they are involved in experiences of the real world instead of passively listening. In the HCI course of the Master Program in Informatics, of the Pontifical Catholic University of Peru, the professors proposed the active-learning approach for the teaching of several topics related to HCI such as usability, user-centered design, human factors, accessibility, ergonomics and internationalization that are nowadays relevant concepts for the design process of graphical user interfaces. For this purpose, we received the committed collaboration of a local financial entity, the BBVA Continental Bank, which voluntarily agreed to establish a set of requirements for the design of GUIs that support the implementation of new features in ATMs. This proposed exercise has allowed the students to apply what they learned in real work situations, and to consolidate the discussed concepts from practice.

Keywords: Human-Computer Interaction · Education · Software Engineering · Software development process · User Experience

1 Introduction

Nowadays, the concept of User Experience (UX) has achieved relevance not only in academia but also in the industrial context [1]. Given the current competitive market and the diversity of available options to obtain software products, the final users are more concerned about the quality attributes that these applications can have, than on the functionality. In the past, the approach of software developers was to focus efforts on the technology, the project resources, and the programming techniques. However, the emergence of innovative methodologies such as Lean UX, Agile UX, and User-Centered Design (UCD) has impacted on the UX awareness in the IT industry [2].

A new philosophy which involves the final users in the entire process of software development is reaching an extensive acceptation because of the benefits that this approach provides. The advantage of a participatory design based on the user's opinion

W. Karwowski et al. (Eds.): AHFE 2019, AISC 963, pp. 131–142, 2020.
https://doi.org/10.1007/978-3-030-20135-7_13

allow to decrease the number of errors that the software product can present at the end of the development process, and therefore, the costs associated to the modifications that are required [3]. The importance of these topics at present has led to incorporate them in the course of Human-Computer Interaction (HCI) as part of the curricula of the Master program in Informatics at the Pontifical Catholic University of Peru (PUCP). However, instead of teaching these theories and concepts in a traditional way, by pure expository classes, we decided to change the conventional paradigm and to propose a learning-by-doing scenario for our students. Under this approach, they were able to apply all the concepts learned in the classroom in a real case study. To carry out this teaching experience, we received the support of a financial company in Peru, the BBVA Continental Bank, which voluntarily agreed to collaborate with the formulation of the students' assignments.

In this paper, we report a teaching experience in which the active learning approach was employed to impart knowledge on the main topics of HCI. The purpose of our research is to establish the preliminary foundations for a future proposal of a teaching methodology of the HCI course. The paper is structured as follows. In Sect. 2, we defined the method that guided the educational experience as well as the way in which the course was structured for this purpose. In Sect. 3, we describe the requirements and the case studies that were formulated for the students. In Sect. 4, the results of the educational practice are presented and discussed, and finally, in Sect. 5, the conclusions and works derived from this research are established.

2 Theoretical Overview

2.1 Learning-By-Doing Approach

Learning-by-doing is an innovative educational approach developed from the theoretical basis exposed by the American philosopher, psychologist and educational reformer, John Dewey, which established that the learning process is an iterative process [4]. Professors must look at learning as a cycle of experiences where lessons are planned and executed based on observation and reflection from their own and their learners' previous experiences and interests [5]. In this theory, Dewey theorizes that the learning must be relevant and practical, and not just only passive and theoretical.

Current pedagogical practices hold that students learn and retain information best when they are involved in real-world projects that give them direct experiences that can control to meet their objectives [6]. Professors should present real-life problems to the students and then guide them to solve problems by providing them with a hands-on activity to learn the solution. This methodology requires teachers to come up with engaging class projects that lead their students to the required skills and knowledge through practical application.

Learning-by-doing is a theory based on the processes that enable students to learn while they are engaged in solving problems [7]. Nowadays, this learning methodology has evolved to become "Project-based learning". Some educational institutions have a system in which the students work in real projects linked to relevant companies in the local industry. Project-based learning is a teaching approach that engages students in hands-on activities and experiences that help them generate and apply new skills and concepts learned in the classroom [8]. This methodology, in fact, represents a challenge for professors, since the best projects involve students working in groups, professors must also know how to mix according to personality, learning style or other criteria.

2.2 Structure of the HCI Course

The HCI course of the Master Program in Informatics at the Pontifical Catholic University of Peru (PUCP) is part of the first semester of studies as can be appreciated in Fig. 1. The purpose of this subject is to develop the necessary skills in the students for the design, evaluation and implementation of software systems from the perspective of a user-centered methodology, which involves the indispensable participation of the final users during the entire process of development, for the obtaining of a product that meets the expectations of the stakeholders.

Semester I	Semester II	Semester III	Semester IV
[CIC602] **Programming Languages** 3 Credits	[INF649] **Research Methodology and Ethics** 3 Credits	[INF618] **Thesis Seminar 1** 3 Credits	[INF628] **Thesis Seminar 2** 6 Credits
[INF647] **Human-Computer Interaction** 3 Credits	[INF650] **Technological Innovation, Development and Entrepreneurship** 3 Credits	[INF653] **Software Project Management** 3 Credits	
[INF639] **Software Construction Process** 3 Credits	**Elective Course** 3 Credits	[MAT836] **Statistics and Experimentation in Computing** 3 Credits	**Mandatory Mention Course** 3 Credits
[INF648] **Machine Learning** 3 Credits	**Mandatory Mention Course** 3 Credits	**Mandatory Mention Course** 3 Credits	**Mandatory Mention Course** 3 Credits

Fig. 1. Distribution of the Master Program in Informatics at PUCP

Given the current relevance that the field of Human-Computer Interaction has reached during the last years, the course is mandatory for the two specialization areas that the program offers: (1) Computer Science, and (2) Software Engineering. The purpose of the HCI course is the teaching of the User-Centered Design (UCD) methodology for the design of high-quality graphical user interfaces. In order to establish a learning-by-doing approach, the professors reorganized the concepts imparted in the course, in a way in which each topic could be applied in a specific phase of an assigned project. The final structure of the course content is shown in Table 1.

The teaching experience was performed during the semester 2018-2, in which the course had the enrollment of thirty students. Considering this number of students, the professors established six teams of five members each one. The teams were requested to achieve the creation of a specific interactive prototype, and the organization of the content allows them to develop the designs, at the same time in which the multiple

Table 1. Structure of the topics in the HCI Course

Unit	Topic	Week
1. Introduction	1.1. The field of Human-Computer Interaction 1.2. Historical background	1
2. Human Factor	2.1. The users 2.2. The computers 2.3. The interaction	2
3. Interaction Design	3.1. Interaction devices 3.2. Analysis of requirements 3.3. User-Centered Design 3.4. Design techniques: Personas, Empathy map and Card sorting 3.5. Prototyping	3, 4, 5, 6, 7
4. Usability	4.1. Paradigms and principles of usability 4.2. Usability Engineering 4.3. Usability evaluation: Inspection methods and Testing methods	8, 9
5. Communicability	5.1. Semiotic Engineering 5.2. Communicability Evaluation Method 5.3. Semiotic Inspection Method	10, 11
6. Accessibility and Globalization	6.1. Accessibility 6.2. Internationalization	12

prototyping techniques and general aspects of the UCD methodology were studied. At first, the students developed low-fidelity prototypes as a base to subsequently develop a more advanced version in the form of wireframes.

The main goal of the projects was to design interfaces according to all the aspects established by the UCD approach and some guidelines of HCI, but not to develop a functional implementation of the system. In this sense, the teaching experience consisted of presenting the students with a scenario to be solved over the elaboration of certain graphical user interfaces that meet the requirements proposed. The course is formed by fourteen weeks; however the week 13 is destined for the presentation of the final result of the project, and week 14 is reserved for the final exam.

3 Design of the Learning-By-Doing Experience

To involve students in real projects, in which they could apply the learned concepts, the professors looked for the support of an IT company from the local industry. The BBVA Continental Bank of Peru voluntarily agreed to participate in the formulation of the students' project. The case studies were formulated according to real needs and requirements of the bank. The purpose of the projects was to design usable interfaces for Automated Teller Machines (ATMs) to support special and uncommon features.

In a previous work [9], we have developed a similar experience with students from the University of Cauca (UNICAUCA) - Colombia. In the same way as our last work, the bank provided all the related information in order to students could develop their proposals according to requirements and complementary information. The bank established that: (1) the ATM interfaces had to fulfill the users' needs and expectations, (2) an interaction design process had to be followed, and (3) the final prototypes had to be tested with real users. The ATM interfaces were designed by the students employing a user-centered process according to the developed and discussed topics in class.

The BBVA Continental Bank established the scenarios described as follows. A different case study was assigned to each team of six students, which was responsible for the proposal of ATM interfaces that meets the requested conditions.

- *Interactive advertising*: The bank needs to display offers in the ATM screen to its customers in a proactive way, without forcing the users to consult the information. The advertising must be received in a usable and friendly way, without interrupting the experience of the use of the ATM.
- *Cheque deposit*: The bank needs the ATM to be used to collect cheques that belong to its financial entity. The money collected from the cheque could be deposit by the users to any of their bank accounts.
- *Withdraw with QR for non-customers*: The bank needs the ATM to be used to withdraw money in an experience which involves biometrics and current technologies such as Quick Response (QR) code.
- *Advanced withdraw*: The bank needs the ATM provides the users the possibility to withdraw money with a more detailed personalization to the current one, offering the opportunity to choose the quantity and denomination of the money bills, without impacting on the agility and efficiency of the process.
- *Delivery of cards with QR*: The bank needs the ATM to be used by the users to receive a credit card in an end-to-end experience in which the interaction with an office is not required.
- *Quick withdraw*: The bank needs its customers to be able to withdraw money in the ATMs in a more agile and efficient way than the regular and established process. The ATM usage time must be optimized, and the users must be able to receive the money that they frequently withdraw from the ATM.

4 Results of the Teaching Experience

In this section, the result of one of the teams is reported. According to the topics that were reviewed in the classroom, students adopted the user-centered design process in the following techniques: (1) Personas, (2) Empathy map, (3) Storyboards, (4) Paper prototyping, (5) Heuristic evaluation and (6) User testing. Next, each of the techniques used in the design process as well as the results obtained by one of the teams will be described.

4.1 Personas

Personas is a method to describe the target users of a tool, software product or system, establishing a clear picture of their expectations and the way in which they are likely to use the technological innovation [10]. Persona is a conceptual model of a targeted user group(s) that can serve to promote the shared understanding throughout the process of analysis, design, development, and implementation. Past studies have argued that conceptualizations of users' mental models could be used effectively and successfully to create a shared understanding among multiple individuals over a single topic or domain and thus a better system [11]. As aggregated conceptual models, personas should facilitate shared understanding in project communications by creating a mutually understood context. In Fig. 2, we can appreciate the application of Personas in the development of the case study related to *Interactive advertising*.

Proyecto:	Publicidad en ATMS	
Nombre del Usuario	**Estudios**	**Ocupación**
Mariano	Bachiller en Ingeniería de Sistemas	Programador en una empresa desarrolladora de software
Photo	**Detalles Demográficos**	**Breve descripción de la persona**
	**Edad:** 25 **Sexo:** Hombre **Estado Civil:** Soltero, vive con sus padres **Lugar:** Comas **Discapacidad:** Ninguna **Ingreso:** 1500-2700 Sole **Horario de trabajo:** 9am-6pm, 40 horas semanales **Tiempo de Experiencia:** 3 años	Sus hobbies son bailar, tomar, probar comida nueva y es fanático del futbol. Sus principales gastos son comida, bebida, ropa y transporte.
Información Bancaria	**Información sobre las operaciones**	**Opinión sobre la publicidad**
Es cliente de los bancos Interbank y Cencosud. Posee tarjeta de crédito y débito. Emplea más la tarjeta de débito ya que en la de crédito los intereses son muy altos.	Acude aproximadamente una vez por semana a los ATMS de Globalnet. Sus operaciones frecuentes son las de retiro y depósito y tienen una duración promedio de 2 minutos.	Opina que la publicidad debe ser selectiva y de su interés ya que desea realizar operaciones rápidas. La publicidad que lo cautivará debe estar orientado en beneficios del banco como cuentas nuevas de depósitos a plazos fijos, promoción en cuanto a tasas de préstamo, ofertas de productos inmobiliarios exclusivos del banco. Él considera que la publicidad debe ser fácil de remover ya sea al inicio o al final.

Fig. 2. Application of the technique Personas by the students

4.2 Empathy Map

Empathy map is one of the main techniques applied in the Design Thinking process [12]. By displaying user attitudes and behaviors in a map, the UX team gets proper insight and guidance for developing prototypes accordingly to the final user's needs. The method allows to create the degree of empathy for a specific person [13]. The focus is on understanding other individuals by looking at the world through their eyes. When the stakeholders understand the user, they will be able to understand how small changes in design can have a significant impact on users. In Fig. 3, it is possible to appreciate the empathy map performed by one of the teams related to the case study of *Cheque deposit*.

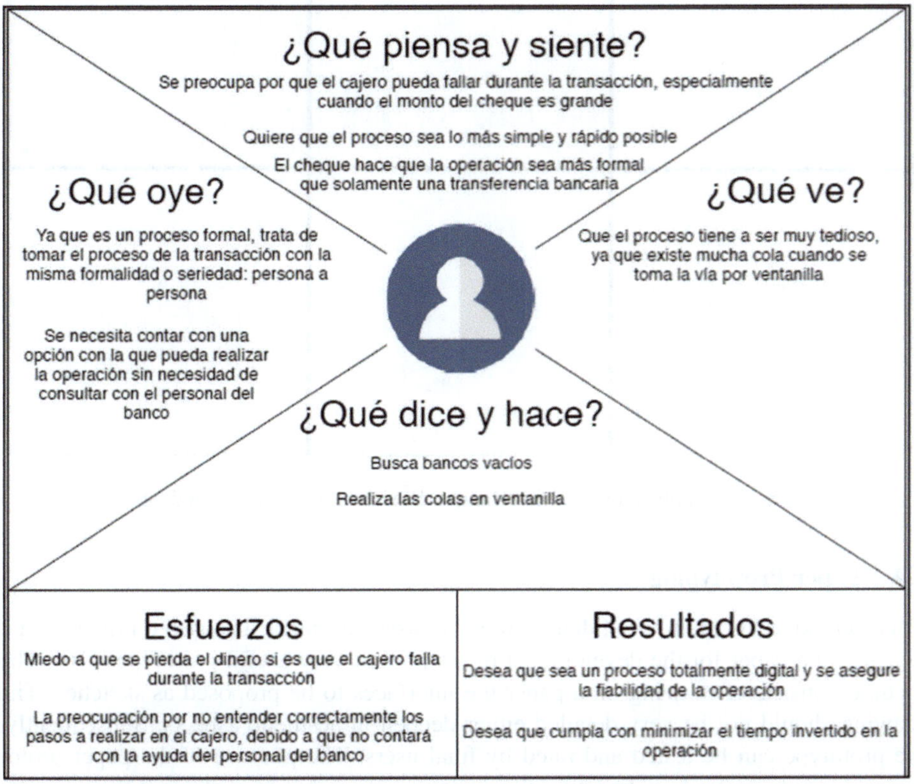

Fig. 3. Application of the technique Empathy Map by the students

4.3 Storyboards

The technique of storyboarding was originally conceptualized on the cinema industry [14]. The method consists of a series of drawings or images organized in a sequential format (storyboards), that according to the user-centered design methodology, represent how the new tool, system or software product will be used for the achievement of a certain task. Storyboards can be cataloged as effective instruments to represent in a graphical and informal way, the problem, the technological solution and the scenario of use. Usually, the storyboard is accompanied by an explicit narrative that helps the stakeholders to understand the situations, scenarios and reasons by which the proposal that is being designed is actually required [15]. In Fig. 4, it is possible to notice the storyboard performed by the team in charge of the case study related to *Withdraw with QR for non-customers*.

Fig. 4. Application of the technique Storyboarding by the students

4.4 Paper Prototyping

This technique to obtain low-fidelity graphical user interfaces, is based on the use of a pencil and a paper for the development of simple but very useful prototypes [16]. The method consists of drawing on a paper the interfaces to be proposed as sketches. The drawing should not be very detailed either deepen in aesthetic aspects. Subsequently, the prototype can be tested and rated by final users. The purpose of the paper prototyping is not to verify how beautiful the design is, but rather to evaluate if the users are able to perform their tasks with the proposed interfaces. This technique does not establish the incorporation of technological advances, just only the functionality of the system and the way in which the information and the interaction is communicated.

In Fig. 5, it is possible to appreciate the prototypes that were performed by the team which developed the case study related to *Advanced withdraw*.

4.5 Heuristic Evaluation

The heuristic evaluation is an inspection method in which specialists judge if each element of a graphical user interface follows recognized usability principles [17]. The advantage of this technique against other approaches is that the execution of the heuristic evaluation is usually less expensive than other assessment techniques in which a representative number of users is required. According to Nielsen, a maximum number of five experts is enough to identify most of the usability problems that are present in a

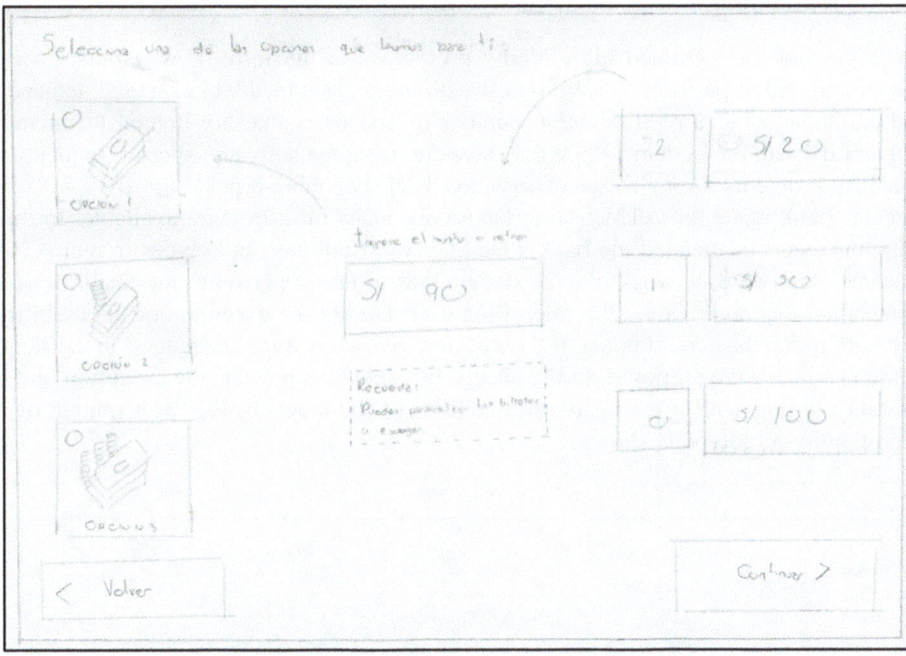

Fig. 5. Application of the technique Paper Prototyping by the students

graphical user interface [18]. However, the identification as well as the recruitment of usability experts can represent a difficult process. In this specific teaching experience, the students used the ten usability heuristics proposed by Nielsen, to carry out the revision of their design proposals. In Fig. 6, it can evidence some of the problems that were identified by the team whose case study was related to *Delivery of cards with QR*.

ID	Área editable Definición del Problema	Comentarios/Explicación	Ejemplos de Ocurrencia	Heurísticas Incumplidas	Screenshots
P1	Flujo de navegación no queda claro	No se puede identificar el orden de las pantallas al no estar definida la navegabilidad	Todas las pantallas	H4	SS1.jpg
P2	Metafora de pin no queda clara	Se muestra un pin de color azul y los demas en gris pero no queda claro que quiere indicar con esto	En la pantalla numero 3	H2	SS2.jpg
P3	No se muestra cabecera con el nombre del banco	En la pantalla no se muestra el nombre del banco, a diferencia del estandar mantenido por todas las pantallas del cajero del banco	Todas las pantallas	H4	SS3.jpg
P4	No permite personalizar los montos de retiro rápido	Se muestra la lista de montos rapido con metafora de eliminar mas no muestra ninguna forma de editar dichos montos	En la pantalla numero 1 y 3	H2, H7	SS4.jpg
P5	No se utiliza metafora para Menu principal	Se deberia colocar el icono de home para indicar que se vuelve al meno principal y no unicamente texto.	En la pantalla numero 6	H2	SS5.jpg

Fig. 6. Application of the technique Heuristic Evaluation by the students

4.6 User Testing

Once the students obtained high-fidelity prototypes in the form of wireframes, they were requested to perform a usability test with users. This method in specific demands the participation of a representative number of end users that are invited to interact voluntarily with the system. While the users are using the software, specialists identify usability problems through user observation [19]. For this purpose, the BBVA Continental Bank made the ATM testing laboratory of its infrastructure available. In this way, the students attended the bank's facilities to install their prototypes in real ATM systems. Afterward, the task of each student was to take a person to the facilities as a companion that could cover the role of the user. During the execution of the usability test with users, they could notice if their design intentions were understood in the same sense in which were proposed. In Fig. 7, it is possible to appreciate the execution of the usability testing with users inside the facilities of the bank. In Fig. 8, a sample of a high-fidelity prototype is shown.

Fig. 7. Execution of the usability testing with users performed by the students

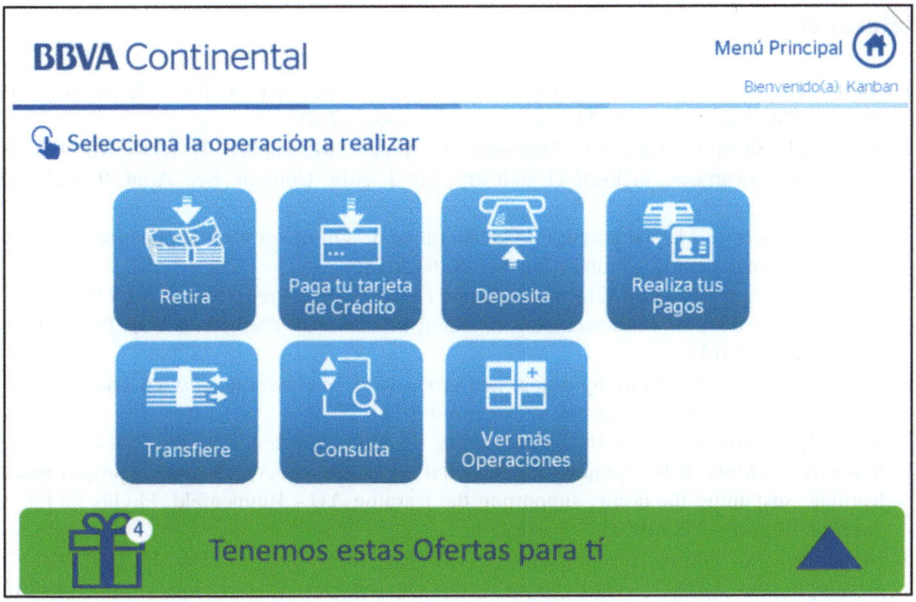

Fig. 8. A sample of the screen of the final prototype developed by the students

5 Conclusions and Future Works

Usability and UX are nowadays two essential concepts that can cause a significant impact on the acceptance by the users of the final product. In this sense, these aspects must be considered in the curricula of a Master Program in Computing or related areas. The course of Human-Computer Interaction can be the perfect scenario for these topics in which the user-centered design methodology could be studied in detail. According to the teaching experience that we describe in this research, learning-by-doing is an effective methodology that can be applied in the classroom especially for these concepts. The students have a more in-depth learning when they are involved in a real scenario which challenges them to think about innovative ideas, and to apply all the concepts learned in class in the solution proposal. Some future works could involve the application of the teaching methodology to related courses or in similar scenarios at other educational institutions. With this work, the authors are consolidating the base for a more detailed proposal of the teaching of the HCI course in the field of Software Engineering.

Acknowledgments. The authors thank all the participants involved in this teaching experience that has allowed the development of the present research work. This study is highly supported by the *Section of Informatics Engineering* of the *Pontifical Catholic University of Peru* and the "HCI, Design, User Experience, Accessibility & Innovation Technologies Research Group (HCI-DUXAIT)". HCI-DUXAIT is a research group of the *Pontificia Universidad Católica del Perú* (PUCP).

References

1. Santoso, H.B., Schrepp, M.: Importance of User Experience Aspects for Different Software Product Categories, pp. 231–241. Springer, Singapore (2018)
2. Ashraf, M., Khan, L., Tahir, M., Alghamdi, A., Alqarni, M., Sabbah, T., Khan, M.: A study on usability awareness in local IT industry. Int. J. Adv. Comput. Sci. Appl. **9**, 427–432 (2018)
3. Mirri, S., Roccetti, M., Salomoni, P.: Collaborative design of software applications: the role of users. Human-centric Comput. Inf. Sci. **8**, 6 (2018)
4. Dewey, J.: Experience and Education. Kappa Delta Pi, Free Press, New York (1997)
5. Aubrey, K., Riley, A.: Understanding & Using Educational Theories. SAGE Publications Ltd, London (2016)
6. CRISS: Applying 'Learning by doing' and group work in class. https://www.crissh2020eu/applying-learning-group-work-class/. Accessed 5 Feb 2019
7. Anzai, Y., Simon, H.A.: The theory of learning by doing. Psychol. Rev. **86**, 124–140 (1979)
8. Soloway, E., Marx, R.W., Krajcik, J.S., Guzdial, M., Palincsar, A.: Motivating project-based learning: sustaining the doing, supporting the learning AU - Blumenfeld, Phyllis C. Educ. Psychol. **26**, 369–398 (1991)
9. Moquillaza, A., Molina, E., Noguera, E., Enríquez, L., Muñoz, A., Paz, F., Collazos, C.: Developing an ATM Interface Using User-Centered Design Techniques, pp. 690–701. Springer, Cham (2017)
10. LeRouge, C., Ma, J., Sneha, S., Tolle, K.: User profiles and personas in the design and development of consumer health technologies. Int. J. Med. Inform. **82**, e251–e268 (2013)
11. Sutcliffe, A.: User-Centered Requirements Engineering: Theory and Practice. Springer, Heidelberg (2002)
12. Ferreira, B., Silva, W., Oliveira, E., Conte, T.: Designing personas with empathy map. In: 27th International Conference on Software Engineering and Knowledge Engineering (SEKE 2015), pp. 501–505 (2015)
13. Gray, D., Brown, S., Macanufo, J.: Gamestorming: A Playbook for Innovators, Rulebreakers, and Changemakers. O'Reilly Media, Sebastopol (2010)
14. Truong, K.N., Hayes, G.R., Abowd, G.D.: Storyboarding: an empirical determination of best practices and effective guidelines. In: Proceedings of the 6th Conference on Designing Interactive Systems, pp. 12–21. ACM, University Park (2006)
15. Newman, M.W., Landay, J.A.: Sitemaps, storyboards, and specifications: a sketch of web site design practice. In: Proceedings of the 3rd Conference on Designing Interactive Systems: Processes, Practices, Methods, and Techniques, pp. 263–274. ACM, New York City (2000)
16. Snyder, C.: Paper Prototyping: The Fast and Easy Way to Design and Refine User Interfaces. Morgan Kaufmann, San Francisco (2003)
17. Paz, F., Paz, F.A., Villanueva, D., Pow-Sang, J.A.: Heuristic evaluation as a complement to usability testing: a case study in web domain. In: 2015 12th International Conference on Information Technology - New Generations, pp. 546–551 (2015)
18. Nielsen, J., Molich, R.: Heuristic evaluation of user interfaces. In: Proceedings of the SIGCHI Conference on Human Factors in Computing Systems, pp. 249–256. ACM, Seattle (1990)
19. Murillo, B., Pow Sang, J., Paz, F.: Heuristic Evaluation and Usability Testing as Complementary Methods: A Case Study, pp. 470–478. Springer International Publishing (2018)

Education, Learning and Flipped Classroom

How to Improve Medical Simulation Training: A New Methodology Based on Ergonomic Evaluation

Agnese Brunzini(✉), Alessandra Papetti, Eleonora Brandoni Serrani, Martina Scafà, and Michele Germani

Università Politecnica Delle Marche, Via Brecce Bianche 12, 60131 Ancona, Italy
{a.brunzini,m.scafa}@pm.univpm.it,
{a.papetti,e.brandoni,m.germani}@staff.univpm.it

Abstract. An adequate medical education is the key driver of healthcare quality improvement. Technological innovations have led to consistent improvement in learning outcomes but the systematic measurement of students performance and cognitive workload need further research. The aim of this paper is to propose an innovative method for the Design and Development of new advanced learning models, to be used in the training of medical students, which includes also the analysis of students performance and cognitive ergonomics. A web-based survey, on team simulation training and technology role, has been administered to 180 medical students. On the basis of this analysis, a list of guidelines for the design of medical education training has been proposed.

Keywords: Medical simulations · Training · Advanced learning technologies - cognitive ergonomics · Mental workload · Instructional design · Augmented reality

1 Introduction

European data from the World Health Organization, consistently evidence that medical errors and healthcare related adverse events occur in 8% to 12% of hospitalizations. In this context, 23% of European Union citizens claim to have been affected by medical errors, 18% of them has experienced a serious medical error in the hospital setting, and 11% declares to have been prescribed a wrong medication. Evidence on medical errors shows that 50% to 70.2% of such harm can be prevented through comprehensive systematic approaches to patient safety [1].

An adequate medical education is the key driver of quality improvement. In this context, the safety and innovation challenges have raised the need to step forward from a traditional 'apprentice' learning model to a simulation-based learning model. The increased awareness about clinical risk management, and, nevertheless, ethical issues, impose the shift to a more skilled/hands-on training of healthcare professionals. Technological innovations, such as models, physical medical simulators, virtual, augmented and mixed reality (AR/VR/MR) simulations, have led to consistent

W. Karwowski et al. (Eds.): AHFE 2019, AISC 963, pp. 145–155, 2020.
https://doi.org/10.1007/978-3-030-20135-7_14

improvement in learning outcomes. Students may learn better and retain information more effectively by engaging themselves in an immersive experience. In the last years, several studies have tried to integrate augmented reality and mannequins obtaining simulators in mixed reality. An integrative review of more than 2.500 papers [2] found that 96% claimed AR to be useful in healthcare education. Additionally, authors were able to determine that AR increased the speed at which students learned and made the learning process easier. Several aspects were elicited in the different studies such as decreased amount of practice needed, reduced failure rate, improved performance accuracy, shortened learning curve, easier to capture learner's attention, increased motivation and improved assessment of trainees [2, 3].

Augmented reality supplements the real world with virtual objects, such that virtual objects appear to coexist in the same space as the users' physical reality, making simulations more realistic and immersive [2, 4]. The main use of AR for learning is to provide feedback and to offer immersion in a scenario. A better, effective and immersive medical training will lead to improved clinical performances and consequently to a significant reduction of the risk to the patient.

Thanks to these advanced learning technologies, it is possible to reproduce particular conditions not always verifiable in clinical practice during the training sessions, and to linger on critical issues in emergencies and not. With these simulations, students and practitioners have the opportunity to repeat, in a safe environment, specific situations particularly stressful both in terms of therapeutic and psychological approach. Although most students are very much in favour of this new technologies [4] and recommend using them in the future [5], the VR/AR/MR applications are not as widely accepted as they perhaps should be [4]. The main reasons are the technical and usability challenges with these devices [5], the difficulty in looking at both the real environment and the augmented objects, the teacher resistance and pedagogical issues. Therefore, the main weaknesses of advanced interactive training in healthcare education can be summarized as follows [2]:

- *Lack of learning theories to guide the design* (80% papers did not describe which kind of learning theory was used to guide design and application);
- *Traditional learning strategies applied* (in 64% papers, advanced technology was used as a guidance system or as feedback tool);
- *Mostly applications prototypes reported* (56% papers presented prototypes without studying their impact).

Therefore, few suggestions are given in literature for choosing an advanced learning model that is better for healthcare education. Further research in this area should be taken to clarify the appropriate model, instructional designs and how to effectively use VR/AR/MR for healthcare education [2, 3]. In fact, sometimes, AR applications are evaluated only taking into account their technical efficacy, but it is important to verify that the AR app satisfies also the real purposes of education. For this aim, several Instructional Design models are available, and they are composed mostly of five basic phases: Analysis, Design, Development, Implementation, and Evaluation [3].

In this paper we will focus on the first two phases: the pedagogical context, the users and the specific needs will be analysed and then, with the results of the Analysis phase, several guidelines for the Design of an advanced learning model will be defined.

Another key aspect in medical education is the assessment of Cognitive Workload. Indeed, during simulations, students can feel discomfort, fatigue and stress. Commonly recognized stressors include technical complications, time pressure, distractions, interruptions, and increased workload [6].

A more efficient medical training will reduce the risk of erroneous events and wrong approaches which may generate excessive levels of anxiety and acute stress to the clinicians, thus compromising their performance. For this reason, during the Design of an instructional model, also cognitive workload and stress level of the trainees should be considered. In this way, the Cognitive Ergonomics Analysis becomes an important aspect to be included in the Design phase.

The principal reason for measuring mental workload (MWL) is to quantify the mental cost of performing a task in order to predict operator and system performance [7], and it is a fundamental construct for exploring the interaction of people with technological devices [8]. Several reviews attempted to organize the significant amount of knowledge behind measurement procedures [9]. Generally, researchers agree in classifying MWL into three main broad categories [7]: *self-assessment, task performance* and *physiological measures.*

Examples of common performance parameters are response, reaction time, accuracy and error rate, estimation time, objective speed and signal detection [10].

The physiological measurements take into account physiological responses of the operator's body that are believed to be correlated to MWL. Their scope is to interpret psychological processes by analyzing their influence on the state of the body.

Nevertheless, a systematic method for the cognitive ergonomics analysis is still not included in the design of simulation-based training. Therefore, the aim of this paper is to propose an innovative method for the Design and Development of new advanced learning models, to be used in the training of medical students, which includes also the analysis of cognitive ergonomics.

2 Method

In the Analysis phase of the Instructional Design Models, it is necessary to define the nature of the problem, identify its origin and propose some possible solutions. This phase may include specific research techniques such as user analysis, pedagogical context analysis, and analysis of specific needs. These results are the inputs to the Design phase [3].

In this paper we propose the Analysis of the context of the simulation-based training in medicine, in order to define a list of guidelines for the Design and Development of innovative learning technologies. The Analysis is carried out through a web-based survey. Users involved in the questionnaire should include medical students of different years of study, in order to investigate attitudes and possible resistances.

2.1 Web-Based Survey

The web-based survey proposed in this paper addresses the benefits and limits of simulation, technology's opportunities and barriers, team dynamics, and personal feelings and attitudes. It is divided in two main sessions: the first one is about the team simulation and the second one is about the role of technology in the simulation-based training. For each session, several questions have been prepared to allow the users to quantitatively evaluate different aspects of simulations.

An introductive session is included to collect personal data of the users such as sex, year of birth, degree course. Other general questions are expected to investigate if the user has previously participated in clinical case management group simulations, which kind of previous group training experiences the user had and if she/he participates in health volunteer activities.

Team Simulation Training. In the web-based survey, the simulation is explicitly defined as a process based on the reproduction, through models, of a system or environment in which the participants act in order to acquire or implement the skills necessary to face the simulated context. The questionnaire session relative to the group simulation training consists in two sections, described hereunder in detail:

- Relevance of the simulation
  - Simulation is an effective teaching method;
  - I would not recommend simulation-based training to my colleagues;
  - Simulation succeeds in transmitting a greater motivation in the learning phase;
  - Simulation can help to deal with situations that cause anxiety or fear;
  - Simulation allows to develop greater empathy with the patient;
  - Simulation is an inadequate tool for training group decision-making skills;
  - Simulation training methods help to understand the importance of team working;
  - Simulation does not provide me with clear information for understanding the actions to be performed during clinical practice;
  - Simulation confuses me in the identification of the significant theoretical elements, among those acquired by studying, for the practical resolution of the simulated case;
  - I can keep more attention during simulation training;
  - Simulation allows me to face a good discussion on the simulated case;
  - Simulation helps me to formulate a workable solution for the problem;
  - Simulation can help me to develop critical thinking (process of forming a judgment through the analysis and objective evaluation of information).
- Relevance of inter-professional education
  - Learning together with other professional figures is important for the improvement of collaboration in the workplace;
  - Simulation carried out together with other professional figures is an effective context for learning;
  - The opportunity to learn together with other professionals should be a priority in my training;
  - Learning shared with other professionals will improve my ability to understand clinical problems;
  - Interprofessional learning opportunities will not have a positive impact on the outcome of my patients;
  - Communication within the group is as important as technical skills;

- It is not necessary for members of the team, providing immediate assistance to the patient, to announce their actions aloud;
- Team members should paraphrase or repeat the instructions received to clarify what they understood;
- Safety in the delivery of care increases if all the members of the team share information regarding patient management;
- Frequent summaries of patient test results are useful for keeping team members' attention to patient needs;
- Within the team, establishing and knowing "who-does-what" is essential for improving the quality of the care provided;
- It is important that team members ask for assistance if they need help completing a task.

Role of Technology. Technology is defined, in the web-based survey, as the application of IT and telematic devices. The questionnaire session relative to the role of technology in the simulation-based training is composed by two sections, described hereunder in detail:

- Personal attitude to the use of technological devices
  - I am familiar with the use of the following devices:
    - Personal Computer
    - Tablet
    - Smartphone
    - Virtual Reality glasses
    - Gloves with haptic feedback for gaming
  - I am familiar with simulation video-games (e.g. "The Sims") and serious games (i.e. video-games with educational purposes);
  - I use technological products that support health and lifestyle (e.g. smartwatch, activity tracker, etc.);
  - I feel ready to work in a high-tech environment;
  - I find it stressful to work in a high-tech environment.
- Relevance of technology in the context of simulation
  - Technological devices (i.e. virtual reality glasses, gloves with haptic feedback, …) are a valuable tool for learning during training;
  - Multisensory interaction (tactile, visual and auditory) through technological devices such as sensorized gloves, glasses and earphones, encourages learning in the simulation;
  - Providing feedback through technological devices promotes learning in the simulation;
  - A high degree of immersion during the simulation has a positive effect on:
    - Learning
    - Psychological component
  - In the context of learning during simulation, I would like to be supported by high-tech devices for:
    - Team working
    - Decision process

- Practice with the physical simulator (dummy)
- Understanding of human anatomy and of physiological and pathological processes

For each question, the users are invited to answer accordingly to a 1 to 5 scale (where 1 is for 'Totally Disagree', 2 is for 'Disagree', 3 is for 'Undecided/Neutral', 4 is for 'Agree' and 5 is for 'Totally Agree').

3 Results and Discussion

3.1 Web-Based Survey

In this section, the main outcomes of the survey are reported. In total 123 students of the sixth year of the degree course in Medicine and Surgery answered. It means a response rate of 68%. It is a representative sample, which accurately reflects the members of the target population. Missing data rates among those questionnaires were low ($\leq 1\%$): only 1 questionnaire was excluded.

Most of the study sample were female (57%). The mean age of the sample was approximately 24 years. Table 1 summarizes the characteristics of participants. It emerges that the majority of students participated to group simulations in clinical case management. Although they declared to be ready to work in a high-tech environment, their familiarity with technological devices is mainly limited to traditional ones (i.e. PC and smartphone).

Table 1. Sample characteristics (score 1 = totally disagree to 5 = totally agree).

	% of respondents				
Gender:					
• Female	57%				
• Male	43%				
Participation to:					
• Residency	52%				
• Professional training activity	90%				
• Training course (e.g. BLS)	90%				
• Working experience	5%				
• Health volunteer activities	19%				
Familiarity with:	*1*	*2*	*3*	*4*	*5*
• Personal Computer	2%	0%	7%	36%	55%
• Tablet	2%	5%	14%	33%	45%
• Smartphone	2%	0%	5%	24%	69%
• Virtual Reality glasses	21%	19%	24%	26%	10%
• Gloves with haptic feedback for gaming	57%	24%	12%	5%	2%
• Simulation video-games and serious games	29%	24%	14%	17%	17%
• Products that support health and lifestyle	31%	26%	17%	14%	12%
Feeling:					
• Ready to work in a high-tech environment	2%	2%	24%	29%	43%
• Stressful to work in a high-tech environment	43%	43%	10%	5%	0%

Analyzing their opinions about the relevance of simulation (Fig. 1), it emerges that it is an effective teaching method that transmits a greater motivation in the learning phase and supports the formulation of a workable solution for the problem, clarifying the actions to perform. Conversely, the simulation does not seem to foster the development of empathy with the patient. About the efficacy of simulation to support the development of decision-making skills, controversial opinions exist.

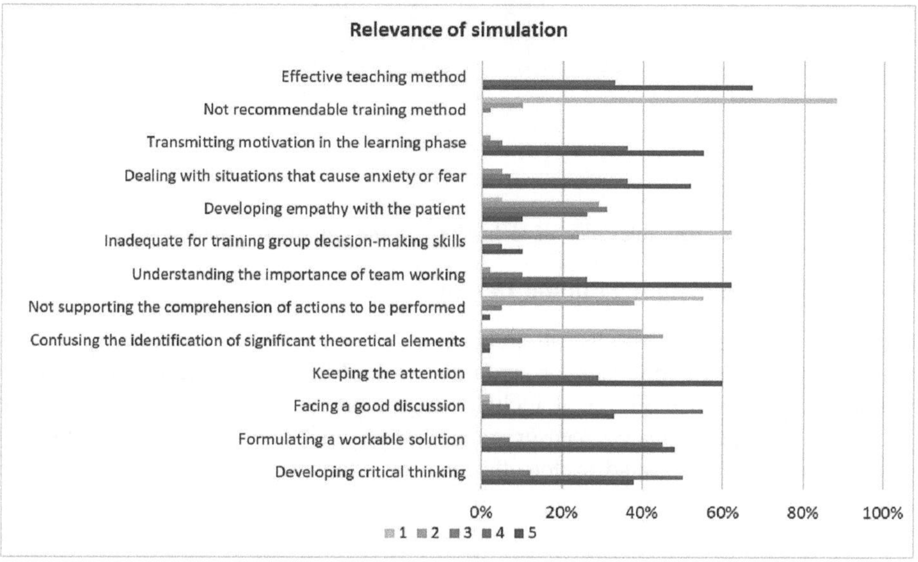

Fig. 1. The relevance of simulation.

The analysis shows that learning together with other professional figures is important for the improvement of collaboration in the workplace (97% agree and strongly agree). However, its priority in the learning process is controversial (7% disagree and strongly disagree). Going into more detail of inter-professional education, the most of respondents considers essential that team members ask for assistance if they need help completing a task (97% agree and strongly agree) and share information regarding patient management (95% agree and strongly agree). On the other hand, the importance to paraphrase or repeat the instructions received to clarify what team members understood is perceived less (26% disagree and strongly disagree).

Observing the relevance of technology in the context of simulation (Fig. 2), it emerges a greater distribution of responses. Almost all the students consider technology a valid support to improve the practice with the physical simulator (dummy) and understand the human anatomy and physiological and pathological processes. It is mainly related to the higher degree of immersion that can be reached. However, it seems to be more suitable for developing specific individual skills rather than team working or decision-making process.

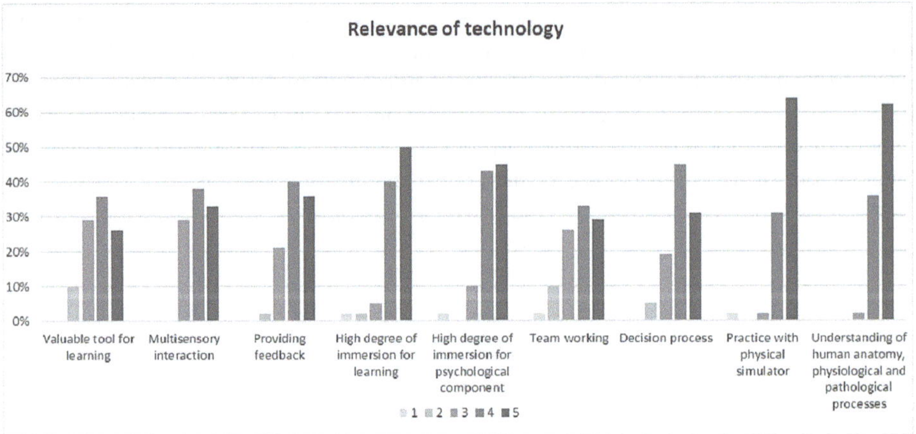

Fig. 2. The relevance of technology in the context of simulation.

3.2 Medical Simulation Training Design

According to the survey results, an architecture to define innovative training courses and new learning methodologies for junior and senior healthcare professionals has been designed. As shown in Fig. 3, it includes high-fidelity simulation practices, mixed reality devices, performance measurement system and wearable sensors aimed at achieving a single integrated system that can be adapted real-time according to the student's actions.

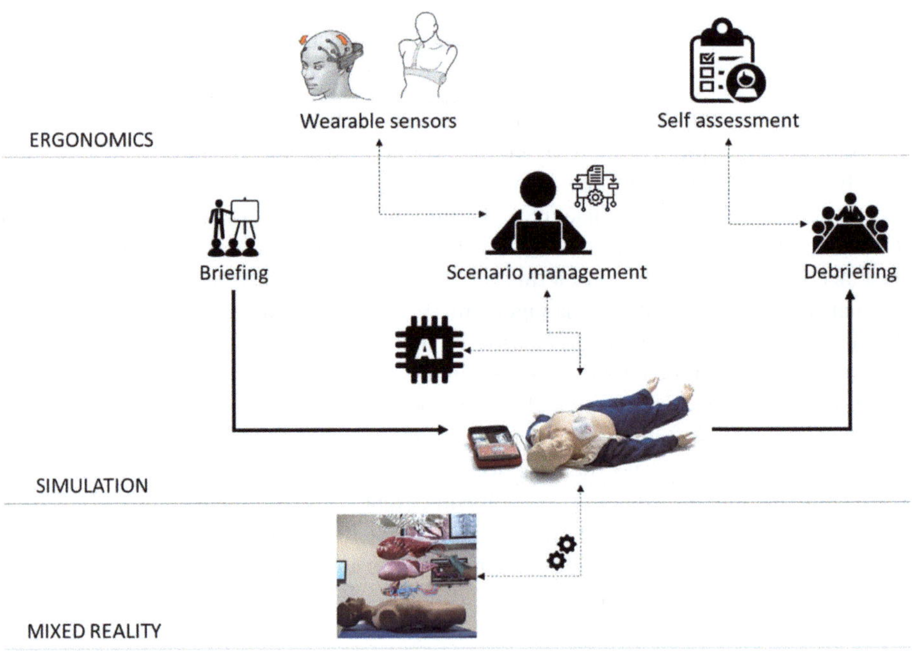

Fig. 3. Training architecture based on mixed reality simulation and ergonomics analysis.

The first step is the briefing session, which aims to review the essential skills, present the learning objectives and environment, give an overview of learner roles and expectations and foster students engagement during the simulation experience. To provide clear and explicit instructions in advance was found to be beneficial to improving learner performance.

Then, the scenario-based simulation allows reproducing in high-fidelity particular conditions not always verifiable in clinical practice during the training sessions and to linger on critical issues in emergencies and not. Moreover, the use of smart dummies supports the development of new scenarios (in addition to standardized ones), implements intelligent algorithms to give immediate feedback and enables the real-time management by moderator.

Mixed reality devices inject a greater realism into scenario-based training. They increase the degree of immersion by reproducing realistic environment (e.g. emergency room, operating room), patient characteristics (e.g. realistic appearance, injuries) and reactions (e.g. movements, pallor) and sensorial feedback (e.g. vital parameters).

Wearable sensors allow monitoring physical and mental stress in students, which could inhibit the learning process and compromise their performance, during more or less complex medical procedures. They mainly collect specific biometric signals (e.g. heart-rate variation, respiration, perspiration, pupil dilation, salivary cortisol, etc.) and/or track the students' movements.

Thanks to the information collected by dummies, sensors and direct observation, the teacher can supervise and modify the clinical case, including patient reactions and feedback, and provide cues according to the student's performance and feeling.

Subjective and multidimensional assessment tools allow evaluating the workload, feeling and performance perceived by students during the simulation session. It is one of the key inputs of the debriefing. The latter is an important component of full-scale simulation because it allows to encourage discussion about knowledge, attitudes, self-confidence, self-efficacy, anxiety and skill performance and to make sure that all learning objectives were covered.

It is worth to specify that the proposed architecture is suitable for transversal (e.g. emergency room/first aid) and specialized skills (e.g. pericardiocentesis, rachicentesis) and learning techniques such as high-fidelity models, task trainers, and low-fidelity models.

The efficacy of the proposed architecture is strictly related to the setting and scenario definition. The adoption of a structured approach that clarifies, formalizes and analyses the casual relationships between objective and subjective variables is essential. For this aim, a set of guidelines is proposed in Table 2. They support:

- Selection of ergonomic indicators to be used to monitor students physical and mental stress;
- Characterization of clinical scenarios for the diagnosis and management of different conditions;
- Characterization of each task to be performed by students;
- Use of probabilistic models to define the diagnostic hypothesis from the map of signs/symptoms and to derive the consequences of the therapy chosen by the student;

- Consideration of clinical risks associated with different treatment options and patient management;
- Management of feedback to be given by physical and virtual systems;
- Definition of parameters that allows evaluating the performance.

Table 2. Design of scenario-based simulation.

Aspect	Description
Environment characteristics	Detailed definition of the simulation setting in terms of room characteristics, equipment and actors. To be specified real and virtual components
Ergonomics indicators	Selection of which indicators and sensors to be used to monitor students physical and mental stress (e.g. heart-rate variation, respiration)
Patient characteristics	Detailed characterization of patient specifying relevant demographic information (e.g. gender, age), physical characteristics (e.g. obesity), anamnesis and signs/symptoms
Pre-scenario	Description of tasks executed before the scenario objective of the simulation that are relevant for the clinical case (e.g. anesthesia)
Tasks	Identification of the sequence of actions to be performed by the student in order to contextualize following items
Verbal interaction	Specification for each task if a verbal interaction between students and patient is expected. In that case, possible patient answers should be planned. Verbal interaction is important to consider empathy skills and interprofessional collaboration
Physical interaction	Specification for each task which physical interaction between students and patient is expected (e.g. palpation, needle insertion)
Side effects	Identification of possible side effects (predictable events) that could occur during each task
Complications	Identification of possible complications (unpredictable events) that could occur during each task
Dummy feedback	Identification of feedback to be provided by dummy (e.g. fluid leak from a paracentesis puncture site)
MR feedback	Identification of feedback to be provided through mixed reality (e.g. patient movement, sounds)
Quantitative feedback	Identification of quantitative feedback to be provided (e.g. heart rate, SPO_2) really (i.e. monitor) or virtually (i.e. augmented reality)
Success indicators	When relevant, definition of indicators that measure the success in each task
Time	When relevant, definition of the time to be taken to perform each task
Quality indicators	When relevant, definition of indicators that evaluate the quality of the performance in each task
Errors	Definition of possible errors that could be made during the execution of each task, specifying the severity level

4 Conclusions

The paper investigates the key elements of medical student education, highlighting the relevance of simulation and technology to increase students engagement and to support the definition of actions to be performed to manage a clinical case. Based on the main outcomes, a new architecture, more student-centric, able to improve the clinical educational path according to an adaptive learning approach is proposed. It is focused on scenario-based smart simulators, mixed reality, human factors and performance assessment. In this way, students and practitioners will have the opportunity to repeat, in a safe environment, specific situations particularly stressful both in terms of therapeutic and psychological approach, by reviewing in detail their own performances. Moreover, an immersive learning environment and the scenarios adaptation according to different variables, indicators and events allows improving both the learner technical skills and his/her ability to make decisions in a stressful situation. It will have positive relapses on safety, quality and education in healthcare.

Future works will focus on the development and experimentation of the proposed architecture.

Acknowledgments. We thank our colleagues from the Faculty of Medicine and Surgery who provided insight and expertise that greatly assisted the research.

References

1. World Health Organization – Regional Office for Europe. http://www.euro.who.int/en/health-topics/Health-systems/patient-safety/data-and-statistics
2. Zhu, E., Hadadgar, A., Masiello, I., Zary, N.: Augmented reality in healthcare education: an integrative review. PeerJ **2**, 1–17 (2014)
3. Garzón, J., Pavón, J., Baldiris, S.: Augmented reality applications for education: five directions for future research. In: Proceedings of 4th International Conference on Augmented Reality, Virtual Reality and Computer Graphics – AVR, Part I, Ugento, Italy, pp. 402–414 (2017)
4. Herron, J.: Augmented reality in medical education and training. J. Electron. Resour. Med. Libr. **13**(2), 51–55 (2016)
5. Chaballout, B., Molloy, M., Vaughn, J., Brisson III, R.: Feasibility of augmented reality in clinical simulations: using Google glass with manikins. JMIR Med. Educ. **2**, 1–6 (2016)
6. Arora, S., Sevdalis, N., Nestel, D., Woloshynowych, M., Darzi, A., Kneebone, R.: The impact of stress on surgical performance: a systematic review of the literature. Surgery **147**, 318–330 (2009)
7. Cain, B.: A review of the mental workload literature (2007)
8. Longo, L.: Formalising human mental workload as a defeasible computational concept (2014)
9. Xie, B., Salvendy, G.: Review and reappraisal of modelling and predicting mental workload in single- and multi-task environments. Work Stress **14**(1), 74–99 (2000)
10. International Encyclopedia of Ergonomics and Human Factors - 3 Volume Set. Taylor & Francis. https://www.taylorfrancis.com/books/e/9780849375477

Professional Portfolio of the Preschool Teacher: What Is the Real Content?

Petra Trávníčková(✉) and Barbora Petrů Puhrová

Department of School Education, Faculty of Humanities,
Tomas Bata University in Zlín, Štefánikova 5670,
76001 Zlín, Czech Republic
ptravnickova@utb.cz

Abstract. Quality in education is directly linked to teachers' professional qualities. Teacher quality develops over a career in terms of personality, knowledge, skills and approaches. A teacher's professional portfolio is a comprehensive tool which provides the opportunity to acquire information not just on professional activities, but also professional identity, which may significantly affect career performance. A professional portfolio has a particular structure, such as a CV, career development plan, lesson planning, learning processes, teaching reflections, self-assessment and co-operation with colleagues, parents, the public and others. The professional portfolio is gradually becoming part of the job for teachers in nursery schools in the Czech Republic in order to improve teaching quality. If, however, this tool is perceived as just another obligatory document, its content also changes. So what is the real content of the professional portfolio? In this study, this content is structured on the basis of an analysis of the content of a preschool teacher professional portfolio. Since this type of portfolio is currently in the phase of being introduced to the nursery school environment, data from the portfolios of students who are just setting out on their journey to becoming a professional is used. Specifically, they are ending their third year having completed a month's work experience in nursery school. Because the analysed document is very large, attention is mainly focused on self-reflection aspects.

Keywords: Professional portfolio · Preschool teacher · Preservice student · Professional identity

1 Theoretical Definition

The professionalisation of nursery school teachers as an ever more-discussed topic in the Czech Republic. One of the tools with which this professionalism can be investigated is the professional portfolio. The meaning of the term "professional portfolio" differs in academic literature. In this study, professional portfolio is perceived to mean, "an inventory of portfolio products collected over time whose continuous generation is important for itself" [1]. In foreign studies, the issue of defining the term is similar. We can, for instance, encounter the terms "professional teaching portfolio", "professional development portfolio", "professional practice portfolio" and even the relatively new term of "eportfolio". A professional portfolio is a thoughtful, organised, and continuous

W. Karwowski et al. (Eds.): AHFE 2019, AISC 963, pp. 156–162, 2020.
https://doi.org/10.1007/978-3-030-20135-7_15

collection of a variety of authentic products that document a professional's progress, goals, efforts, attitudes, pedagogical practices, achievements, talents, interests, and development over time [2]. The professional portfolio can be termed a tool for professionalising nursery school teachers, and it is currently also used in working with two-year-old children [3]. Nursery school teachers, and also elementary school teachers are often also monitored in regard to their professional career. The development of one's professional career can also be monitored, for example, through this portfolio [4].

The objective of this report is to clarify what the content of teaching students' self-assessment is as novice teachers. This research could also provide a more in-depth response to the question of to what extent teaching students' professional identity crystallises as novice teachers. This alternative can be perceived as a suggestion for future investigation. As also defined by Berliner, the phases of the path to profes-soinalism are: beginner – novice, advanced beginner, competent teacher, proficient teacher, expert. For the beginners who are the subject of this study, it particularly applies that they rely on their own experience, and emulate the teaching they them-selves have undergone.

Attention is focused on the process of professional identity development, which can be monitored through the portfolio, specifically in the self-assessment, teaching self-reflection aspects. One can differentiate three phases in teachers' professional identity development: a starting identity (identifying the preconcepts of teaching students), a transformed identity (reconstruction of preconcepts to new theory), and professional identity (subsequent reflection and self-reflection) [5]. A secondary objective of this study is to ascertain how early childhood preservice teachers consider themselves and the management of their teaching during their teaching work experience. It has been argued that reflection skills are crucial for teacher development, and improvement of student learning [6].

2 Methodology

The main objective of this study is to ascertain what the content of the self-reflection aspect of professional portfolios is for students of teaching for nursery school, novice teachers.

This sub-objective follows from the main objective:

1. To describe how teaching students/novice teachers consider themselves in regard to the education process.
2. To describe what teaching students/novice teachers pay attention to in their self-assessment.

This research was undertaken in winter 2018 and focuses on the professional portfolio of nursery school teachers in the context of self-assessment. In undertaking this qualitative research, I have used the content analysis research method. Teachers' professional portfolios were analysed. The content analyses were performed using open coding, with categories which were interpreted subsequently created.

Research participants characteristics: The research participants were five students in Year 3 of Teaching for Nursery Schools, bachelor's course. As part of their work experience, which lasts 1 month, they undertook a professional portfolio. This portfolio involved having to implement 10 teaching performances. This means that the student herself has to manage and organise the education process with the children. Two participants already have a job in nursery school arranged, and three are going to apply for a master's course. The selection of participants was available. In the following text, these participants will be named P1–P5. In research looking at self-efficacy of teaching students, it has been shown that this is quite high, which is a condition for suitable work with a professional portfolio. As such it is fair to support students in working with the portfolio [7].

A total of 5 professional portfolios were available, and 50 self-assessments were analysed; 10 self-assessments from each participant.

Research Limits. Data from the research cannot be generalised for the whole population. In the next period, however, further research will follow on from these results, which will focus on other professional portfolio categories.

3 Interpretation of Research Findings

From open coding, a number of categories were generated which will be further interpreted (Table 1).

Table 1. Category

Research findings category
Plan versus improvisation
Organisation as foundation
Fortune favours the prepared
The fewer children, the fewer problems
Teacher as player
Feeling as a tool of self-assessment

Plan Versus Improvisation. Planning is undoubtedly a very important part of the work of a nursery school teacher. In this regard, however, we encounter in self-assessments that participants consider a plan to be a kind of dogma which has to be followed in order to achieve their objective. As such, the self-assessment also includes whether the participant followed the plan they set themselves.

P1: *"I think in general this time my prepared activities went according to plan."*

P3: *"Since it didn't take place according to the plan, this objective could not be met."*

As such, fulfilling the plan is indicative of whether the objective was fulfilled. Participants also state that their plan was not always ideally thought through, and thus so-called blind spots occur which they try to fill through improvisation, which not everyone perceives as a simple solution.

P2: *"It all began according to plan, though I also improvised a little too…"*

P1: *"Today's output was totally not according to plan... I thought the kids would want to investigate the natural materials more, and as such I hadn't planned an alternative solution."*

In general, one can see that participants pay great attention to following their plan. As has been noted, planning is a natural part of the work of nursery school teachers. Another natural part, however, is improvisation, which is very difficult for novice teachers due to a lack of experience. An inability to improvise can then explained as a failure as a teacher.

P1: *"I didn't manage the situation; I could come up with a more appropriate solution at the time... I see it as a personal failure."*

Organisation as Foundation. Like planning, organisational abilities are a part of a nursery school teacher's competencies. This category was also heavily focused on by the participants. Essentially in every self-assessment, the participants mentioned organisational matters. In particular, these regarded, e.g. dividing children into groups (P3), spatial organisation of children (P4, P5) and organising the process in terms of sequence (P2).

P3: *"You can also divide up the children, as there were a lot around the parachute. I'd try it with half first, and then the other half."*

P4: *"I hadn't realised that when I manage to make a circle with the kids and I put pictures and pegs in the middle, then suddenly it won't be a circle any more. Children charged the pegs and pictures because I had poorly explained what they were going to do and each child wanted that peg."*

P5: *"Because each child didn't have their place marked with a sign they chose, the children tended not to move around. They sat where they should."*

P2: *"Next time, I could first introduce the activity on the carpet, and only then move to the desks. Then all the children would see everything."*

Fortune Favours the Prepared. It isn't a good idea to underestimate preparation for the education process. Even participants who often found that their preparation was not thoroughly thought out in terms of thinking and preparing the environment were aware of this. According to the responses, preparation can be divided into practical material preparation (P1) and preparation in terms of the theoretical part of the work (P2).

P1: *"Next time I'd prepare the paints in advance."*

P1: *"I think that I should clear up the play area in advance so as not to cause delays in activities with children. I had to clear it and bring aids which the children used as an opportunity to focus on their own activities."*

P2: *"Since I already know which children are in the class, I let them all guess..."*

P3: *"I know of a child with specific learning needs in the nursery school, which I included in my preparation."*

The Fewer Children, the Fewer Problems. In teachers' self-assessments it is frequently seen that success or failure of the education process is directly affected by numbers of children. The responses suggest that participants work better with smaller children numbers. This is logical from an organisational and disciplinary perspective too, which are given here as reasoning.

P1: *"There weren't many children in school today, so I was able to use this organisational form without chaos in class resulting."*

P2: *"Overall, my output was calm, which can certainly be attributed to the small number of children, who were calm over the whole period of output."*

Participants assess their output positively when fewer children are present. This is an understandable phenomenon since these are novice teachers.

Senior Teacher as Player. In this context we think of the senior teacher as someone working in a nursery school for a longer period with some experience of working with children. Her role in working with the student is as mentor, partner who helps the student to get a grasp of the nursery school environment. It is natural then that she is found in the self-assessment category in novice teachers' professional portfolio.

Her role differs, however. It may happen that the participant does not agree with the teacher's opinion (P4). The teacher is also involved in student performances, and thus affects student plans which are important for them (P3). Naturally, their role is mainly positive because they give participants feedback, advice for preparing activities, recommendations for next time, and so on (P5, P1, P2).

P4: *"Because everyone was in St Nicholas mood in class, the teacher scared the kids with devils. One boy was so afraid by the teacher that he couldn't stop crying until we told him no-one was going to be taken away."*

P3: *"I'd originally had exercises planned at the start, but on the basis of the situation which arose, with the teacher needing to practice a performance with certain children in the play area, I went straight to making St Nicholas models."*

P5: *"The teacher told us that this was the first time the kids had worked with a parachute. They hadn't tried it with them before, so I was a little worried how the kids would respond, but they managed it very well."*

P1: *"At that moment, the teacher intervened and told the boys to sit down and I worked just with the girls."*

P2: *"In the end, even the teacher praised me in feedback for responding quickly to a game which was on my topic, if not science-focused."*

Feeling as a Tool of Self-assessment. In participant testimony, participants often used feelings in their assessments of their performance in the education process. It is natural that a good feeling from your work is an important factor in satisfaction here.

P5: *"I had a very good feeling from my performance."*

P2: *"Overall I'd give my performance a good assessment, I have a good feeling from it and I definitely know what I want to do again, add or remove next time."*

P3: *"I have generally positive feelings from my overall performance, and next time I would think in particular about preparing individual aids."*

P1: *"On the other hand, I have the feeling that the model was too easy for some of the older children, because they had no great problems with the folding."*

4 Conclusion

Criteria are seen in the professional portfolios of novice teachers or final-year students, according to which they assess their success in regard to the education process. Because these are novice teachers, organisation, planning and preparation are very important to them. In contrast, improvisation and changing activities for understanding is very difficult for them. In terms of organisation, they particularly state that it is difficult for them to keep children in a planned position, e.g. in a circle. Another problematic aspect of organisation is the sequence of steps. This means which activity is undertaken first, which second, and why.

Hand in hand with organisation is children's discipline, which is also mentioned here. The objective of novice teachers is also very frequently to include all children and hold their attention. We can see a shift in their path when they try to apply other, relatively unconventional organisational forms, such as activity centres. (Activity centres are inspired by the Step by Step programme.)

It is also shown that novice teachers perceive working with smaller groups of children to be an advantage. Here we can see a relationship to organisation and discipline, which is important for them in this phase because they use this to measure success. The senior teacher is a fundamental point in the process of introducing novice teachers to practice. They are partners who give the novice teacher advice, recommendations and help to keep control of various situations. Sometimes, however, they also intervene where the novice teacher does not feel it is necessary, and does not entirely agree with the intervention.

Finally, one must mention what is essentially the most important in teachers' further work, this being their own satisfaction and good feeling from work. An assessment of feeling is found amongst all participants. It is clear, then, that this is important to them. Whether it is a good or bad feeling, these emotions are naturally linked to all the teacher's steps from the start of his or her professional career.

The results show that the content of self-assessment in the professional portfolio includes a number of categories which it would be appropriate to look at in the phase of preparation for the profession. This study serves as an initial research report for deeper investigation of the issue.

References

1. Lukášová, H., Svatoš, T., Majerčíková, J.: Studentské portfolio jako výzkumný prostředek poznání cesty k učitelství: příspěvek k autoregulaci profesního učení a seberozvoje. Univerzita Tomáše Bati, Zlín (2014)
2. Spencer, J.S.: Creating your own professional portfolio. Interv. Sch. Clin. **36**(4), 195–201 (2001)
3. Wiegerová, A., Gavora, P.: Preschool attendance of very young children: parents' choice and outcome satisfaction. J. Lang. Cult. Educ. **6**(2), 36–47 (2018)
4. Wiegerová, A., Navrátilová, H.: Crucial moments in professional careers of preschool and primary teachers. Adv. Intell. Syst. Comput. **19**(785), 259–269 (2019)
5. Spilková, V.: Současné proměny vzdělávání učitelů, Brno (2004)

6. Foong, L., Binti Md Nor, M., Nolan, A.: Individual and collective reflection. Australas. J. Early Child. Early Child. Aust. **43**(1), 43–51 (2018)
7. Gavora, P., Wiegerová, A.: Self-efficacy students in a preschool programme: the construction of a research instrument. New Educ. Rev. **47**(1), 125–138 (2017)

Measuring Student Engagement and Commitment on Private Academic Institutions Using Fuzzy Logic Expert System Metrics Applications

Evangelos Markopoulos[1(✉)], Jarno Einolander[2], Hannu Vanharanta[2,3], Jussi Kantola[2,3], and Ari Sivula[2]

[1] HULT International Business School, 35 Commercial Road, Whitechapel, London E1 1LD, UK
evangelos.markopoulos@faculty.hult.edu
[2] University of Vaasa, Wolffintie 34, 65200 Vaasa, Finland
jarno.einolander@gmail.com, ari.sivula@gmail.com, hannu@vanharanta.fi, jussi.kantola@uwasa.fi
[3] Poznan University of Technology, Plac Marii Skłodowskiej-Curie 5, 60-965 Poznań, Poland

Abstract. Even though academic knowledge is provided to academic institutions under very specific academic standards in teaching and research, the instruction's management can impact significantly the student engagement and commitment on receiving and utilizing such knowledge. To analyse this challenge, a Fuzzy Logic, expert system-based software application has been developed and applied on a private academic institution. In this research the institution participated with 40 undergraduate students, from 24 different countries from two different semesters on the same course. The technology measures the student engagement and commitment via the co-evolute methodology for knowledge elicitation. By utilizing this approach, the management of academic institutions can make development analysis based on concrete bottom-up results. The collective analysis of the test results clearly identifies where students see the needs for greatest development and how they view their current state of engagement.

Keywords: Student engagement · Commitment · Expert system · Private academic institution

1 Introduction

Astin [1] defined the term engagement in academic context as 'the amount of physical and psychological energy that the student devotes to the academic experience'. Later, Kuh [2] described student engagement as the time and effort students devote to educational activities that are empirically linked to desired outcomes and what institutions do to induce students to participate in these activities. Commonly, student engagement have been used to describe students' commitments in learning, their identification and

W. Karwowski et al. (Eds.): AHFE 2019, AISC 963, pp. 163–173, 2020.
https://doi.org/10.1007/978-3-030-20135-7_16

belonging at their educational institution, as well as, describing their participation and activities in the school environment to accomplish desired outcomes, is associated with academic, social, and emotional learning outcomes [3].

2 Student Engagement and Retention

An increasing issue and major concern in universities around the world is the low student retention which has become one of the most serious topics in higher education. Low student retention rates have a negative effect on both the students and the academic institutions who devote huge amounts of resources to revert students from dropping-out with incomplete education. Often dropout students have accrued large debts from their studies but leaving without a degree doesn't make it easy to get a well-paying job to pay them [4]. As higher education has a strong connection to the social, economic and political growth of the whole society its important this issue to be addressed collectively and not individually as it impacts the majority of the universities and not specific ones.

Student engagement or involvement within the educational institute can greatly influence student success and persistence. The term student engagement has become more popular in education during the recent decades, probably resulting from an increased understanding that certain intellectual, emotional, behavioral, physical, and social factors have effect in the learning process and social development [5]. Student engagement describes the time and effort students dedicate to activities that are related to the desired outcomes of the educational institutions and on what these institutions do to encourage students to participate in these activities [2]. According to Harper and Quaye [6], engagement is not just involvement or participation; it also necessitates feelings and sensemaking as well as being active. Studies have shown that students' psychological attachment to the university, that is, their commitment, can be a significant predictor of retention which affects many other attitudes and behaviors.

Based on previous research results, people who are committed to their studies are likely to perform better and are more likely to follow them through. Frequently research on retention has focused on academic abilities to predict the students' retention. However, research has also suggested that academic goals, institutional commitment, self-confidence, social support, as well as, institutional selectivity, financial support, and social participation have a positive effect on student retention. Students who are unable to develop these factors are more likely to drop out. Previous studies have indicated that the strongest factors on how students feel, is related to their academic skills, academic self-confidence and academic goals [7]. In addition, previous studies have shown that students who are committed to a particular university are more likely to graduate than those who do not show commitment to any particular institution.

Tinto [8] argues that students are more likely to stay and graduate when the institute, (1) expects them to succeed, (2) provides academic, social and personal support, (3) gives feedback regularly on their performance, (4) regards them as significant members of the institution (e.g., in frequent and high-quality interaction with the academics, the staff and other students), and most importantly (5) fosters learning.

An important factor in the student's commitment is also the level of motivation for their studies. According to Skinner and Pitzer [9] engagement is a result of motivation which promotes engagement if students needs for relatedness, competence, and autonomy are satisfied. Failure to fulfil these needs would lead to disengagement [9], which generates higher risks of later drop-outs [10]. Motivation that comes from students' desire to participate in the learning process has a positive impact on students' academic performance, learning strategy, adaptation and well-being [11]. Motivation in primary education, secondary education and higher education influences academic performance through study effort as a mediator [12].

3 Student Engagement in Private Academic Institutions

Private institutions operate under different management practices, strategies and standards from the state (government) institutions. The students pay higher tuition fees, and have expectations that could differ from what students expect in state institutions primarily in operations, support, administration management, infrastructure, and to an extend in the teaching quality, career development, student life, activities, etc. Many private academic institutions can be considered highly multicultural therefore the student engagement and commitment need to be approached from a cultural dimension as well, in both studying and learning. This cultural diversification extends also to the financial background of the students, were many might have secured professional careers while others might enter directly into their family business. Such students expect teaching and learning to be more practical, interactive, enjoyable, less research driven, more case driven and hands-on. This on the other hand can be a challenge for the academics who follow traditional academic teaching and research practices not quite aligned with the student's expectations and interests. Such cases, and not only, impact heavily the degree of student engagement and commitment, and become quite significant issues and concerns for the development of the strategic management and leadership on the institution's operations.

To further analyse this challenge, a Fuzzy Logic, expert system-based software application has been developed and applied on a private academic institution. In this research the institution participated with 40 undergraduate students, from 24 different countries from two different semesters on the same course. The technology measures the student engagement and commitment via the co-evolute methodology for knowledge elicitation.

4 Evaluating Student Engagement

The evaluation method used in this case-study utilizes a generic, Internet-based application environment called Evolute. The Evolute supports various purpose specific fuzzy logic based evaluation instruments [13–15]. The evaluation instruments are ontology-based, and they are used to acquire perception and collective understanding of different organizational resources. The instrument used in this study is called Helix Academic v2, and it is based on well-known models of student persistence and retention [16, 17]. The main models used in the creation of the framework for the ontology and the statements for the instrument application was Tinto's Student Integration Model [18, 19] and Bean's Student Attrition Model [20, 21]. The evaluation instrument uses self-evaluation in the assessment of different concepts through semantic entities, where in this case are statements. After the data collection, the system

Table 1. Helix Academic features.

Feature	Description
Goal progress	The degree students feel that their studies are moving forward helping them to accomplish their goals
Competence (personal agency beliefs)	The degree students feel about their performance and competencies to study effectively and the attainability of their goals
Autonomy (centralization)	The degree students feel about the freedom from coercion and their influence on their study environment and decision making
Routinization	The degree on how students feel whether their studies are exciting and challenging
Social Integration (relatedness)	The degree of student participation and integration to a social group related to studies
Goal setting	The degree students feel about course requirements being on the right level
Responsive environment	The degree students feel the study environment is responsive to promote effective learning
Learning support	The degree students feel their university is providing the support they need for their studies
Learning resources	The degree students feel the university is providing conditions and resources for learning
Distributive justice	The degree students feel they are treated fairly and recognized for their efforts
Teaching quality	The degree students feel they are receiving quality teaching at their university
Stressors	The degree students feel there are things in the study environment that may cause stress
Institution commitment	The degree of student loyalty to the specific university and their intent to graduate from it
Emotional attachment	The degree of students feeling emotional attachment and connection to this specific university
Utility	The degree of how students recognize the future value, usability and utility of their studies and performance
Student satisfaction	The degree on how students feel about various facets of satisfaction towards the university as a whole
Development	The degree on how the students feel a sense of accomplishment and personal development
External commitments	The degree of personal binding variables external to the study environment (family, community etc.)

computes and visualizes the meaning of the knowledge input collected from the students. Such a fuzzy logic-based system reminds an expert's task of evaluating and reasoning based on linguistic information.

Helix Academic analyzes the goal commitment (personal goals) of the students. This is the degree students committed to the goal on obtaining their degree and how responsible are for their studies. Table 1 presents the key features for feedback categorization in Helix Academic.

5 Case-Study

A case study was carried out at the HULT International Business School, in London, UK. 110 students who attended a course on Creating Problem Solving in two semesters were asked to participate and access codes to the expert system were given to all. In the end, 40 students (36.4 of the participants) responded and completed the statements of the research application. The mean age of the respondents was 21 years of age. Sixty-five (65%) percent of the respondents were male, and 35% female. All in all, there were 24 different nationalities in the case group.

The results of the case study are reported in the following figures. Figure 1 presents the results according to current state values, Fig. 2 is based on target state values, and Fig. 3 is based on creative tension.

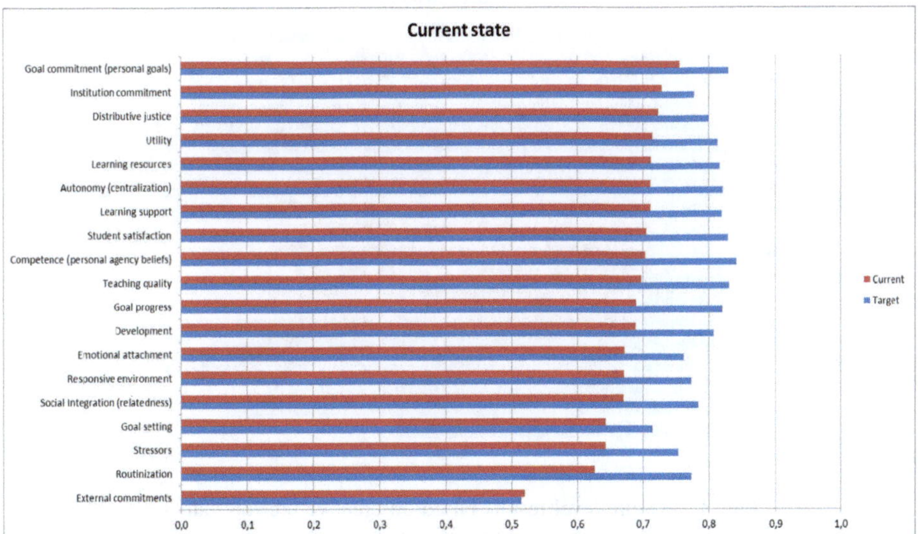

Fig. 1. Feature-level results based on current state values.

The results based on the creative tension show that the highest creative tension i.e. the gap between the state of collective current feelings and the target state, in the test group was in the feature of routinization. This feature assess the students feelings on

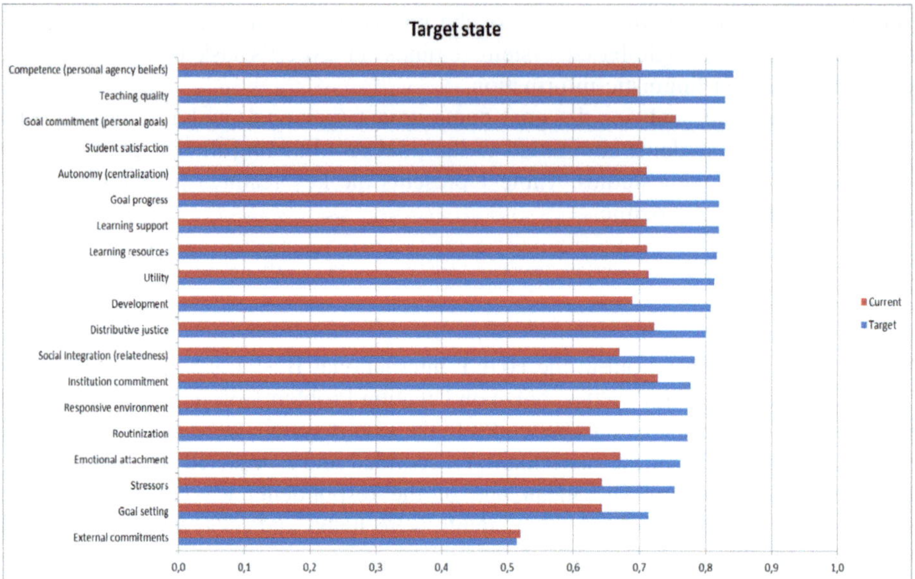

Fig. 2. Feature-level results based on target state values.

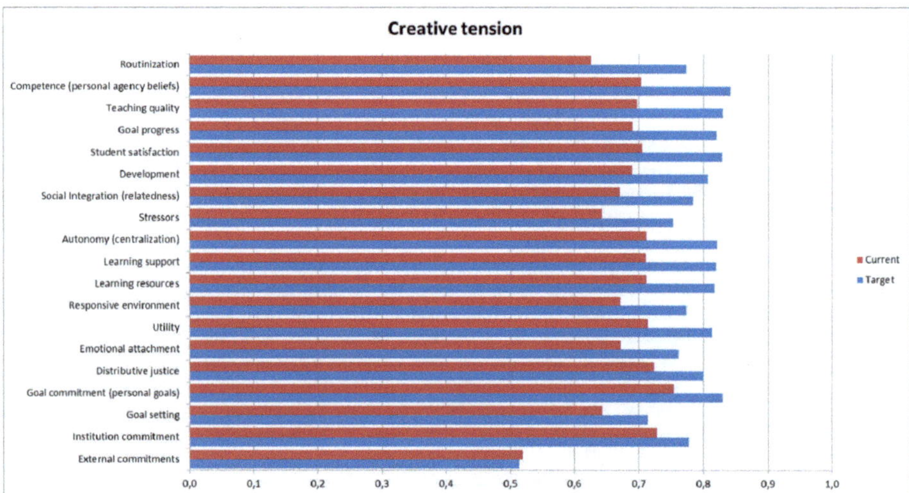

Fig. 3. Feature-level results based on creative tension values

how existing and challenging their studies are. Based on this result the respondents feel that the studies should provide more variety and be more challenging and stimulating. The next highest creative tension was in the feature of competence (personal agency beliefs). This feature describes the state of competence and performance of the students, as they perceive it and its match between the study environment's requirements.

This shows that the respondents collectively feel they wish they would feel more competent regarding studying and would like to advance their study related skills and capabilities. The third highest creative tension was in teaching quality. This feature assesses e.g. the quality, consistency, professionalism, and interestingness of teaching staff and methods. The next highest creative tension was in the feature of goal progress describing the feeling of progress towards to accomplish the study related goals. In addition, the features of development and student satisfaction were at a relatively high level.

Interestingly these features were not regarded as highly based on target state evaluations. The target state indicates where the respondent would like to see improvement. In the target state evaluation, respondents express their own desires and feelings, according to their own situation and knowledge. Also, by looking at the results of the target state evaluations, it is possible to see how the respondents value each feature, in other words, which ones they think are the most important ones.

Figure 4 presents the results according to current and target state values against the average values from other institutions. The darker red and blue bars illustrate the research results from this case study while the lighter bars represent average values from all research conducted with the instrument.

According to the Fig. 4, it seems that the HULT case group scores higher in all of the features than the mean of all the case studies conducted with the research instrument, both in current and target states. Only in the feature of external commitments the HULT case had lower average scores.

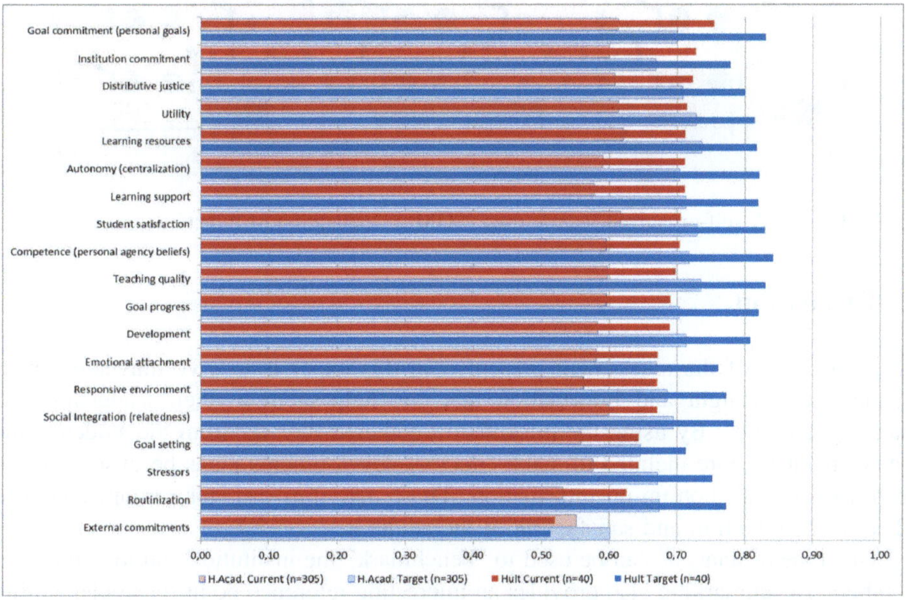

Fig. 4. Feature-level results compared between HULT and other institutions.

In the later analysis, the results were subjected to Self Organization Map (SOM) analysis. The SOM maps were used to confirm the results. Figure 5 illustrates the six highest features based on target state evaluations. The color of the nodes in the component planes visualizes the value distribution of the variables. Warm colors (red, orange) indicate high values, while bluish colors indicate low values. Under each feature component, the color scales are presented as values. Based on SOM maps the results were similar to the results in Fig. 2. This kind of analysis was used in order to confirm the previous results (c.f. [22]).

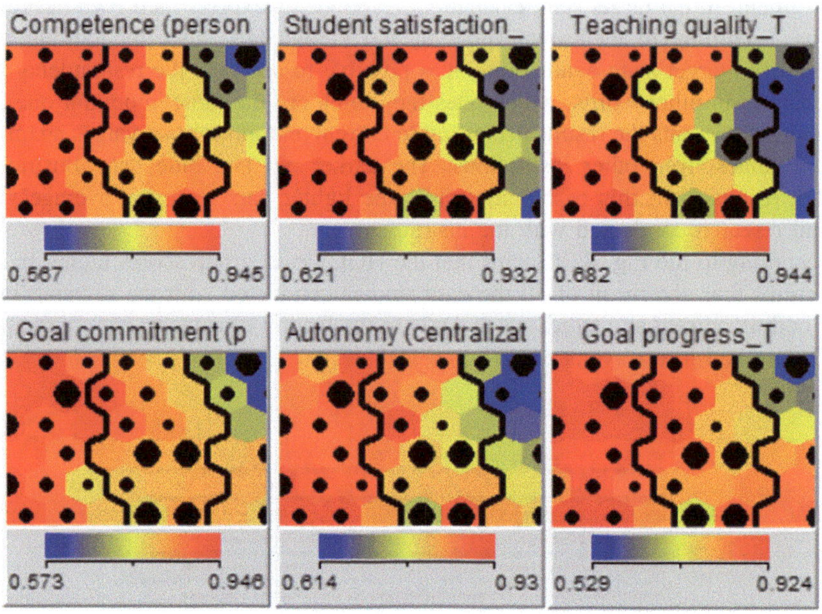

Fig. 5. A self-organized feature map (SOM) visualization of target state results

6 Discussion

The main goal of the assessment presented in this paper is to give university management new insight and information that can help them in their leadership and planning activities. By using the information gathered directly from the students, the management is more inclined to make effective developments plans because they can be based on such bottom-up information. These activities can, for example, help to advance the retention and satisfaction of the students.

This type of analysis can be used to “benchmark” the institution and to compare it to others. Benchmarking also provides an interesting reference point for implementing and managing change. Comparisons to other groups and subsequent analysis can uncover good and useful practices utilized by other institutions.

Based on the results, the respondents feel a high degree of commitment to their goals to graduate from their current educational institute. The respondents feel this way much more highly than the long term average from multiple cases in different conventional universities. When looking at this case, the differences between current case results and long-term average values, the feature of goal commitment has also the biggest difference between them. These results are kind of expected in this case because HULT students aim higher, primarily due to their background.

Also, for example, the respondents feel a high level of distributive justice is prevalent in their educational institute, as well as they recognize the utility and the value of their studies more highly than the multiple case average. Only the external commitments are regarded at lower levels which seems correct based on the characteristics of the respondent group.

7 Conclusions

The research created data, information, and knowledge that can be used to support students in their commitment to complete their studies as well as to help the university, the academics and the administration to find out the drivers behind the so-called drop-out problem and promote positive education performance. This is a new way to respect students' thinking for their studies, and it is also a modern way to lead and manage academic organizations.

The collective analysis of the test results clearly identified where students see the needs for greatest development and how they view their current state of engaging factors. As indicated in the research results, the respondents have creative tension in each feature which indicates that HULT has room to improve and narrow the distance from the current state to a desired, future state. This however cannot be considered critical as HULT students are above average from other institutions on both their current state (how they feel now) state and on their expectations for their future state (where they want to be). This indicates that HULT students are very active, have achieved a lot but want to achieve more. This creates space for improvement which can be achieved with a strategy based on the analysis of the results.

8 Areas of Further Research

HULT International Business School is an academic institution which operates in six cities and three continents. With undergraduate, post-graduate and summer programs offered and delivered in London, Boston, San Francisco, Dubai and Shanghai, the institution can be considered one of global scale. This research studies the engagement and commitment of the students in the London Undergraduate campus only and involved 40 students from 24 different nationalities. The research will extend on understanding the results from each nationality group or individual in order to identify the students with the most or least engagement and commitment based on the country they come from and their educational culture which then can be related to social, financial and other elements that affect their engagement and commitment on their studies. Furthermore, the research will extend to the rest of the HULT campuses as students in Europe might have different engagement and commitment drives from students in the United States, Asia or Middle-East while they all study in the same institution which provides the same education culture and operations processes.

The geographic distribution of the HULT campuses, the internationality and its global operations, can be considered ideal on studying student engagement and commitment in distance learning programs via e-learning. In this case, students also compose international classes while the programs can be delivered from different parts of the world under one institution.

As the future of education will be border-less, distance-less and available to all from all, it is crucial to understand the student's drives and needs while students operate in classes and environments by themselves or within cultural/national groups. Understanding the student's culture, backgrounds, goals and expectations in the international arena can not only contribute towards designing and executing effective, efficient and affordable educational programs, but also on creating processes and strategies towards a unified educational philosophy and culture.

References

1. Astin, A.W.: Student involvement: a developmental theory for higher education. J. Coll. Stud. Pers. **25**(4), 297–308 (1984)
2. Kuh, G.D.: What student affairs professionals need to know about student engagement. J. Coll. Stud. Dev. **50**(6), 683–706 (2009)
3. Christenson, S.L., Reschly, A.L., Appleton, J.J., Berman, S., Spanjers, D., Varro, P.: Best practices in fostering student engagement. In: Best Practices in School Psychology, vol. 5, pp. 1099–1120. National Association of School Psychologists Bethesda, MD (2008)
4. Vedder, R., Gillen, A., Bennett, D., Denhart, M., Robe, J., Holbrook, T., Neiger, P., et al.: 25 Ways to Reduce the Cost of College. Center for College Affordability and Productivity (NJ1). ERIC (2010)
5. Student Engagement. Retrieved from http://edglossary.org/student-engagement/
6. Harper, S.R., Quaye, S.J.: Beyond sameness, with engagement and outcomes for all. In: Harper, S.R., Quaye, S.J. (eds.) Student engagement in higher education, pp. 1–15. Routledge, New York and London (2009)
7. Lotkowski, V.A., Robbins, S.B., Noeth, R.J.: The Role of Academic and Non-academic Factors in Improving College Retention. ACT Policy Report. American College Testing ACT Inc, ERIC (2004)
8. Tinto, V.: Promoting student retention through classroom practice. In: Enhancing Student Retention: Using International Policy and Practice, an International Conference Sponsored by the European Access Network and the Institute for Access Studies at Staffordshire University Amsterdam, pp. 5–7 (2003)
9. Skinner, E.A., Pitzer, J.R.: Developmental dynamics of student engagement, coping, and everyday resilience. Handbook of Research on Student Engagement, pp. 21–44. Springer, Boston (2012)
10. Archambault, I., Janosz, M., Morizot, J., Pagani, L.: Adolescent behavioural, affective, and cognitive engagement in school: relationship to dropout. J. Sch. Health **79**(9), 408–415 (2009)

11. Lumsden, L.S.: Student Motivation To Learn. ERIC Digest, 92. ERIC Clearing-house on Educational Management (1994)
12. Vansteenkiste, M., Zhou, M., Lens, W., Soenens, B.: Experiences of autonomy and control among Chinese learners: vitalizing or immobilizing? J. Educ. Psychol. **97**(3), 468 (2005)
13. Kantola, J.: Ontology-based resource management. Hum. Factors Ergon. Manuf. Serv. Ind. **19**(6), 515–527 (2009)
14. Kantola, J.: Organizational Resource Management: Theories, Methodologies, and Applications. CRC Press (2015)
15. Kantola, J., Karwowski, W., Vanharanta, H.: Managing managerial mosaic: the Evolute methodology. In: Electronic Globalized Business and Sustainable Development Through IT Management: Strategies and Perspectives, pp. 77–89 (2011)
16. Einolander, J., Vanharanta, H., Chang, Y.S., Kantola, J.: Comparing university students' commitment – a multicultural case study. Theor. Issues Ergon. Sci. **17**(3), 267–283 (2016)
17. Einolander, J., Vanharanta, H., Kantola, J., Paliszkiewicz, J., Madra-Sawicka, M.A.: Comparative study of student engagement in two Polish Universities. Zeszyty Naukowe Politechniki Poznanskiej. Organizacja i Zarzadzaniev (Sci. J.S Pozn. Univ. Technol. Ser. "Organization Manag.") **76**, 87–98 (2018)
18. Tinto, V.: Dropout from higher education: a theoretical synthesis of recent research. Rev. Educ. Res. 89–125 (1975)
19. Tinto, V.: Leaving College: Rethinking the Causes and Cures of Student Attrition. University of Chicago Press (1987)
20. Bean, J.P.: Dropouts and Turnover: The Synthesis and Test of a Causal Model of Student Attrition. ERIC (1979)
21. Bean, J.P.: Student attrition, intentions, and confidence: interaction effects in a path model. Res. High. Educ. **17**, 291–320 (1982)
22. Kantola, J., Vanharanta, H., Laukkanen, I., Piirto, A.: Revealing asymmetries in safety culture through proactive vision. Chem. Eng. Trans. **31**, 319–324 (2013)

The Evolution of Pre-school and Elementary School Teachers' Career Trajectories – Career Beginnings and Ambitions

Adriana Wiegerová(✉) and Beáta Deutscherová

Department of School Education, Faculty of Humanities,
Tomas Bata University in Zlín, Zlín, Czech Republic
b.deutscherova@seznam.cz

Abstract. The professionalization of nursery school teachers is currently a much-discussed topic within the Czech Republic. In this study, the authors present interim results of their research into the career trajectories of nursery school teachers, and compare these to the trajectories of junior level elementary school teachers. The research sample comprised thirteen nursery school teachers (3 of whom are currently Czech School Inspectorate inspectors), and also ten junior level elementary school teachers. This sample included five nursery school principals and five elementary school principals. In the qualitative design of the research, two research methods were used, specifically the writing of a themed text, and interviews. By analysing and processing the data obtained, categories were produced which capture different evolutionary phases in the career trajectory of nursery school teachers, elementary school teachers and inspectors.

Keywords: Profession · Career · Career trajectory · Career change · Career event

1 Introduction

There is a wide spectrum of expectations, opportunities and challenges linked to the professional career of a teacher which help shape their learning and education process as they crystallise over the course of their teaching career. The profession of teacher consists of synchronic, reflecting and diachronic dimensions which reflect their career journey from joining the profession to retirement. This is a process which takes a number of decades. It is a decision of professional direction, followed by a long differentiated period during which an individual acquires practical teaching experience in particular through responding to professional challenges and acquiring a mastery of the profession.

In the ideal case, a career is an invigorating source of personal satisfaction and stimulus to further professional development. It is an "interaction of work roles and other life roles over a person's lifespan, including how they balance paid and unpaid work, and their involvement in learning and education" [1].

Relevant institutional and development factors in a given place and time most influence the course of a teacher's career. This process begins after completion of

W. Karwowski et al. (Eds.): AHFE 2019, AISC 963, pp. 174–183, 2020.
https://doi.org/10.1007/978-3-030-20135-7_17

education and professional preparation, when work and life experiences build up, allowing an individual to achieve the pinnacle of their professional career and personal successes. A teacher's professional career is undoubtedly significantly modified by individual life stages of the individual in the profession, as well as their successive development phases. Initial experiences at the start of teaching may prove relevant predictors of their approach to their further career, although only teachers themselves are able to decide whether to enter the next phases of their professional maturity with full commitment, or whether they are resigned to their future career advancement.

For effective teaching, teachers' own knowledge of their professional development phases is significant for us [2]. This leads to two fundamental and mutually correlating reasons. This is the relevancy of investigating teachers' development, which is not just "research for research's sake" but also a contribution to help teachers teach better and pupils learn better [3].

Specialist texts dedicated to career or managing one's own career are mainly based on two areas:

- The first is based on the psychology of the organisation, personnel psychology, management (e.g. [4–6]),
- The second is based on career advice or career education, or even the psychology of choosing a career or education (e.g. [7–9]).

Sharing and comparing perspectives can thus be beneficial not just for teachers' professionalisation, but also for teachers and students to have a more effective grasp of their current and future work in the education field. Extensive studies of teachers' careers by [10–14] provided new findings which remain relevant. Nevertheless, since the time these results were published, numerous studies have shown that teachers' work has dramatically changed. Although it might appear that a large number of studies have been published abroad which look at the evolution of teachers' career trajectories, this is not the case in regard to nursery school teachers. The small number of studies (especially within Czech education academia) likely reflects the fact that pre-primary education is a relatively young academic discipline which is not well-established within the pedagogy system. To some extent, the fact that "it is perceived that nursery school teachers do not deserve deeper academic investigation because they are attributed lower status than teachers at other school levels" plays a negative role too. This was confirmed in research by [15] and also [16].

2 Research Methodology

2.1 Research Objectives

It should be stated that the research whose preliminary results we present in this study is a continuation of work on the professionalization of nursery school teachers, and builds on research undertaken by the Department of School Education at Tomas Bata University in Zlín's Faculty of Humanities as an exceptional topic in recent years. Basic findings in the topic were ascertained in research from 2016–2017 [16, 17].

In terms of professional career development for nursery and elementary school teachers, one can suppose on the basis of the above research that there are relevant differences. As such, our objective was to ascertain what the fundamental differences were in teachers' professional careers, why they occurred and what impact they can have on teacher preparation at universities. Another objective was to analyse key career events in the career trajectory of nursery school teachers and junior-level elementary school teachers.

2.2 Research Methods

Two research methods were chosen for this research. The method of a themed written text was used. This is a method of free writing in which a text on a given topic is written. The writing is free, i.e. not controlled by the researcher – the flow of writing is thus not disturbed or otherwise restricted. Themed writing is essentially the qualitative counterpole of the questionnaire, in which respondents answer a number of questions posed by the researcher. The focus of the themed writing is opposite. The research engages participants in considering themselves. The text they produce is a construction of their own subjectivity. The disadvantage of this method is that it is rather unnatural. Writing is not a natural method of expressing oneself.

Participants were explained the purpose of the research and they were given instructions in relation to the content of the themed writing. Participants were contacted either in person, by e-mail or by telephone. The content of the themed writing text was given the following areas: the reasons why the teacher wanted to become a teacher in their particular type of school, and what circumstances contributed to this decision; the teacher's career beginnings; a key moment in the teacher's career; the reasons why they became school principal and the circumstances which contributed towards that decision. Themed writing was undertaken once in the research study. The texts were analysed in order to find the answer to the research questions. The themed writing text type was written self-reflection. The writing was free and unrestricted. Participants' involvement in the research was voluntary on the basis of obtained consent. The themed writing text was anonymous. Some texts were shorter, not exceeding 300 words, while some texts were longer, with a word count of 850. The shortness of the texts was a surprise. We had anticipated broader and deeper reflection. The reason for this may be participants' weak motivation to provide a written report on their career trajectory, or low text skills for grasping the topic. Contact was established with participants on the basis of previous co-operation.

The other research method was the in-depth interview method. One interview was held with each of the participants. The interview was recorded using a dictaphone. Once the interview was completed, this was followed by the stage of producing the transcript. The transcript was produced by converting the audio recording into written form. Transcripts were essential for practical reasons; analysing data using transcripts is more comfortable than analysing audio recordings. Written down, the researcher is able to make better sense of and faster find their bearings in seeking out relevant places and sections in the interview. Another advantage is more comfortable recording of codes and if needed being able to create one's own notes. Transcripts are produced using transcript rules. These rules ensure the transferral to written form is done in a uniform and similar manner. This produces a reliable foundation for data analysis.

There are a number of transcription rule systems in qualitative research, which are different in their approach to capturing data and level of detail. The characteristics of the rules are mainly determined by the objective of the research and the theoretical perspective which the research identifies with. This transcription system was shown to be acceptable and fully functional for the following analysis. The analysis was based on so-called recursive reading. Repeated reading of particular sections of the interview led to the gradual penetration of participants' thoughts and opinions. This provided a summative picture of participants' statements, aiming for an understanding of the interview as a composite whole. For more details see [18].

Upon first reading, the option of denotation, or immersion in the data was apparent. Subsequently, the first interpretive ideas were recorded, relevant sections and found associations marked. Gradually, a set of codes was produced for the next phase of data analysis [19].

The principle of induction was applied in the research. The advantage of the induction principle was in producing a potentially new perspective on the research phenomenon or phenomena. In this way new terms were gradually born and correlations arose with common features between them.

2.3 Research Participants' Characteristics

The research sample comprises a total of twenty teachers from the Czech Republic and three Czech School Inspectorate inspectors. The makeup of teachers was the following: 10 nursery school teachers (5 teachers and 5 principals), 10 junior-level elementary school teachers (5 teachers and 5 principals). In the Czech Republic, junior-level elementary school comprises children aged 6–10/11 years old. These children attend for a total of five years.

Data collection took place within one stage. Participants were middle-aged or older. As such their life experience was apparent. A total of twenty-two women and one man took part in the research. The numbers represented in the sample investigated is representative in terms of gender for the population of teachers in the Czech Republic, as there is a clear higher representation of women than men, essentially corresponding to the actual structure of the Czech school system, known for its feminisation.

3 Research Findings

3.1 Why Be a Teacher…

There are a large number of reasons for teachers to choose their profession, and these may be highly differentiated. It should be noted that one isn't born a teacher; one rather becomes a teacher. The decision to become a teacher is preceded by many processes, obstacles and development phases. Frequently given motives for teachers to choose the profession include, for example, a desire to work with children and influence the young generation; to do creative work; candidates' impression and conviction that it's easier to get into the Education Faculty than any other faculty; a wish to help and take care of others. According to [15], knowing the motive which led teachers to choose a teaching career is important information which tells us about the situations, events, people and family members, peers, teachers and also career decisions which led to the decision. In order to understand the aspects of teachers' career path development, one must begin investigating the period prior to actual study and joining the profession, i.e. the period when nursery school teachers and junior-level elementary school teachers realised that they wanted to work in teaching. The motivation to take on a teaching career is one of the significant determinants for the establishment and development of teaching professionalism. Maintaining the right motivation over the whole period of working as a teacher is not at all easy, yet it is fundamental to maintain work performance quality. The following subcategories are recorded; teaching career motives may differ.

The research findings based on the responses of research participants demonstrate that teaching for them was either: (a) a substitute solution; or (b) they wanted to become a teacher.

(a) *Last resort*

Although the decision to become a teacher may seem unimportant for some, the decision can play a large role in the teacher's future teaching career in terms of their approach and experience, quality of work performance, etc. There were teachers in both nursery schools and elementary schools who wanted to be teachers, and also those who came into the profession because they had no other choice.

I studied vocational teaching at secondary schools specialising in medical, social and teaching subjects, and I worked in that field for 7 years. After maternity leave, I couldn't find a job in my qualification, and because I had studied teaching, if at tertiary level, I was accepted at a small elementary school. (UV)

I hadn't planned to study education. I decided to do so when I found out I wouldn't be able to pass the entrance exam for chemistry at agricultural college. (JM)

(b) *Wish fulfilled*

Another incentive for choosing teaching may be fulfilling a desire to work with children, the opportunity to remain young in spirit, experiencing positive emotions in relations with others, a need to educate, create, continuously self-learn. This desire needs to be fulfilled, and it can be satisfied by starting to study Education, and subsequently the Education student who successfully passes their final exams become a qualified teacher. This desire can represent a purpose in life and personal fulfilment. After satisfying the desire to become a teacher, there are further additional desires in the career which are also driven by incentives, whether internal or external.

"I just wanted to be a teacher in nursery school, already at that time I was saying that I would be a nursery teacher..." (UG)

In childhood when children think of their future career, some participants already knew they wanted to be a teacher. It is true that the profession of teacher is one of the first professions that children encounter in their life. Although a nursery school teacher as perceived by children is not realistic, it can still play an important role in their decision on a future career.

"... so I always wanted to be a teacher (laughter), so it was my dream from childhood and I'm glad I managed to fulfil my dream..." (UH)

In early childhood, a desire to become a teacher is expressed when participants wordlessly play at being a teacher. They remember the powerful experience of playing a teacher at an early age, which is strongly imprinted in their long-term memory.

"I've got one brother, who's 8 years younger, and I often looked after him, from when I first thought about my future job, my goal was clear." (PS)

"I've always wanted to be a teacher. In childhood, I played at being a teacher and forced my younger brother to be my pupil." (MT)

"I looked after my 8-years younger brother. I was interested in being in a group, experiencing events with other friends, competition, shared success, enjoyment from play and friendship. That's why I decided to get into teaching." (UŠ)

There were also teachers amongst the participants for whom school itself fulfilled them. They simply enjoying going to school. This phenomenon is not common amongst children today. It would seem, however, that those involved in this research are from a generation for whom school as an institution was seen positively.

"I liked going to school; I enjoyed finding the answer to why something works this or the other way, and the combination of my rationality and admiration for logic led to my decision to study teaching." (PH)

"Some of my hobbies at school-age were art and pottery clubs, and also numerous sports–swimming, hiking, skiing and skating. All these circumstances probably led to my choice to study at secondary education school." (PG)

"From early childhood, I enjoyed subjects at school like PE and art, and I played the violin. So it was probably natural that I applied for secondary education school." (AL)

Participants were also led to their choice of career by their family, i.e. an example from their own family; where being a teacher was a kind of family tradition.

"... I come from a family of teachers, so the choice was quite easy, I would say it was even a target, and I'd say my parents even encouraged me generally through clubs and interests so I would then be ready for a teaching career..." (PP)

"... I come from a family of teachers, so it kind of followed. Right from the word go, from elementary school, I was thinking about it. I played the piano, I sang in a choir and I enjoyed teaching work, so I just decided to go to secondary education school and onwards..." (JH)

3.2 Adaptation Shock, or "Rocky Beginning"

Adaptation to any new environment can produce various situations which may cause difficulties. Adaptation, or the ability to adapt to the school environment, is an important factor which has an impact both on students, children, teachers and even parents. The more difficult the environment, the greater the justification for regulating the adaptation process in which the student and trained teacher enters. Adaptation is the ability of an organism to adapt to the conditions in which it operates. Some teachers look back on it fondly: *"the beginning of my career was wonderful, and my first class of kids have remained long in my memory,"* while some remember certain difficulties.

"Thrown into practice, basically learning on my feet, intensive self-realisation." (JH)

"The beginning of my career wasn't easy; I didn't know what to do about certain children (and parents)." (PP)

"The beginning of my career was wonderful, and my first class of kids have remained long in my memory." (PS)

"The beginning of my professional career was difficult, because I was working for a very strict principal." (JH)

It's clear from the responses given that the teacher – novice is affected by a certain adaptation shock, in particular in working with their senior teacher, in contact with children and also with parents. These three possible areas of conflict are foreseeable. They represent the foundation for good or problematic communication in nursery and elementary school.

Theoretical and research studies have looked at to what extent teachers get through social, cultural and education problems [20], how universities respond to students' needs [21] and what specific steps lead to adaptation [22]. These studies, however, focus more on adaptation to university conditions. In teaching-focused programmes, one must also bear in mind that adaptation also involves institutions in which students–future teachers do their work experience. Even under the conditions of training institutions, students experience adaptation problems which can be expressed at emotional, and also cognitive and social levels. Disturbing stability in these areas amplifies a more emotional experience of the situation, which can lead to reality shock.

3.3 Gaining the Role of Principal – Major Career Event

People usually achieve a relatively permanent and secure occupational status in middle age. An examination of one's experience and reassessment of one's work plans may lead to a realisation that almost half of one's professional career is over. At some stage in a teacher's professional development, the acquisition of new skills comes to an end or is reduced, and teachers respond to this in various ways. Some settle into routine, others seek out opportunities for self-fulfilment, perhaps through professional growth by acquiring roles which bring new challenges. A school principal is an expert in all areas (curriculum, didactics, school management, school legislation, innovations in teaching, etc.), ideally delegates powers, creates and supports policies leading to improving school quality, seeks out solutions to problems which have a long-term impact, with pupils and children their greatest priority. [23] Organisational and leadership abilities are particularly important for the role of principal.

There are many reasons for becoming a school principal; a higher income, new experiences, personal growth, etc.), and on the other hand there is the risk linked to the job, the level of social responsibility, less time for family and opportunity to exchange attitudes amongst colleagues.

"... I began as a teacher, and, well, then after eight years I was principal because I had worked here as a teacher, so I then moved on to the role of principal when the previous principal took her first retirement..." (UC)

"After a year working at secondary school, I got an unexpected offer from the mayor of the village of principal at my previous elementary school workplace." (CB)

"I got the role of principal by chance when the previous principal left and none of the other employees had the required education." (JH)

"The inspector offered me the position of principal because she knew of my relationship to the place and knew my work from regular inspections at our school." (PP)

The research participants also showed what problems the role of principal can bring, and that they do not always anticipate these. In particular, this applies in terms of social contacts with former colleagues.

"It was a wonderful period, accompanied by the unpleasant realisation that as principal you're always on your own. Your friends become acquaintances, and some of your acquaintances become adversaries, or even enemies." (UH)

"After taking up the role, I became convinced that it was the biggest mistake I'd ever made. I didn't enjoy my job." (MT)

3.4 Another Career Milestone – Becoming an Inspector

As has been mentioned, our participants included three Czech School Inspectorate inspectors. This institution is focused on monitoring and inspection. What was interesting in the presented research was the fact that being an inspector was a kind of imagined objective for many ambitious teachers. In the case of our participants, the inspectors were women who wanted to be school principals, but there ambitions were not fulfilled, and they therefore decided to try being inspectors.

"... our deputy head contacted me to ask if I wanted to go to the Czech School Inspectorate, that I was clever, the town had just done what they'd done, so would I like to try this..." (UG)

"... I met an inspector, and in just an ordinary informal discussion I also began considering it, so of course I looked at the Czech School Inspectorate website career section, and there was an opening for this position. It was the position of special teacher, junior level teacher, so I applied, I fulfilled the essentials that were required and I took part in the recruitment procedure in Prague..." (MT)

"... well, so I was keeping an eye on it before ... whether any recruitment procedure was announced and then it was... I began looking to see what I could do, and the fact that my mum worked at the Inspectorate meant I knew a little about the job..." (UŠ)

The career trajectory of their job can be outlined graphically as follows (Fig. 1):

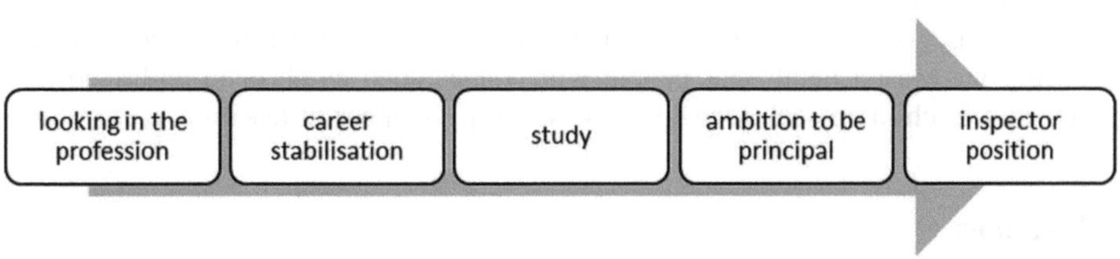

Fig. 1. *Inspector career trajectory.* It is shown that investigating the career of inspector can significantly enrich research. As such, in future we are considering expanding this group of research participants.

4 Conclusion

Teaching is above all a *social category*. It does not exist beyond the human realm. It is shaped and fixed in social acts, i.e. activities, in which people enter into mutual interactions and exchange meanings. These activities include what takes place within the universities, i.e. in the space where humans gradually become teachers. A social act is socially recognised purposeful behaviour, both verbal and non-verbal, direct and mediated, through which and within which one creates new meanings and refines, clarifies and fixes already existing meanings. It is these social acts which are important in the process of professionalising students during the period of undertaking work experience. Social acts, however, are not just about creating and clarifying meanings. On the one hand, social categories construct, and on the other hand they serve as an interpretative framework through which one understands items and phenomena around oneself and acts in accordance.

All people with life experience of school education construct the concept of teaching. However, those who are teachers, or who are on the path to becoming teachers, have a double construction. This second construction is the building of *teacher identity*. The social construct of teacher is individualised, become a part of the "I". The construction of a teacher's professional identity is a continuous, permanent and unceasing process, but the most malleable period is the start of a teacher's professional career. In the first phase a certain preconcept of a teacher is formed, in childhood often done through play in the form of self-presentation as a teacher. Through the impact of new situations (life, academic, professional), this preconcept is deepened.

We believe that it is mainly the influence of the academic and professional situations which students experience during their work experience which deepens their own professional identity. Young people reach an ever deeper grasp of their role, which takes place not just during the process of construction, but also reconstruction, of identity. The processes described take place not just consciously, but also implicitly, unconsciously.

In order for a particular identity to be constructed, social and cognitive processes are not enough. Humans need a certain emotional foundation in order to be receptive to social categories, and in order to engage in the construction process. As such, the *affective side* of the construction process is always important in forming social categories. These are transformed and fulfilled during work experience and higher education study itself.

In the case of teachers, the affective side of the construction process is very powerful, one reason being that people with values such as altruism and a positive relationship to children seek (or should be seeking) a career in teaching.

References

1. Jackson, C.H. (ed.): ELGPN Glosář. ELGPN Tools, no. 2. c2015, Praha: Centrum Euroguidance (2015). http://www.elgpn.eu/publications/browse-by-language/czech/elgpn-glosar
2. Stroot, S., et al.: Peer assistance and review guidebook. Ohio Department of Education, Columbus (1998). http://education.utoledo.edu/par/Stages.html

3. Lukas, J.: Učitel - jeho profesní vývoj a sociální vztahy ve škole: disertační práce. Masarykova univerzita, Fakulta sociálních studií, Katedra psychologie, Brno (2011)
4. Sturges, J.: All in a day's work? Career self-management and the management of the boundary between work and non-work. Hum. Resour. Manag. J. **18**(2) (2015)
5. King, Z.: Career self-management: a framework for guidance of employed adults. Br. J. Guid. Counc. **29**(1) (2001)
6. Noe, R.A.: Is career management related to employee development and performance? J. Organ. Behav. **17**(2) (1996)
7. Vuorinen, R., Watts, A.G. (eds.): Politiky celoživotního poradenství: Evropský nástroj pro tvůrce politik. ELGPN Tools, No. 1. Centrum Euroguidance, Praha (2011)
8. Thomsen, R.: A Nordic Perspective on Career Competences and Guidance – Career Choices and Career Learning. NVL & ELGPN Concept Note. c2015, Oslo: NVL (2014). http://www.elgpn.eu/publications/browse-by-language/english/career-choicesand-career-learning.-nvl-elgpn-concept-note
9. Vaughan, K.: The potential of career management competencies for renewed focus and direction in career education. N. Z. Annu. Rev. Educ. **20** (2010)
10. Lortie, D.: Schoolteacher: A Sociological Study. University of Chicago Press, Chicago (1975)
11. Sikes, P., Measor, L., Woods, P.: Teacher Careers: Crises and Continuities. Falmer Press, London (1985)
12. Huberman, M.: The Lives of Teachers. Cassell, London; Teachers' College, New York (1993a)
13. Huberman, M.: The model of the independent artisan in teachers' professional relations. In: Little, J.W., McLaughlin, M.W. (eds.) Teachers' Work: Individuals (1993b)
14. Fessler, R., Christensen, J.: The Teacher Career Cycle: Understanding and Guiding the Professional Development of Teachers. Allyn and Bacon, Boston (1992)
15. Wiegerová, A., Gavora, P.: Proč se chci stát učitelkou v mateřské škole? Pohled kvalitativního výzkumu. Pedagogická orientace **24**, 4 (2014)
16. Wiegerová, A., Deutscherová, B.: A Comparison of Professional Careers of Early Childhool Teachers: A Qualitative Investigation. E-pedagogium, 4/2018 (2018)
17. Deutscherová, B., Wiegerová, A.: Professionalization of teachers of preschools. In: 9th International Conference of Education, Research and Innovation. Book Series: ICERI Proceedings, pp. 1575–1580 (2016)
18. Pope, C., Mays, N.: Qualitative Research in Health Care, 3rd edn. Blackwell Pub./BMJ Books, Malden (2000)
19. Miles, M.B., Huberman, A.M., Saldaña, J.: Qualitative Data Analysis: A Methods Sourcebook, 3rd edn (2014)
20. Clinciu, A.: Adaptation and stress for the first year university students. Procedia - Soc. Behav. Sci. **78**, 718–722 (2013)
21. Feldt, R.C., et al.: Measurement adjustment to college: construct validity of the student adaptation to college questionnaire. Meas. Eval. Couns. Dev. **2011**, 44–92 (2011)
22. Chang, E.C., et al.: Relation between problemsolving styles and psychological adjustment in young adults: is stress a mediating variable? Pers. Individ. Differ. **42**, 135–144 (2007)
23. Meador, B.: Things a successful school principal does differently (2015). http://teaching.about.com/od/SchoolPrincipals/

Teaching Learning Aspects for Current Technical Education

N. Prasanthi Kumari[1(✉)] and P. Suresh Kumar[2]

[1] Department of Electronics and Instrumentation, University of Petroleum and Energy Studies, Dehradun, India
prasanti@ddn.upes.ac.in

[2] Department of Mechanical Engineering, University of Petroleum and Energy Studies, Dehradun, India
psuresh@ddn.upes.ac.in

Abstract. Technical Institutes and participating stakeholders have to support the process of providing skill based education through inventive practices that blends classroom teaching and experiential-learning resources. Emphasizing to apply the concepts learnt, to develop project prototypes along with the professional software training in the campus are to be introduced as value added courses. Inclination towards student centric learning shall be encouraged. Project based learning will strengthen technical skills and thereby accomplishing quality in Technical Education. Practical knowledge equipped with fundamentals are essential to bridge academic-industry gaps. This paper describes the necessity of education with hands-on, methods that can enrich skills and knowledge that gains troubleshooting skills and employability.

Keywords: Project based learning · Student Centered Learning · Hands on training · Professional software training

1 Introduction

In our country's recent days reforms in education sector are adopted to affect changes in the employability. The quality in the curriculum which was equipped with project-based learning is emphasized. The concept of higher quality, lower cost of production and service are the emerging needs of today's customers [1]. This is a matter of great concern for today's education institutions that the students are not meeting the primary requirement standards of industries. Hence it is very much necessary that the students which are coming out from engineering schools must be directly useful for industries. More than 60% of the eight lakh engineers graduating from technical institutions across the country every year remain unemployed, according to the All India Council for Technical Education (in Times of India) [2]. The thirst and growing desire for attainment of technological competence has paved the way to identify approaches which will help students to get acquainted with the projects out of learnings.

W. Karwowski et al. (Eds.): AHFE 2019, AISC 963, pp. 184–189, 2020.
https://doi.org/10.1007/978-3-030-20135-7_18

2 Excellence in Technical Education

In order to impart high quality technical education, it is vital to maximize the project based learning in the institution. And in order to improve the effectiveness of any institution the curriculum developed is to be project based courses and courses related to professional software tools be identified and the same are inserted and emphasis on the same to be maximized. The quantitative and qualitative analysis are to be kept in the agenda of strategic planning commissions of the university. From the downward look up of the flow chart of the university hierarchy, it is, then, required to: (1) analyze need of TQM concept in technical education, (2) Identify the project based learning of technical institution, (3) Practical based or simulator based theoretical courses and (4) evaluate the factors that they have positive effect and human values in society. In a TQM context the whole institutions need to care about quality improvement and thereby decentralizing the same into departments [3]. In this respect, institutions need cross-functional work groups that will deal with inter-departmental management research problems. concern over acquisition of life-long skills enhancing knowledge, attitude that will help engineering graduates to meet technical and industrial expectations. These skills are imperative to flourish in occupation, higher studies and to involve in research activities [4].

3 Need for Student Centered Learning

The term 'Student-centric' should be principal of our educational shapers. Being innovative is about looking beyond achievement we have and identifying the realizable research concepts for next generation. It is the responsibility of the educational institutions to provide unlimited academic possibilities by an active practical based learning environment for the students. Teaching at Technical schools, mostly, concentrates on delivering lectures and that is not the prime method of Teaching. A student should be able to:

(i) Understand the concept and transform the same into deliverable model, (ii) Reasoning and critical thinking, (iii) Self-learning, (iv) integrating interdepartmental problems, (v) Decision making skills, (vi) trouble shooting, (vii) adaptive to change, etc. [5].

Black board teaching along with simulator based lectures will equip students to exercise the objectives. Introducing professional softwares as credited courses or honor courses will help the students to attain above mentioned goals. Students are to be given some participation in the delivering lectures it is so called active learning. The objectives are multi-dimensional in nature, so for their achievement multiple methods should be used in an integrated fashion. In the light of this, it is important to reflect on Student Centered Learning and focus on practical learning thereby promoting long-standing learning experiences for the students.

4 Learning Methodology and Instructive Schemes

Teaching and learning environment that puts the students' experience as focal point, strong pedagogical methods for faculty and participatory learning schemes for students are to be framed for best yield from the universities. Moreover, teaching/learning have been traditionally paramount in our practices and it is the objective of Teaching, Learning and Assessment Policy and Strategy. Faculty members, should include the aspects of learning approaches in the course plan for the course and to be disseminated so as to make the students to participate in proper learning, which covers wide cooperation in instruction-learning. Appropriate evaluative component shall be designed and encouragements for best performers can be made. Classroom teaching should involve students participation in the process. Sessions like on Quiz, short Recapping of Previous lecture, students discovering innovative projects out of concepts covered, Technology update from technical companies, brain storming fire, Mind mapping techniques and etc., are interleaved with lectures. Below mentioned diagram portrays the retention of learning apex, students they learn from Audio visual aids, black board lectures is about less than <20% whereas discussion and exercises will help better retention. If the learning is done through doing with practical/project model, then the impact is tremendous and chance of exploring newer ideas for (i) upgradation of the existing model, (ii) ways to overcome the defects and (iii) integration with other appliances (Fig. 1).

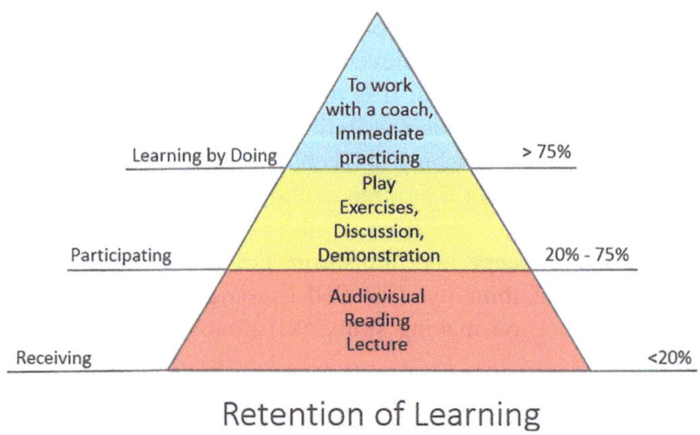

Fig. 1. Retention in practical based learning Courtesy Google Images.

Working with students which in turn makes reduce the gap between learning of student to that of the teaching of the professor. If the classes are aided with practical demonstration of abstract principles, obstacle to imagine will be appreciably reduced. Also the classes if conducted in the laboratory environment the ignition to project based learning will be highly effective.

5 Infra Structure of the University

Quality depends on the institution's infrastructure, faculty's research and development activities and industry-institution interaction etc. [6]. Infrastructure also should be friendly to the student to design and work on the model, if the model has long occupancy for example Students who was working with me to capture the radiations from Jupiter has wide setup. As the frequency of radiation is 24 MHz so the antenna to receive is about 12 m so the setup to be kept in place where usage is zero because the radiations are captured in the night. Similarly, with the case of robot based and drone based projects. Infrastructure plays important role in the nurturing the students with technical education, with topping of quality of projects that puts the students' experience.

6 Interaction with Industry and Society

Though the extracurricular aspects are not very much oriented into awareness camps to use technology; small life hacks students are encouraged to network with industry experts to improve the practical skills in the subsequent semesters, as the student advancing towards final semester. Student need to industrial and domestic applications.

Research and Development: though the research and development does not have much more prominence in the undergraduate level, nevertheless students can be put into the exposure of projects done by academia experts and senior students in the university [7].

6.1 Technical Oriented Extra Curricular Activities

Though the extracurricular aspects are not very much oriented into awareness camps to use technology; small life hacks [8].

6.2 Hands-On Blended Teaching

Teachers should teach the students with explanation through the practical classes for a theory. Although such type teaching covers not all subjects, real world problems can be given to students [9].

7 Knowledge Upgradation for the Faculty

As beginners were good at software tools they can concentrate on basics of the subjects which align to one field. Sequentially they are encouraged to update themselves through enhancing their qualifications attend various professional enrichment programs though industry liaison, like workshops, seminars, conference, hands-on advanced kits or tools. This ultimately leads universities to provide worthy students to the society. Therefore, it is advisable to the faculty to constantly upgrade and to institutes to encourage the faculty according to the field of interest. This kind of upgradation not only to enrich faculty but also students acquire knowledge.

8 Field Visits to Industry

Students are encouraged to visit the industry to have the real look of the procedure and processes related to fabrication and assembly [10]. Students were freely given chance to do internship in the industries it is good, otherwise can be accommodated in the university if sophisticated labs or consultancy work happens from the respective department [11].

9 Working with Faculty in R&D Projects

Students who are in pre-final year are to be encouraged to work with the basics of the building blocks of subsystems in a design or model [12]. If they are encouraged to minute level of design analysis in the consultancy projects which are offered to the department. Students should be given priority to partake in the construction of nutshells of the project which improves the vicinity of the real time problems.

Focus on technical needs of the Society, when the students are allowed to work in the consultancy projects students really understand the needs of the mankind. They can correlate the technology and application of knowledge for the society [13, 14].

10 Conclusion

In this paper, we have briefly examined the various factors that mark the effectiveness of technical education with project-based learning. This type of education is explained with different prevailing factors. A full overview of timeline of life of a student in the technical education institute from the first year to final year. With this experience emerges into a person who cares about the society needs.

Acknowledgments. I sincerely acknowledge the University of Petroleum and Energy studies for giving me liberty to write about the education system.

References

1. Rosiczkowski, J.W., Hahn, W.F.: Total quality management and engineering education at Alfred University. Frontiers in Education Conference. Proceedings of Twenty-Third Annual Conference on Engineering Education: Renewing America's Technology, Washington, DC, pp. 540–543 (1993)
2. www.timesofindia.indiatimes.com
3. Mallesham, P.: Quality in technical education. Indian J. Tech. Educ., **28**(2), 66–70 (2005)
4. Kulkarni, S.: Graph theory and matrix approach for performance evaluation of TQM in Indian industries. TQM Mag. **17**(6) (2005), Emerald Group Publishing Limited
5. Greene, M., Papalambros, P.Y.: A cognitive framework for engineering systems thinking. In: 2016 Conference on Systems Engineering Research, pp. 1–7 (2016)
6. Zemke, S.C.: Student learning in multiple prototype cycles. In: ASEE Annual Conference and Exposition (2012)

7. Accreditation Board for Engineering and Technology, Baltimore, MD, USA, "Engineering criteria 2000" (2011)
8. Isaac, T.S., Kolawole, O.J., Funsho, A.A.G., Adesiji, O.J.: Reviewing engineering curricula to meet industrial and societal needs. In: IEEE International Conference on Interactive Collaborative Learning (ICL), pp. 55–59 (2014)
9. Dodero, J.M., Sicilia, M.-Á., Minguillón, J.: Int. J. Eng. Educ.: Guest Ed. **27** (2011)
10. Sahu, A.R., Shrivastava, R.L., Shrivastava, R.R.: Keys factors affecting the effectiveness of technical education – an Indian perspective. In: Proceedings of the World Congress Engineering, London, 2–4 July 2008 (2008)
11. Gambhir, V., Wadhwa, N.C., Grover, S.: Interpretive structural modelling of enablers of quality technical education: an Indian perspective. Int. J. Prod. Qual. Manag. **12**(4), 393–409 (2013)
12. Senge, P.: The Fifth Discipline: The Art and Practice of the Learning Organization. Doubleday, New York (1994)
13. Mahajan, P.T., Golahit, S.B.: Only education is not enough: a necessity of all-inclusive services for technical education. Int. J. Adv. Res. **5**(1), 1246–1253 (2017). ISSN 2320-5407
14. www.journalijar.com

Music Aesthetics Course Teaching Reform Based on Flipped Classroom Model

Haiyang Qu(✉) and Dahai Xing

South China University of Technology of Art,
Guangzhou University City, Panyu District, Guangzhou, China
{hyqu, xdh}@scut.edu.cn

Abstract. In the practical application process, the flipped classroom teaching mode mainly utilizes the reversal and internalization of current teaching knowledge, changes the traditional teaching mode, makes learning become the main body, turns the teacher into a teaching guide, re-plans the content and time of the classroom teaching and improve the practical teaching efficiency. In this paper, the authors make an in-depth analysis on the structural contradictions in the course of "Music Aesthetics" and uses Schubert's work, "The Devil" as a teaching case to provide teaching examples for the curriculum reform of "Music Aesthetics".

Keywords: Flipped classroom · Music aesthetics · Acoustic structural force · Auditory perception reality · The devil

1 The Ideological Foundation of the Curriculum Revolution of Music Aesthetics

Professor Han Zhongen is a senior professor and a doctoral supervisor of music aesthetics in China. In the past 35 years, he has been involved in the research, teaching activities and discipline construction of music aesthetics. In his article "Listening of music and Speaking of Aesthetics - The problem of music aesthetic teaching forced the subject ontology" [1], he puts forward a profound and systematic exposition on the teaching reform of the music aesthetics course. For example, he believes that it is necessary to observe the positioning problem of music aesthetics through multiple relationships. He said, "Regardless of the ancient or modern, Chinese or foreign, the issues discussed here are always inseparable from the relationship between nature and culture, culture and art, art and music, music and musicology, and musicology and music aesthetics." Therefore, he focuses on the following aspects:

1. The issue of cultural separation through the relationship between nature and culture is observed, that is, how culture can achieve separation from nature through imitation.
2. The origin of art through the relationship between culture and art is extracted, that is, the sighing problem in parallel with the imitation of the origin of art.
3. The problems of aesthetics through the relationship between art and music are highlighted, here, the art form is the key to trigger the aesthetic occurrence directly.

W. Karwowski et al. (Eds.): AHFE 2019, AISC 963, pp. 190–199, 2020.
https://doi.org/10.1007/978-3-030-20135-7_19

4. The problems of aesthetics through the relationship between music and musicology are generated, that is, how the meaning of different levels reflects the emerging problems.
5. The perceptual ontology problems and the corresponding aesthetic problems of music are explored, which are mainly about the acoustic structural force and the abstract pure formal language, the auditory sensory reality and imagination intermediary as the super-structural existence, the meaning of music and a kind of existence existing in its own name.

These five aspects comprehensively summarize the scope of the music aesthetics discipline, especially the fifth point, which is related to the sound structure force and other issues, is the core of the other four points. Aiming at and around the audio-sensory reality presented by the sound structure force, due to the intervention of sound ontology, audio-ontology and emotion ontology, the purposeless purposive existence that touches the meaning of music is the cornerstone of the positioning of music aesthetic discipline.

Based on the perceptual experience of the sound structure force of the work, this paper transforms the auditory image into imaginative conceptual language and even humanistic meaning expression, and on the basis of seeking practical facts, constructs a kind of beautiful, practical facts unique to the listener with the discipline language conforming to its own existence.

2 Curriculum Design of Acoustic Structural Force Combined with Flipped Classroom - Taking the Devil, a Schubert's Work as an Example

2.1 Video Recording Before Class

Auditory Sensory Reality. The so-called auditory sensory reality, only literally, is a fact presented by listening and the corresponding listening actions [2]. However, once it is placed in the middle of a particular art and aesthetic field, it is far from enough to understand it literally. As far as art is concerned, it is necessary to clarify the premise of its listening object, that is, what kind of sound to listen to? Concerning aesthetics, it is essential to clarify the precondition of listening sense and listening behaviour, that is, relying on what to listen to such a sound? If these two premises are missing, there will not be such an auditory sensory reality with an artistic attribute and an aesthetical sense; then it means that there is no difference between the sound of art and the sound of non-art. Therefore, the purpose of this explanation is to explore the presentation methods and means of the content displayed in the work "The Devil" through the auditory perceptual representation of the artistic attributes and aesthetic nature of the acoustic structural force in musical works.

Schubert's art song "The Devil", a perfect unification of poetry, music, and piano accompaniment, is a narrative song with accurate characterization and strong dramatic colour. It was created in 1815, belong to his early art songs. Someone once commented on this song. "In Schubert's whole life, even if he only had this song, it would put him in the ranks of outstanding musicians" [3].

The content of the work is rich in fairy tale flavour and dramatic effects, telling a tragic story. On a dark night, a father was holding his sick son in his arms and rushing home by a horse. The tree demon in the forest continually lured and intimidated his son with sweet talks, trying to take the child's life. The son was terrified and asked his father for help, while the father always tries to soothe the child and continued his way. However, by the time they got home, the child was already dead in his father's arms.

The overall structure of the work is to develop the four character images presented in the poem through the thematic motivation contained in the introduction. The auditory perception is expressed as follows:

Intro: rapid influx and force field polymerization;

Narrator: An uncomfortable lyricism;

Father: Anxious and growing anxiety;

Son: A cry of terror and haste;

Devil: Slightly lingering humour;

End: Sighs progressive and kinetic energy release;

The overall situation: the perfect interaction between dynamic theme and lyric motivation through the whole rhythm running, which is from tension and urgency to clarity and completion through the intermediary, so that the contrast and interaction between the dynamic theme of grief and the sentimental melody, indicates the coordination and harmony between the two basic modes of tonality music and represent different presentations of a sensory reality. Grief and sadness mean that people's emotions and moods exist through the joint sound.

2.2 Further Description of the Core Motivational Pitch Organization

Intro: rushing in and converging the force field, just like the devil knocking on the door and urging it, is to achieve a kind of structure-based spreading drive, constantly generating its basic sentence structure kinetic energy and forming the overall sound potential field.

Driven by the triplet rhythm pattern of the continuous octave of the right-hand piano accompaniment, the two themes motivation of the left-handed, are one tight and one loose, and the lengths are varying. The partial-focus balance and the symmetrical tension are the basis for the development of the whole movement. The dynamic theme is to form an arch-shaped arc pattern with a triplet scale in g minor, which becomes the prototype of the melody jumping into the dynamics, as if the sand in the wildness blown by the night wind or lifted by the horseshoe. The rhythm of the sentimental motivation is elongated by the small second-degree long-sounding consonant of the upward progression, increasing emotional tension and helplessness. Therefore, these two core motivations can be regarded as the structural engines of the whole movement, that is, the reason why the extensional drive is the core structural force. The specific description as follow (Fig. 1).

Fig. 1. Ex. 1.

Narrator: With the main comparison of lyric tunes, the reason why the sensuous rhetoric of lyric is not comfortable is that some audio narration is more or less stagnant. The sound rhetoric comes mainly from such aspects.

The blocking caused by the hidden lines of the second interval with minor fluctuations and the palpitation condition with slightly convulsive and spasmodic created by the comparison of excessive fluctuations. The second interval like example 2 - (1) A long zigzag journey after many climbs to the top and back. See example 2 - (2) Up four hops and progressive uplink sound pattern through the follow-up of the modular architecture. See example 2 - (3) (Fig. 2).

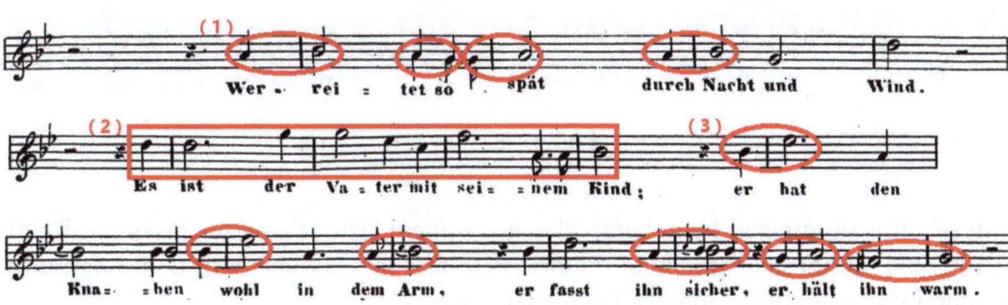

Fig. 2. Ex. 2.

Father: Its main role is to promote the development of musical emotions with connected tunes. The reason why the emotional rhetoric of giving a steady and growing anxiety is that the father's acoustic narrative is low and rapid. The sound rhetoric mainly comes from its relatively low sound zone, giving people a sense of stability. The repeated four-degree jump and the continuous progression of the semitones highlight the inner anxiety and desire. See example 3 (Fig. 3).

Fig. 3. Ex. 3.

Son: It is mainly composed of progressive shouting tunes. The reason for giving the emotional rhetoric of panic and hasty shouting is to feel that its sound is high and in screaming style. The sound rhetoric comes mainly from such aspects.

1. Repeating a certain tone(the tone is D, see example 4), again and again, seems to have a sense of immobility.
2. Based on the second-half-tone advancement (see example 4, from C-D and C bD), the model advances seem like constantly screaming for help under extreme fear (Fig. 4).

Fig. 4. Ex. 4.

The devil: It is dominated by a dance-like atmosphere. The reason for giving a slightly lingering and humorous perceptual rhetoric is to feel the expansion of audio narration. The sound rhetoric comes mainly from such aspects.

1. The three appearances of the devil are all displayed in the major, giving a sense of harmony, happiness and warmth.
2. The theme, which is full of contradictory dynamics, changes its normal state and uses the dancing three-beat rhythm to depict the joyful feeling. See example 5 (Fig. 5).

Fig. 5. Ex. 5.

The wavy melody lines highlight the devil's gentleness and patience, while the two downward jumps at the end of the phrase reveal the devil's sinister intentions. See Example 6 (Fig. 6).

Fig. 6. Ex. 6.

Finale: The gradual progression of sighs and the release of kinetic energy are mainly derived from such aspects.

1. The second-half-tone progression continued to develop upwards and eventually grew to a climax on the highest tone G, which seems like that there was inner grief swelled in the heart and kept accumulating until it broke out. See Example 7 (Fig. 7).

Fig. 7. Ex. 7.

1. In the last sentence, the accompaniment of the previous triplet also came to an abrupt end. Instead, the chords of the column chords and the melody voices created a deep atmosphere and a premise of the fate of the son. See Example 8 (Fig. 8).

Fig. 8. Ex. 8

In conclusion, the structural drive here is mainly between the dynamic thematic motivation of piano accompaniment and the second-order advancement motivation of melody development, as well as the structural characteristics presented by each other and the structural interaction formed between each other. There is no doubt that line movement here is the most aesthetic artistic attributes of the intuitive perceptual style. To a considerable extent, it is of central significance.

2.3 In-Class Experience and Problem Discussion

Listening, Speaking, Singing, and Writing of Music at the Same Time. By watching the teacher's video multiple times, students can understand the information conveyed by the video, and have sufficient time to think and query through all kinds of data, so that the students will have a more in-depth understanding of this work. However, music is a language after all, and it is a way to provide emotional communication. Music cannot form an operational means of communication just by understanding, without practical internalization. Therefore, the performances and practices in the class, including the chorus, solo-singing, role-singing and various audio versions, are all external stimulation that motivates the students' musical language perception. "The stronger the external stimulus, the greater the motivation, the more active the student's performance, and the better the language learning effect" [4]. Therefore, it is the most efficient way to learn music and the best way to memorize when listening, speaking, singing and writing in the meantime. Consequently, the key to further thinking and promotion is the discussion on the subject of music aesthetic.

Problem Discussion. Given Schubert's work "The Devil", I give the following questions:

1. Aiming at the aesthetic problems of music: Why does the same work produce a completely different auditory experience after being transferred? Why do the equal tones of the same work have their own differences in emotional overtones? Voice description with the core tonal expression: Can the absolute pitch position control the acoustic structure and auditory sensory reality? Is the tonal expression mainly in the tone or in the pitch?

2. Aiming at the cultural problem: Is it possible to feel the difference between the style of the times of this work and the one you live in, the national style of the work and yours, while feeling the movement trend of the structure, the division of the climax, the degree of tension and relaxation and the density of the volume in the work? For example, if we are asked to express the connotation of the work, "The Devil", in contemporary times, what means and music forms will we use? Why is that? This way can cultivate students' imagination and creativity.
3. Aiming at the musical art problem: When feeling, understanding and further aesthetic judging the work, can we use the principle of agreement of music works, as well as the principles of structure, stability, order and emotion to further feel the difference and enlightened relations between this work and other works and feel the aesthetic study of music study in this work. For example, a comparative study with works of the contemporary composer like Schumann, or with other narratives like Chopin's piano narratives should be done. The settings of these questions are based on the subject orientation of the music aesthetics course and the acceptance of reality, which provide an implementable plan for the reform of the course.

2.4 Post-class Practice and Literature Reading

After completing the pre-class and in-class learning, students need to select a work for aesthetic analysis of the sound structure force and upload it to the online teaching platform, to master and understand the essence of this lesson and make practical work operation. The teacher will evaluate students' assignments, like advantages discovery and finding out the deficiency, and thus accumulate the teaching experience, which is helpful to propose referential foundation to the next.

The extensive literature reading involves the discussion of conceptual interpretation, structural description, presentation and other issues mainly through literal reading, topic generalization, key tips and meaning interpretation. Students can raise questions on all the related musical works, music literature, music culture, music anthropology, etc. caused by this work. The feedback will be sent to the teachers through the online teaching platform, and they will explain deeply and analyze the problems with high frequency. Such a music aesthetics course will truly become an innovative course in which students learn how to think, how to apply what they have learned and then explore creatively.

3 Review and Thinking

The online flipped classroom breaks thinking mode reflected by traditional teaching method, which is mainly based on convergent thinking, paying attention to the common understanding of the theory, rational analysis and seeking homogeneity in solving problems and pursuing the common correct answers. The formation of this mode of thinking is closely related to the monotony of the teaching method, which will inhibit and hinders the development of students' artistic personality and creativity to a certain extent. As the current curriculum of music aesthetics, the boundaries of the nature of the discipline are vague, the subject of the subject is not accurate, the language of the subject is missing, and the level of teaching objectives is mixed. At present, music aesthetics, however, has such problems as the fuzzy boundary of disciplines, the inaccurate orientation of subject objects, lack of subject language orientation, and mixed levels of teaching objectives. In this case, the online flipped classroom has become a beacon, pointing out the direction of this discipline.

3.1 The Online Flipped Classroom Is Conducive to Efficient and In-Depth Learning

Focusing on music works and building a considerable literature information network around them are conducive to efficient and in-depth study. The detailed analysis of the overall sound structure force of the works, from the whole to the part, from practice to theory, complete the internalization process of teaching content, through the change of teaching methods, making music a truly language tool that can communicate, think, and create.

3.2 Online Flipped Classroom Promote the Production and Development of Music Teaching Resources

The flipped classroom, which absorbs the main advantages of online open courses, pays attention to the collection, utilization, research and development of online teaching information resources. Therefore, the flipped classroom is quite beneficial to the production and research and development of teaching resources. In other words, it not only promotes the utilization of online teaching resources but also encourages the production, especially for music, which appeals to the art of hearing. The reasonable integration and usage of relevant audio-visual video and audio have a very positive effect on the students' learning and thinking, which is beneficial to the expansion of the scale and scope of the students' academic vision.

3.3 The Online Flipped Classroom Is Helpful in Building a New Teacher–Student Relationship

The flipped classroom has changed the teacher–student relationship in the past. No matter whether students watch teaching video before class or interact with teachers and classmates in class, it takes students as the centre and focuses on students' subjectivity, which is student-oriented classroom teaching. Students will gain more initiative in the whole teaching process, and more control in the progress of self-study teaching videos, and can also put forward their own ideas to communicate with teachers or classmates, thus forming a new type of teacher–student relationship. Teachers transform from knowledge transfers and managers to mentors, while students become more proactive in learning knowledge. Flipped classroom teaching mode promotes the role transformation of teachers and students, which makes students' learning more subjective and creative.

To achieve better results in the course study, on the one hand, we must make rational use of the new method of online flipping classrooms without a doubt. On the other hand, the design and layout of the music aesthetics course based on the correct teaching concept is also the fundamental guarantee for students to explore the essential meaning of music and music beauty.

References

1. Han, Z.: Music listening and aesthetic speaking—subject ontology is forced by music aesthetic teaching problem. J. People Music (July 2018)
2. Han, Z.: A philosophical terminal based on aesthetics and aesthetics. J. Art Music (1) (2016)
3. Shen, D.: The symphony of poetry and music—analysis of Schubert's art song "The Devil". J. Voice Yellow River (2) (2012)
4. Gong, Z.: Language acquisition law and language teaching. J. Soc. Sci. Hum. Med. Univ. **12** (3) (May 2010)

Human Factors in Training, Education, and Learning Sciences

Curriculum Infusion Through Case Studies: Engaging Undergraduate Students in Course Subject Material and Influencing Behavior Change

Ellen J. Bass[1(✉)], Holly A. Foster[2], Douglas W. Lee[1], Susan E. Bruce[3], and R. Reid Bailey[3]

[1] Drexel University, Philadelphia, PA, USA
ejb96@drexel.edu, douglas.w.lee2980@gmail.com
[2] The University of Southern Mississippi, Hattiesburg, MS, USA
holly.foster@usm.edu
[3] University of Virginia, Charlottesville, VA, USA
{sbruce,rrbailey}@virginia.edu

Abstract. This study investigated infusing health promotion topics into an engineering course via problem-based case studies and lecture to assess student learning and self-reported behavior. Junior-level systems engineering students in two sections participated: one section with 52 students and one with 36. One section received a celebratory drinking case; one received distracted driving case and a lecture about hazardous drinking. Student ability ratings related to the course subject matter generally improved with both cases. The lecture appeared to enhance health promotion knowledge. Students self-reported behavior change with both cases. Case studies as a form of curriculum infusion for health promotion topics show promise. The use of case studies overall was well-received by students and coupled with lecture material can increase student health promotion knowledge and behavior change.

Keywords: Curriculum infusion · Health promotion · Systems engineering education

1 Introduction

Curriculum infusion (CI) is a pedagogical approach that integrates health-based information into academic courses that may not traditionally focus on health-related topics [1]. It has the potential to leverage the socially-situated experiential model of learning by incorporating relevant examples that actively engage students [1, 2]. Integrated curriculum could be introduced through problem-based case studies that provide an opportunity to learn, retain, and think critically about the content [3].

The goal of this study is to investigate the use of integrated problem-based case studies to provide health promotion information in an undergraduate engineering course, a discipline generally void of health promotion content. To evaluate the health promotion content, one case study involved the analysis of university-specific survey data regarding an annual high-risk celebratory drinking event [4]. The second involved

W. Karwowski et al. (Eds.): AHFE 2019, AISC 963, pp. 203–214, 2020.
https://doi.org/10.1007/978-3-030-20135-7_20

the analysis of national data on texting and driving. Two sections of a junior-level system engineering course were selected based on the course content requiring data analysis and on the pedagogy, which already included several case studies.

This study addresses the hypotheses that infusing health promotion topics into academic courses will increase student interest, engagement, and overall learning of health-related material in a non-health promotion-related academic course. It also addresses hypotheses related to the academic course content with regards to student ability in the academic content. The study also compares the learning of high-risk drinking related information via lecture and through a case study.

2 Methods

The study institution's Institutional Review Board for Human Subjects gave approval.

2.1 Participants

A required junior-level systems evaluation course included case studies as part of its curriculum and was offered with one section of 52 students and another with 36. Participation in the case study analyses was part of the graded homework assignments. Students were assigned teams to analyze the data. The health promotion lecture in one section was part of the scheduled class period. Participation in the pretest, posttest, and the course evaluation was optional and not part of the students' final grades.

2.2 Instruments

Celebratory Drinking Case. The celebratory drinking (CD) case involved a university-specific high-risk drinking practice in which approximately 20% of seniors participate. The case described social norms marketing [5] and highlighted a social norms intervention at the university. The case included specific health promotion content on general safety information, local emergency room (ER) procedures, risks of mixing alcohol with energy drinks, identifying the signs of alcohol overdose, and handling a situation when someone is ill from alcohol. The latter text stated to closely monitor the person for four signs of alcohol overdose and to call for help when needed. The case text described a survey about student drinking norms and related behaviors associated with the high-risk drinking event and student perceptions of others' drinking behaviors. De-identified survey data from 1,335 respondents were included. Students were to analyze data to identify the value of potential social norms marketing interventions by considering: (a) is the CD event worthy of university resources? (b) how many students attempt it? (c) does when a student learns about it impact participation? (d) are there misperceptions associated with it such that social norms marketing may help? (e) are there positive and negative consequences of participating in it? and (f) are there sub-populations for whom the answers to the questions above change?

The assignment requirements included submission of a presentation tailored to a student affairs client (not a statistician). For the statistical analysis requirements, the teams needed to provide backup documentation for all statistical tests conducted.

Distracted Driving Case. The distracted driving (DD) case involved a national sample. The National Highway Traffic Safety Administration (NHTSA)'s General Estimates Systems (GES) data identify highway safety problem areas, provide a foundation for regulatory and consumer information initiatives, and form the basis for cost and benefit analyses of highway safety initiatives [6].

The case included access the GES data and to determine whether DD is deserving of investment by auto insurance companies. To reduce the negative effects of DD, the instructions required students to identify where resources should be targeted with questions including: (a) what types of distracted driving are the most prevalent? (b) are crashes involving distracted driving more severe and do different types of distraction lead to different crash severities? (c) are injuries from crashes involving distracted driving more severe and involve more people who are injured? (d) do crashes involving distracted driving occur at different speeds than non-distracted driving? (e) are there any sub-populations for whom the answers to the questions above change? and (f) how may potential underreporting of crashes affect the meaning of the results?

The assignment required similar deliverables as the CD case.

Pre and Posttests. A pretest and an identical posttest were administered one week prior to each case study and immediately following assignment completion respectively. To address academic content, students provided ability ratings with respect to the learning objectives. The ratings addressed evaluating data and applying basic data-cleansing methods, identifying when to use various statistical tests, and managing time effectively as a team. To test health promotion knowledge, students were asked about local emergency room (ER) procedures, the risks of mixing alcohol with energy drinks, and identifying the four signs of alcohol overdose.

End-of-Course Evaluation. Seven questions were added to the standard end-of-semester course evaluation. Students compared the CD (or DD) case study to other case studies in the course and provided level of agreement with the following:

- I learned something new about alcohol or drinking (distracted driving) in the Celebratory Drinking (Distracted Driving) case.
- I have changed my drinking (driving) behavior because of things I learned from doing the Celebratory Drinking (Distracted Driving) case.
- I have talked to others about things I learned about drinking (distracted driving) from the Celebratory Drinking (Distracted Driving) case.

Students rated their level of agreement with three statements comparing the CD/DD case to the other cases by filling in the appropriate blank. The Celebratory Drinking (Distracted Driving) case was (Much Less, Less, About the Same As, More, Much More): (a) Interesting; (b) Engaging; and (c) Relevant to me personally.

Students rated their use of outside resources (Many Fewer, Fewer, About the Same As, More, Many More) for the statement "Compared to other cases in this course, I researched ___ outside resources for the case." This measure attempted to address any additional effort students were willing to exert.

2.3 Protocol

Students could voluntarily participate in a pretest on Survey Monkey. Faculty randomly assigned five to six students to each group (10 groups for the CD case and 7 for the DD case). A health promotion professional presented health promotion knowledge including answers to the posttest to the DD group on the assignment day. The CD group could obtain this information from the case. After the assignment, students were invited to participate in the voluntary posttest administered via Survey Monkey. The pre- and posttests were coded to identify responses for paired analysis while maintaining anonymity. The anonymous and voluntary end-of-course evaluation was administered through the standard method as with all university courses.

2.4 Independent Variables

There were two independent variables: health promotion topic (CD and DD) and timing of responses (pretest vs. posttest). The topic variable differentiated the case topics and the delivery mode of the risky drinking material (CD: case; DD: lecture).

2.5 Dependent Variables

Systems Evaluation Learning Objective Ability Ratings. Pretest and posttest ability ratings identified change in self-reported ability regarding course learning objectives.

Health Promotion Learning Objective Knowledge. For the health promotion knowledge concerning the signs of alcohol overdose, the number of correct selections was collected in the pretest and posttest periods. For how to deal with an intoxicated friend, the number of correct responses was collected pretest and posttest. In the course evaluation, ratings using a five-point scale addressed level of agreement regarding learning something new about hazardous drinking or distracted driving and about talking to others about the new knowledge.

Behavior Change. The end-of-course evaluation collected ratings using a five-point scale with respect to level of agreement with statements regarding changing drinking or driving behavior based on new knowledge learned.

Case Study Measures. The end-of-course evaluation collected ratings regarding the case studies with respect to level of interest, engagement, personal relevancy, and use of outside sources as a proxy for willingness to work harder.

2.6 Data Analysis

The Wilcoxon signed-rank test compared pretest and posttest responses within each health promotion topic group. The Mann-Whitney test compared pretest and posttest scores across the groups and the course evaluation scores across the groups. A test of proportions compared correct answers across the groups for the signs of alcohol overdose and for helping an intoxicated friend.

Where n is the number of subject pairs, Z for the Wilcoxon Signed Rank test was calculated as [7]:

$$z = \frac{V - \left(\frac{n(n+1)}{4}\right)}{\sqrt{\frac{n(2n+1)(n+1)}{24}}} \quad (1)$$

Where n_1 and n_2 are the number of subjects in each group, Z for the Mann-Whitney test was calculated as:

$$Z = \frac{W - \left(\frac{n_1 n_2}{2}\right)}{\sqrt{\frac{n_1 n_2 (n_1 + n_2 + 1)}{12}}} \quad (2)$$

Effect sizes associated with the Mann-Whitney and Wilcoxon Signed Rank tests were calculated as:

$$r = \frac{z}{\sqrt{N}} \quad (3)$$

where z is the standardized value of the test statistics from the Mann-Whitney test (W) and the Wilcoxon Signed Rank test (V) and N is the total number of subjects [8].

Effect sizes from the tests of proportions were calculated using Cohen's h:

$$h = 2\sin^{-1}\sqrt{p_1} - 2\sin^{-1}\sqrt{p_2} \quad (4)$$

where p_1 and p_2 are the two proportions being compared [9].

3 Results

3.1 Sample

48 and 50 students in the CD group and 36 and 34 in the DD group completed the pretest and the posttest respectively. 46 students in the CD group and 33 in the DD group completed both the pretest and posttest, and remain for analysis. There were 52 end-of-course student evaluations from the CD group and 35 from the DD group.

3.2 Systems Evaluation Learning Objectives

Addressing health promotion information in a case does not interfere with the ability of the students to learn the subject matter of the course. From the pretest to the posttest, self-reported ability ratings in both groups indicated significant improvement in all academic learning objectives except for using the Wilcoxon signed-rank test statistic and managing time effectively while working in a team (Fig. 1). For the Wilcoxon signed-rank test statistic, students in both groups reported median ratings of 3 in the posttest, perhaps due to not considering the use of the test during the case study analysis and thus not gaining experience. For team skills, students in both groups reported median ratings of 4 in the pre- and posttests. There may not have been enough focus on team skills to raise self-reported knowledge from the relatively high base value.

After completing the cases, the DD group tended to rate their abilities higher than the CD group for several dependent variables (Fig. 1). This difference did not appear to be due to a systematic bias toward higher ratings in the DD group as a comparison of pretest scores across the two groups only showed a statistically significant difference with respect to the Wilcoxon signed-rank test statistic ratings (Fig. 1). Perhaps the national automotive data set provided a richer experience for data analysis as there were more samples and potential factors as compared to the CD set.

3.3 Health Promotion Learning Objectives

Knowledge of the Best Options for Dealing with an Intoxicated Friend. In the pretest, many knew the correct option for dealing with an intoxicated friend (pretest in Fig. 2). There were no significant differences between the pre- and posttests for the CD group. For the DD group, there was a significant improvement in knowledge for the correct option (pretest vs. posttest: DD in Fig. 2). Compared to the CD group, the DD group had significantly higher posttest knowledge ratings for the correct option for dealing with an intoxicated friend (right side in Fig. 2).

Knowledge of the Four Signs of Alcohol Overdose. At the start of the study, 13% in the CD group (median pretest score of 1) and no one in the DD group (median pretest score of 0) knew all four signs of alcohol overdose. A Mann-Whitney test indicated that more students in the CD group had prior knowledge of the signs of alcohol overdose ($p = 0.005$; $W = 1027.5$). In the posttest, 17% of the students in the CD group and 30% in the DD group knew all four signs of alcohol overdose. While there was no statistical difference indicating learning between the pre- and posttests in the CD group, the difference in the DD group was significant ($p < 0.001$; $V = 14$). Thus, the original knowledge advantage for the CD group did not transfer to the posttest.

Taking Ill Friends to the Emergency Room. In the pretest, most students knew to take an ill friend to the ER if alcohol related illness from celebratory drinking is suspected. 89% (41 of 46) in the CD group and 79% (26 of 33) in the DD group indicated agreement or strong agreement with the statement "If you or your friends are hurt or ill from alcohol, it is important to go to the (local) Emergency Room (ER)." A Mann-Whitney test comparing the agreement levels with the statement found no significant difference between the groups.

In the posttest, 87% (40 of 46) in the CD group and 97% (32 of 33) in the DD group indicated agreement or strong agreement with the statement. There was a trend toward higher posttest levels of agreement for the DD group ($W = 601.5$, $p = 0.071$). A Wilcoxon signed-rank test for the DD group indicated a significant increase in the level of agreement with the statement between the pre and posttests ($V = 9$, $p = 0.010$).

Question		Group	Pretest Mean	Pretest SD	Pretest Median	Posttest Mean	Posttest SD	Posttest Median	Pretest v. Posttest	Pretest v. Pretest	Posttest v. Posttest
If given an unknown data set, please rate your ability to evaluate data quality and apply basic data-cleansing methods		CD	3.17	0.80	3	3.72	0.69	4	**p < 0.001**, V=36, r=0.355		**p= 0.021**,W=554,r=-0.230
		DD	3.09	0.77	3	4.06	0.56	4	**p < 0.001**,V=9.5,r=0.429		
If given a data set, please rate your ability to identify where (and if) ___ could be used to analyze those data	one of the many forms of t-tests	CD	3.48	0.86	4	3.94	0.61	4	**p < 0.001**,V=24,r=0.332	.	**p = 0.003**,W=499, r=-0.291
		DD	3.67	0.78	4	4.36	0.60	4	**p < 0.001**,V=7,r=0.369		
	one of the many forms of ANOVA	CD	3.15	0.79	3	3.63	0.57	4	**p < 0.001**,V=46,r=0.343	.	**p = 0.014** W=541.5 r=-0.243
		DD	3.39	0.83	3	4.00	0.75	4	**p<0.001**,V=28.5,r=0.332		
	the Wilcoxon signed-rank test statistic	CD	2.59	0.96	2.5	2.74	0.77	3	.	**p = 0.039** W=559.5 r=-0.223	**p = 0.017** W=538 r=-0.247
		DD	3.06	1.03	3	3.18	0.73	3	.		
	one of the many forms of tests of proportions	CD	2.76	0.99	3	3.78	0.89	4	**p < 0.001**,V=50,r=0.464		
		DD	3.06	0.79	3	4.03	0.73	4	**p < 0.001**,V=15,r=0.402		
	a contingency table/chi-square test statistic	CD	2.57	0.89	3	3.41	0.86	3	**p < 0.001**,V=52,r=0.438		
		DD	2.61	0.66	3	3.58	0.90	4	**p < 0.001**,V=35,r=0.386		
	the Kruskal Wallis test	CD	1.61	0.68	1.5	3.24	0.82	3	**p < 0.001**,V=0,r=0.575		**p = 0.011** W=515.5 r=-0.273
		DD	1.82	1.01	1	3.73	0.94	4	**p < 0.001**,V=0,r=0.482		
	the Mann Whitney test	CD	1.37	0.61	1	3.43	0.75	3	**p < 0.001**,V=0,r=0.595		**p = 0.039** W=565.5 r=-0.216
		DD	1.46	0.67	1	3.79	0.86	4	**p < 0.001**,V=0, r=0.538		
Please rate your ability to manage time effectively on a team		CD	4.09	0.46	4	4.13	0.50	4			--
		DD	4.00	0.66	4	4.03	0.47	4			

Fig. 1. Self-reported ability ratings for course learning objectives (1 = no ability; 2 = little ability; 3 = moderate ability; 4 = good ability; 5 = excellent ability)

	Pretest				Posttest				Pretest v. Posttest: DD	Posttest CD v. DD
	CD	%	DD	%	CD	%	DD	%		
Give coffee	46	100	32	97.0	46	100	33	100		
Let sleep it off alone	43	93.5	32	97.0	45	97.8	33	100		
Make throw up	43	93.5	27	81.8	44	95.7	33	100	**p=0.010**,Z=-2.57,h=-0.881	
Sit and watch person*	41	89.1	26	78.8	44	95.7	31	93.9		
Give food	31	67.4	19	57.6	29	63.0	32	97.0	**p<0.001**,Z=-3.82,h=-1.419	**p<0.001**,Z=-3.55,h=-0.957
Give water	13	28.3	4	12.1	17	37.0	22	66.7	**p<0.001**,Z=-4.53,h=-1.199	**p=0.009**,Z=-2.60,h=-0.604

*Correct answer

Fig. 2. Options for dealing with intoxicated friend (CD: n = 46; DD n = 33)

Emergency Room Confidentiality for Alcohol-Related Visits. About two-thirds of the students knew about confidentiality at the ER at the start of the course. For the pretest, 67% (31 of 46) in the CD group and 64% (21 of 33) in the DD group indicated agreement or strong agreement with the statement "The (local) Emergency Room (ER) respects confidentiality for an alcohol-related visit and does not contact parents, administration or police." A Mann-Whitney test found no significant differences between the pretest levels of agreement of the groups.

In the posttest for the CD group, 13% (6 of 46) strongly agreed and another 27 agreed with the statement (72% (33 of 46) total); for the DD group, 52% (17 of 33) strongly agreed and none agreed (79% (26 of 33) total). This difference between the groups was significant ($W = 516.5$, $p = 0.010$). While there was no significant difference in agreement between the pre- and posttests for the CD group, there was for the DD group ($V = 12$, $p = 0.003$).

Risks of Mixing Energy Drinks and Alcohol. Most students knew about the risks of mixing energy drinks with alcohol at the start of the course. For the pretest, 83% (38 of 46) in the CD group and 85% (28 of 33) in the DD group indicated disagreement or strong disagreement with the statement "Drinking alcohol mixed with an energy drink poses no additional risks compared to drinking alcohol alone." There was no significant difference between the two groups.

For the posttest, 87% (40 of 46) in the CD group and 76% (25 of 33) in the DD group indicated disagreement or strong disagreement with the statement. There were no significance differences for either group from the pretest to the posttest. There was a trend towards a difference between the groups ($W = 598$, $p = 0.077$) in posttest responses. It is not clear why the DD group performance declined from the pretest.

Self-Reported "Learning Something New about the Case Topic." The cases appeared to support learning about health promotion topics. In the end-of-course evaluation about three-quarters of the students reported that they learned something about the topics. 71% (37 of 52) in the CD group indicated strong agreement or agreement that they learned something new about risky drinking, and 77% (27 out of 35) indicated strong agreement or agreement that they learned something new about distracted driving. There was no statistical difference in the ratings between the two groups.

Self-Reported Discussion of Case Topic Learning with Others. There was evidence that the reach of the health promotion knowledge stimulated conversation about the topics beyond classroom boundaries. 50% (26 of 52) in the CD group indicated strong agreement or agreement that they talked to others about what they learned about drinking; in the DD group, 49% (17 out of 35) indicated strong agreement or agreement that they talked to others about what they learned about distracted driving. There was no statistical difference in the ratings between the two groups.

3.4 Behavior Change

Self-Reported Impact on Student Behavior. Some students self-reported that as a result of completing the case studies, they changed their behavior. For the CD group, 8% (4 of 52) indicated strong agreement or agreement that they changed their drinking behavior as a result of completing the CD case, while 17% (6 out of 35) reported changing their driving behavior as a result of completing the DD case. The DD group provided significantly higher ratings than the CD group (W = 663; p = 0.024).

3.5 Case Study Measures

Self-Reported Interest. The CD group rated their case as interesting. Specifically, 54% (28 of 52) in the CD group rated their case "More" or "Much more" interesting than other cases, while 37% (13 out of 35) in the DD group rated their case "More" or "Much more" interesting. There is also a trend (W = 1086.5; p = 0.100) for students to rate the CD case as more interesting as compared to the DD case. This may have been due to the university-specific focus of the CD case.

Self-Reported Engagement. Some students rated the cases engaging. Specifically, 44% (23 of 52) of those who participated in the CD group rated it "More" or "Much more" engaging than other cases, and 40% (14 out of 35) of the DD group. There was no statistical difference in the engagement ratings between the two groups.

Self-Reported Personal Relevance. Some students rated the cases as personally relevant. Specifically, 52% (27 of 52) of those who participated in the CD group rated it "More" or "Much more" personally relevant than other cases, as did 40% (14 out of 35) of the DD group. There was no statistical difference in the personal relevance ratings between the two groups.

Self-Reported Use of Outside Resources. Compared to other cases, some students reported that they tended to use more outside resources for either case. 15% (8 of 52) of those who participated in the CD case rated using "More" or "Many more" outside resources than for other cases, and 29% (10 out of 35) rated using "More" or "Many more" outside resources for the DD case. There was no statistical difference in the outside resource ratings between the two cases.

4 Discussion

This study investigated infusing health promotion topics into an academic course via case studies to see if increased student learning of and interest in health promotion-related material can occur while not interfering with student ability in the academic content. The study also sought to compare the learning of health promotion-related information via lecture and through a case study. A study was conducted in two sections of the same engineering course where students completed different case studies and gained knowledge related to risky drinking through different pedagogy.

The results support that the case studies on the health promotion topics did not interfere with student learning of the course subject matter. Ability ratings related to the subject matter generally improved with both cases. This is especially important if the academic faculty are not be willing to "sacrifice" students learning the topics they are teaching to "make room" for health promotion information.

With respect to learning the health promotion knowledge, many (three-quarters) self-reported learning something new via the case studies and about half reported discussing this new knowledge with peers outside of the classroom. The measured learning of specific declarative health promotion knowledge was superior with the course lecture as opposed to the case study. The CD group learned the health promotion information through the text of the case study, while the DD group received an in-person discussion from a health educator. The finding that the case study was less effective may have been due to the case study design as the health promotion information was not a focus of the analysis and was not highlighted in the materials. The CD addressed in the case study focuses on a practice among seniors, while the students in the class were juniors. A case study involving a CD more prevalent among all students may provide different results. Future work should investigate better methods for including the health promotion knowledge in the cases and for supporting more active learning.

Faculty teaching academic courses do not want to introduce cases that deter student interest and this concern was not an issue herein. This study supports the hypothesis that there will be an increase in interest and engagement in topics of health promotion. For example, over half of the CD group reported that the case was more interesting and personally relevant than other cases in the course.

A positive finding was self-reports of changing behavior based on the cases. While the percentage of the students was not high (8% for the CD group and 17% for the DD group), any positive behavior change is considered successful, especially considering the limited time period during which the case study occurred. Additionally, the CD addressed a practice among seniors. It is possible their work with the case study may influence their decision to engage in the practice the following year.

Research on the use of case studies in academic courses with the infusion of health promotion topics would be of value to the academic and student affairs communities.

This exploratory study does have limitations that should be addressed in future work. It included confounds between the case study topics, delivery methods of celebratory knowledge delivery (live presentation and case study narrative), and size and type of the case study datasets. Future work should uncouple these confounds to determine better case study designs.

Another limitation was the small sample size used. New experimental designs will be required in future work. A larger population would allow for more power and certainty in the overall results. A post-hoc analysis indicates a minimum sample size of 450 participants is needed to achieve power of 0.8 on four outcome measures. Barriers are associated with achieving a large sample: most notably that upper level engineering courses are rarely this size.

No follow-up occurred after the semester. Future work should include a plan to contact students the following year to address future learning and behavior change and longer-term follow-up.

In conclusion, case studies as a form of curriculum infusion for health promotion topics show promise. The use of case studies was well-received and was coupled with a positive impact on student self-reported behavior change.

Acknowledgements. This study was funded in part by the University of Virginia Learning Assessment Grant program for a project entitled "Curriculum Infusion: Evaluating Alcohol Education Delivered via Systems Engineering Case Studies." The data collection for the celebratory drinking case was supporting in part by the Virginia Alcoholic Beverage Control Board. The authors thank the students who volunteered to complete the pre- and post-tests as well as the course evaluations.

References

1. Lederman, L.C., Stewart, L.P., Russ, T.L.: Addressing college drinking through curriculum infusion: a study of the use of experience-based learning in the communication classroom. Commun. Educ. **56**(4), 476–494 (2007)
2. Kolb, D.A.: Experiential Learning: Experience as the Source of Learning and Development. Pearson Education, Upper Saddle River (2015)
3. Kreber, C.: Learning experientially through case studies? A conceptual analysis. Teach. High. Educ. **6**, 217–228 (2001)
4. Foster, H.A., Bass, E.J., Bruce, S.E.: Are students drinking hand over fifth? Understanding participant demographics in order to curb a dangerous practice. J. Alcohol Drug Educ. **55**(3), 41–60 (2011)
5. Berkowitz, A.D.: The social norms approach: theory, research and annotated bibliography. Higher Education Center, Newton (2004)
6. National Highway Traffic Safety Administration, FARS/NASS GES coding and validation manual. U.S. Department of Transportation (2016)
7. Bellera, C.A., Julien, M., Hanley, J.A.: Normal approximations to the distributions of the Wilcoxon statistics: accurate to what N? Graphical insights. J. Stat. Educ. **18**(2), 1–17 (2010)
8. Fritz, C.O., Morris, P.E., Richler, J.J.: Effect size estimates: current use, calculations, and interpretation. J. Exp. Psychol.: Gen. **141**(1), 2–18 (2012)
9. Rosnow, R.L., Rosenthal, R.: Effect sizes for experimenting psychologists. Can. J. Exp. Psychol./Revue canadienne de psychologie expérimentale **57**(3), 221–237 (2003)

The Importance of Didactic Toys in Kindergarten Educational Process

Barbora Petrů Puhrová(✉) and Hana Navrátilová

Tomas Bata University in Zlín, Stefanikova 5670, 760 01 Zlín, Czech Republic
{petru_puhrova, hnavratilova}@utb.cz

Abstract. The aim of this study is to identify preschool children's strategies in choice and play with didactic toys. The research was based on participatory observation of children with the set of original didactic toys. In total, 77 children aged 2 to 6 years from four different kindergartens were observed, with children always in small groups. The data obtained in the form of video recordings and field notes was then transcribed and analyzed using qualitative methods. The results point out the teacher's influence, usually determined what the toy serves or what task have to accomplished. We also point out the possible conditionality of the child's interest in the toy. Findings highlights the child's strategy in the play, especially the modification of the toy that arises just when the child has the opportunity to independently explore the toys. The importance is to support the didactic thinking of kindergarten's teachers including pre-service teachers.

Keywords: Didactic toys · Kindergarten · Observation

1 Introduction

Preschool teachers are significant actors of play initiation, support and development of children in kindergarten. An inspiring environment is one of the basic assumptions of play opportunities for children planned and implemented solely according to their imagination. And yet, play and learning are not mutually exclusive at all [1]. The essence of child play developing the child's personality is based on terms like creativeness, adaptation, exploration, experiment, learning, socialization, enculturation and skill [2]. The teacher is a partner and assistant to the child in this. Utilization of play means not only giving instructions for activities, but also providing a natural space for various activities and forms of play between which the child moves freely. This space is incited by the kindergarten teacher in a spontaneous but still targeted manner by preparing the environment for play and inspiring the child's independent creative activity. Despite the number of empirical studies, the potential of play as a learning strategy still remains largely unused.

1.1 Games and Toys

Play is a basic phenomenon in the child's life, and yet there is no clear definition or accurate specification of it [3, 4]. The definition, causes, value, meaning, nature and influence differ largely and they even are controversial [5]. Although forms of play are

W. Karwowski et al. (Eds.): AHFE 2019, AISC 963, pp. 215–222, 2020.
https://doi.org/10.1007/978-3-030-20135-7_21

similar, they are hard to define [6]. Play is supposed to be a spontaneous activity of a child which is representing a reflection of its life, bringing satisfaction and expressing the current needs [7]. Rogers [8] defines play in a wider sense as a natural activity springing out of children cognitive need and the necessity to become mainly part of the social world around them. In the context of play, teachers use didactic games with the main share of a didactic objective governed and reflected by the teacher's direct or indirect guidance, oriented towards the development of individual skills and the ability to cooperate, explore, try or create. The purpose of didactic games is development or fixation of a certain skill [9]. Benefits of didactic games include their stimulating charge, for they evoke interest, inspire creativeness, spontaneity, cooperation and competition, making children use various knowledge and skills and engage their life experience [10]. The play of preschool children is certainly connected with toys, considered the instrument of play, for toys, according to Rogers [8], stimulate and extend play. Didactic toys and their use in the process of learning thus support motivation for learning, communication, cooperation and individual cognitive, affective and social levels of the child's development. A toy is at the same time a possible space for social-centered play with the peers [11].

1.2 Preschool Teacher's Position in Children Play

The learning of children in kindergarten is implemented through various educational programs or concepts of the given kindergarten, with the teacher as the principal figure or the activator and implementer of the educational process. Fehessa and Pyle [12] emphasize the importance of the role of the teacher in play, especially as the teacher is aware of the purpose of the play. Effective play-based learning begins if and when the teacher creates conditions for facilitative and scaffolding learning. Involvement of the teacher in a child's play is a supportive tool for the child's learning. The teacher performs the key role in the creation the conditions of a child's play [13].

- Professional knowledge of the teacher is required in the following areas:
- Pedagogical with a focus on the content and management of the educational process;
- Didactic in the specification of the particular taught subject, including knowledge of the environment;
- Curriculum with the knowledge of planning, implementation and evaluation of the key documents;
- Value with an accent on the knowledge of objectives, purposes and values of education and upbringing;
- Pedagogical-psychological in the sense of knowledge of the child and their characteristics;
- Professional development in the sense of knowing and developing oneself [14].

Teachers' interventions in play can take several forms, from assistance in problem solving, asking questions and rectification of undesirable behavior to involvement of children in thematic games. One of the significant teacher's roles is development of understanding of the specific knowledge and skills of the children and their need to develop [15].

To plan a game for a group of children that is acceptable for them and meets their current needs and interests is not an easy task for a kindergarten teacher. Part of the teacher's planning of the opportunities for involvement of the child in play, following a certain target of the child's development, is prior consideration of the desirable framework for play and setting of an environment that is comfortable for the involved child to play in. An inevitable component should be fixed rules of the game, serving for setting certain boundaries for the play and for the teacher's expectations. The teacher's preparation of play should, however, be first of all focused on the children themselves. To plan a game means first of all to create encouraging atmosphere [16]. Through observation of children, the teacher acquires significant information to specifically organize the playroom and activities for the children. Certain freedom of the structure in the preparation of play then allows the children:

- To choose from several variants of the game;
- To negotiate these options;
- To make use of their own ideas;
- To be physically and mentally active overall;
- To experiment, explore and investigate.

1.3 Research Participants

The research involved 77 preschool children (age 2 to 6 years) from four particular kindergartens in the Czech Republic. Detailed figures of participating children per age and gender are mentioned in the Table 1 below. We observed smaller groups (with approx. 6 children per group) from each preschool classroom during the play with the original set of didactic toys. These toys were specially designed and manufactured by the preservice preschool teachers based on their decision of a didactic aim to be fulfilled in a selected field of children development (social, cognitive, motoric etc.).

Table 1. Data of children participating in the study

Age of child	Boys	Girls
2 years	1	1
3 years	3	2
4 years	10	9
5 years	8	17
6 years	14	12
Total	36	41

1.4 Procedure

The main research method of data collection was based on using the participatory observation of the children. This method belongs to actually preferred practices in research with young children in pedagogical and psychological sciences. Participatory observation helps the researcher (as well as the teacher) get to know better the children and thus profit from valuable awareness of their independent and group behavior [17]. Video recording enabled to observe the different moments in repeated replay, returning to the recorded play situations with the option to identify progressively what one cannot notice in the course of a single observation. Despite the possible limitation in camera influence on children behavior, our data suggest that the children acted in play as usually. They were more interested in presented toys than in video camera in the corner of their playroom.

All kindergartens participating in the research allowed us to realize the observation in a separate playroom with enough space which the children knew as a standard area for play with their peers. The small groups of preschool children had a possibility to use and play with given toys as much time as they wanted. The mean length of one recording was 30 min. Data were collected in four state kindergartens in the Zlín region during three months. The data obtained in the form of video recordings and field notes were transcribed and subjected to qualitative analysis.

2 Results

We used a set of didactic toys as a means for observation of pre-school children in play. The observations focused on the toy selection and subsequent play with the toy. At the same time, we focused on the factors that conditioned the toy selection.

2.1 Toy Selection

Part of the approach to the playing child in the kindergarten may be the very moment of the toy selection by the child. In the first stage of the research, the children were attracted by the researchers entering the classroom and the new toys they brought in. Most children were interested in the toys, and the final selection was governed by the visual attractiveness of the toy (color, shape, similarity to another toy, originality of the toy). The comments of the children on their choice were interesting: I visit dancing and painting lessons, there we always create something new; I also like creating and drawing, which clearly distinguished between the children who simply wanted to play without purpose and those who followed a goal of creation or reconfiguration of something.

The very beginning of the research result description reveals that the children mostly do not name the toys, pointing to them with their fingers and saying give me this, that… In another observation, one girl said: Hey, come and look at this funny game. One boy said: Look here! We have a strange thing, come and see the strange thing.

The children did not feel like naming the toys accurately, they even didn't need to ask for it. They instead asked about the purpose of the toy: I want to see what it is; I want that box (Meaning the glove puppet theatre). Neither in the course of the toy selection nor during the play did the children show interest in the name of the toy, which is documented by a sad statement of one of the girls at the end of the play: I could not figure out at all what it was.

As the purpose of the study was verification of a set of toys and whether they fulfilled their didactic goal and a description of the way the kindergarten children played with them, the research workers intentionally avoided interference in the play and explanations of how to play with the toys while observing the children. We observed that many children first wanted to investigate the offering of toys before selecting one, and thus they moved among the toys, sitting down at each of them for a couple of moments, shorter or longer, and then leaving them and approaching others. Basically, the group showed spontaneous immediate selection, i.e. a direct approach to the selected toy. The selection was affected not only by the selecting child but also by the other children around them. Thus there were numerous moves from one toy to another. After a while, the observed children devoted more time to a single toy. B: The football is totally good, I will not give it up so soon. It is really good. What was interesting was observation of the children who focused on a single toy from the beginning and spent most of their play time with it.

2.2 Pleasure of Toy and Shadow of Teacher

The child was pleased with the toy when they could reconfigure it into something else. The first choice of a toy was commenced by the toy discovery after opening the lid of the box, accompanied by a surprised look in the face of the child, voice expressions of surprise, body language, jumping around etc. However, we also noticed unpleasant surprise, disappointment of the child's expectations, as well as doubt and misunderstanding. Fontana [18] confirms the role of emotions in the early childhood age, where pleasure is to be perceived as desirable by itself. The children expressed themselves spontaneously if they were indulged in the game and perceived it emotionally, then their emotionality was expressed not only by voice but also by body movements.

The children were used to asking their teacher for help whenever they needed it, and also at the moment when they needed to boast of their achievement. The teacher as the originator and inspirer of the child's play in preschool education is an ever more frequently discussed theme. More emphasis in the recent theory of play is placed on playful pedagogy, including both the teacher and the child as active actors in the process of common learning [19]. In child's play, the adult takes up the role of encourager of the child's self-control, with the pedagogical strategies focusing on stepwise decentralization of the child. In child's play the teacher should propose solutions for controversial situations and resolve conflict situations among the children, but should not intervene if it is not yet necessary. If the teacher is able to understand and appreciate the spontaneity of the event helping to understand the child's intentions, then the teacher can assist in the development and extension of play.

2.3 What Conditions the Child's Interest?

When observing toy selection, we identified factors significantly affecting the children's interest.

Choice of a Toy or a Peer. Observing peer relations and strategies brought the evidence of mutual influence of child and his/her peer on play in the group. Children mostly learn social skills in cooperation but also and above all by observation and imitation of others. Actions working for one child are tried by other children on themselves and on others. We observed children subordinating to other children and being voluntarily influenced by them:

G[1]: I do not want to play with this anymore.
G2: Neither do I.

I'd Have to Learn This First. Toy selection was also conditioned by the child's conviction that they would be able to understand the principle of the game. As the children did not receive any direct instructions about how to play, some children simply did not choose any toy and others developed their own way, thus creating modifications. Other children verbalized the choice conditioning on the basis of whether they could play or not. This approach corresponded to the initial inquiry of the children about the rules and limits of the toys, therefore we believe this is an adopted mechanism given by the approach of the teacher or the parents.

B: I will try to play with this.
G: We will play with this if we can figure out how.

Partner in Play. The observed groups of children revealed that regular kindergarten attendance led them to a certain degree of socialization and adaptation to play in groups.

G: K, come here!
G2: L, play with me.
B: Hey, let us play football.
G, B: Let us take the theatre! F, come with me (Waves at his friend with a glove puppet in his hand).

Toy selection by the child is conditioned by multiple factors. The research clearly shows that the quality, appearance and didactic goal of the toy is not the most important for a child. The circumstances, phenomena and real children entering the toy selection in our study show that it is more than necessary to see the issue of use of toys in kindergartens as a complex phenomenon.

2.4 Course of Play Activities with a Toy

In addition to toy selection, we also observed subsequent progress of play with the toy. We mainly focused on modifications of the toy and of play. The children obviously first perceived the chosen toy as a whole, but then they began to focus on its uses and the individual elements it consisted of. Discovery of the principle of play was mainly interesting in the case of the toys where the children themselves had to identify what the toy might be used for.

[1] G = girl; B = boy.

G: Look, animals (opening the box with the game). I will take the dolphin.

(turning the engraved animal) Magnet! (Beginning to look at the other animals) They are magnets!

The girl first tries to join the animals together with their magnets, but then she discovers the boards with landscape motifs typical of individual continents where the animals can be placed.

In accordance with the research interest in possible toy modification invented by the preschool children, all the toys were offered to children in simple form (closed boxes with no examples of final solution to be fulfilled). The track for marbles was presented as a plain board where the particular paper obstacles needed to be placed to create many possible forms of lanes for the marbles. Participating researcher intentionally avoided interfering with the play unless the children asked for help (for example when they need someone to read the instructions they've discovered in the box). This strategy was found as strongly confusing for many children who are used to wait for teacher's instructions in play. In this particular case, a gender-conditioned difference in play was manifested. While all boys immediately placed the obstacles in the track with various levels of difficulty, the girls did not pay attention to a "boy" strategy. The girls preferred to modify this toy according their actual needs and interests. In this case, the track became quickly a home for invisible guinea pigs.

Some children preferred activities other than play. They were focused and interested as long as they could create something. Many of the offered toys allowed for modifications, and when the children completed them they began to look for new ones. If they could not find any other modification, they lost interest in the toy.

The course of the play with presented toys was visibly influenced by peer interactions. Particularly in peer groups of older preschoolers (5–6 years of age), the children often changed toys and play buddies or they moved from one toy to another. However, these quick shifts in play with more toys at the same time led to less detailed concentration on each toy. The children didn't focus on one toy to find out all opportunities to play. For other children in the group, this interrupted their play and distracted their attention. One of the boys wanted his friend to play the hide-and-seek game with him, while the latter was concentrated on another game. Another case was represented by children who tried to obtain one of the toys for themselves alone. One of the observed boys did not want to lend the marble track to anybody and invented various rules to prevent others from playing with it. The research by Fleer [20] showed that children not liking the offered scenario or partner in the game chose to abandon the game altogether. We also observed such cases, but at the same time, our children tried more active solutions than retreat, asserting their interests and views of the game and use of the toy.

3 Conclusions

We found out that it was not enough to concentrate on the toy alone, but in accordance with the finding of [21], it was possible to observe impact of play on the development of social, emotional and educational skills of the playing children. The child's development is mainly the task of the teacher – a professional whose didactic thinking must be supported. The research based on children's participation opened currently supported ways of analyzing and interpreting children's own perspectives represented in their micro social worlds as compared to previous predomination of adult's views and understanding of children play (based on play with didactic toys). In accordance with [12], we found out that kindergarten teachers were not yet open or prepared to adopt teaching through play and implement it in the educational program of the kindergarten.

References

1. Nutbrown, C., Clough, P.: Early Childhood Education. History, Philosophy and Experience. Sage, Los Angeles (2014)
2. Ashiabi, G.: Play in the preschool clasroom: it's socioemotional significance and the teacher's role in play. Early Child. Educ. J. **35**(2), 199–207 (2007)
3. Nutkins, S., McDonald, C., Stephen, M.: Early Childhood Education and Care: An Introduction. Sage, Los Angeles (2013)
4. Theobald, M., Danby, S., Einarsdóttir, J., Bourne, J., Jones, D., Ross, S., Carter-Jones, C.: Children's perspectives of play and learning for educational practise. Educ. Sci. **5**, 345–362 (2015)
5. Brooker, L., Blaise, M., Edwards, S.: The SAGE Handbook of Play and Learning in Early Childhood. Sage, London (2014)
6. Hartl, P., Hartlová, H.: Velký psychologický slovník. Portál, Praha (2010)
7. Tomanová, D.: Úvod do pedagogické diagnostiky v mateřské škole. Univerzita Palackého v Olomouci, Olomouc (2006)
8. Rogers, S.: Rethinking Play and Pedagogy in Early Childhood Education. Routledge, New York (2011)
9. Koťátková, S.: Hry v mateřské škole v teorii a praxi. Grada, Praha (2005)
10. Průcha, J., Walterová, E., Mareš, J.: Pedagogický slovník. Portál, Praha (2009)
11. Blakemore, J.E., Centres, R.: Characteristics of boys' and girls' toys. Sex Roles **53**(9/10), 619–633 (2005)
12. Fehessa, E., Pyle, A.: Conceptualising play-based learning from kindergarten teachers' perspectives. Int. J. Early Years Educ. **24**, 361–377 (2016)
13. Martinsen, M.: Structural conditions for children's play in kindergarten. J. Nord. Early Child. Educ. Res., Tidsskrift for Nordisk Barnehageforskning **10**, 1–18 (2015)
14. Tomková, A., Spilková, V., Píšová, M., Mazáčová, N., Krčmářová, T., Kostková, K., Kargerová, J.: Rámec profesních kvalit učitele: Hodnoticí a sebehodnoticí arch. Národní ústav pro vzdělávání, Praha (2012)
15. Rice, M.: What is the teacher's role in supporting play in early childhood classrooms? http://www.ttacnews.vcu.edu/2014/02/what-is-the-teachers-role-in-supporting-play-in-early-childhood-classrooms/ (2014)
16. Bruce, T.: Developing Learning in Early Childhood. Sage, Los Angeles (2004)
17. McLachlan, C., Fleer, M., Edwards, S.: Early Childhood Curriculum. Cambridge University Press, Port Melbourne (2013)
18. Fontana, D.: Psychologie ve školní praxi. Portál, Praha (2003)
19. David, T., Goouch, K., Powell, S.: The Routledge International Handbook of Philosophies and Theories of Early Childhood Education and Care. Routledge, New York (2016)
20. Fleer, M.: Kindergarten in cognitive times: imagination as a dialectical relation between play and learning. Int. J. Early Child. **43**(3), 245–259 (2011)
21. Pyle, A., Bigelow, A.: Play in kindergarten: an interview and observational study in three canadian classrooms. Early Child. Educ. **43**, 385–393 (2015)

Educational Training System for University Students at Initial Levels Through Coaching Strategies and Student Mentoring

Ives Torriente(✉), Rosa Inés Carrión, Eliane Fernández, Paulina Morales, and Gardenia Silva

Universidad UTE, Sede Santo Domingo,
Via Chone Km 4 1/2, esquina Italia, Santo Domingo, Ecuador
{ives.torriente, ccri7013426, efgleo86, paulina.morales, gardenia.silva}@ute.edu.ec

Abstract. The university students of initial levels are marked by the process of transition from one educational system to another superior one; and they bring with them habits and failures that attempt against the adequate academic performance, which generates a high student desertion in the initial university levels. In the current study, reference is made to the application of a system of educational training for university students at the initial levels of the UTE University, through student coaching and mentoring strategies. The research shows as a novelty the design of a training system where outstanding students of the same university specialty perform student mentoring with new students, and also evidences the results of applying student coaching by academic tutors and the department of student welfare.

Keywords: Student desertion · Student coaching · Student mentoring

1 Introduction

In the current context of Higher Education in Ecuador, there is an increase in the number of students entering university studies as a result of the implementation of the Organic Law on Higher Education (LOES), which was implemented in 2010 [1]. According to data from Ecuador's Ministry of Higher Education, the relevant supply of university degrees averages 55.56%, while gross and net enrollment rates are 31.86% and 21.23%, respectively [2]. This growing admission to Higher Education is in line with the contemporary trend in Latin America of the massification of Higher Education.

One of the problems brought about by the massification and increase of university students from the initial levels is student desertion. This phenomenon is complex and has multiple factors that originate it. Some authors affirm that it is the alteration of the student's educational trajectory due to repetition and school failure; considered as an individual phenomenon, and as a collective phenomenon, it is associated with teaching efficiency [3]. On the other hand, other authors establish two groups of factors that originate student desertion: those intrinsic to the process (teacher performance, educational policies, pedagogical relations, methodologies, etc.) and those extrinsic to the process (family situation, motivation to study, cognitive abilities, etc.) [4].

W. Karwowski et al. (Eds.): AHFE 2019, AISC 963, pp. 223–228, 2020.
https://doi.org/10.1007/978-3-030-20135-7_22

This research explores the factors intrinsic to the process, methodologies and pedagogical relationships; and the extrinsic factors in the motivation to study and cognitive abilities. Considering these factors we proceed to establish a diagnosis of which are the elements that most affect student desertion. Once the elements have been identified, we proceed with the analysis and selection of the appropriate coaching and mentoring strategies for the student training system proposed in the model.

2 Student Coaching and Student Mentoring

In today's learning environments, mainly those associated with the business world, coaching is a trend tool with already known results. Therefore there is an incorporation of coaching to the educational field with a positive success. The so called "educational coaching"; is currently theorized and is the subject of an increasingly important number of methodological books in Spanish for the development of academic performance [5]. The research proposes the implementation of educational coaching with the students of the initial levels, where the coaches will be the students of the intermediate levels of the same university careers of the initial students. The intention is that the coaches transmit the fresh experience lived by them to the new income, which act as a kind of personalized tutors who facilitate the adaptation to the new process of education (Fig. 1).

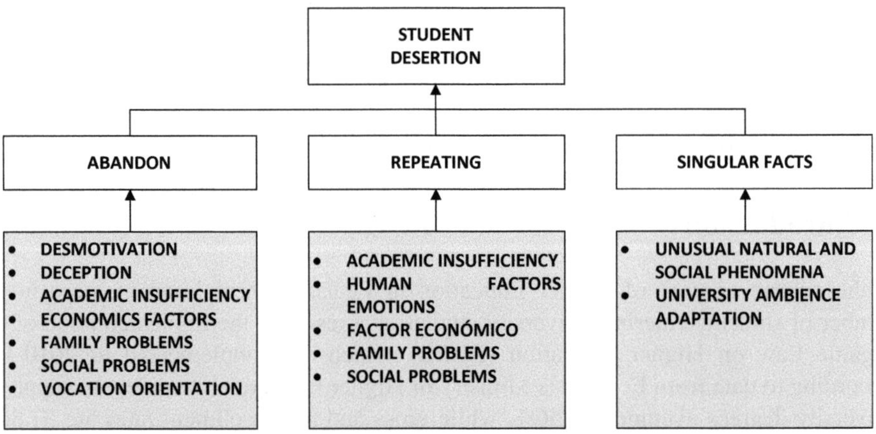

Fig. 1. Criteria's student desertions in the students of initial level of university

For the implementation of the proposal, we took into account the criterion that professional coaching is a personalized and confidential training process through a great set of tools that help to fill the gap between where a person is now and where you want to be [6]. To this end, a training system was established in the summer course for the students who would be the coaches of the new income. The preparation was focused on making coaches facilitators of learning, where their fundamental task is precisely to facilitate the identification and dissolution of those obstacles that limit action and learning.

Mentoring is based on the criteria that it is a process of guidance that is established between a mentor and a protégé. In order to complete the contribution, the term mentor is also defined as a wise and reliable advisor or guide who completes or deepens the skills acquired from the protégé [7]. In the research, mentoring strategies are applied with the students of the terminal levels, those who are already close to leaving the university, with the intention that they complement with their academic experience in the institution, the process of adaptation and retention of the students of the initial levels (Fig. 2).

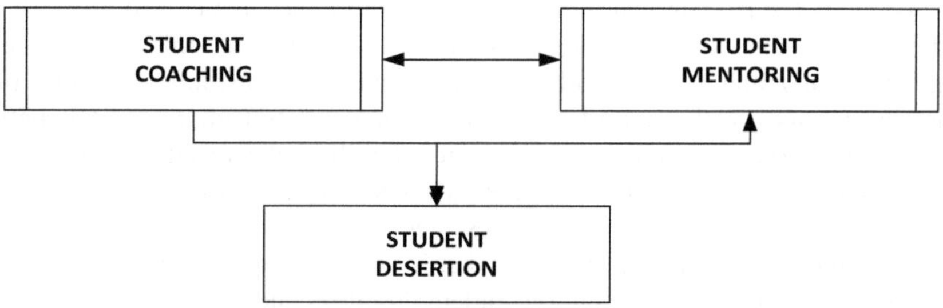

Fig. 2. Model of student coaching and student mentoring versus student desertion

It is necessary to emphasize that mentoring is a process of improvement in various aspects related to the professional career, but also with the improvement of the individual that integrates two parts: a Senior Advisor and a Junior protégé. The relationship that is established has benefits for all parties involved. Moreover, if both strategies are applied simultaneously and in an organized manner, there is a reduction in student desertion from the initial levels.

3 Education System Training

The proposed training system is aimed at the formation and training of three types of coach, the student coach, the teacher coach and the career coach tutor. Of the three types in this research, reference is made only to the student coach, as this is focused on the preparation of the student of intermediate levels to execute coaching strategies with the students of the initial levels. The teaching coach is aimed at preparing the teachers of the subjects of the initial levels to perform coaching actions with their students and the coach of the career tutor is very important because the career tutor is responsible for the implementation, accompaniment and execution of the training system in his career.

In the case of mentoring in the training system is defined as three types of mentoring, student mentoring, career tutor mentoring, career coordinator mentoring. The student mentoring that is the reference in this research lies in the preparation of students who are about to complete the career to professionally guide new income on the important path they are to develop. Career tutor mentoring is essential given the high responsibility of career tutors with the entire teaching and learning process in Higher Education. In the case of the career coordinator's mentoring, the preparation is continuous and profound given that the career coordinator is the maximum authority in the students; formative process.

The training system focuses on a training model for coaches and mentoring, which is based on a comprehensive initial diagnosis of students at the first levels. For the collection of information the technique of the 10 desires is applied in principle, to know the needs, motivations and fundamental aspirations of the subject, as well as to detect aspects that constitute problems or conflicts in the affective life of the examinee. Also simultaneously in the diagnostic process is applied a survey with Likert scale typology, which measures the student's attitudes to sport, culture, academics and research.

After the diagnosis stage, the induction process is carried out; this is the process of linking the family and society with the higher education institution. In this phase, coaches are trained on the importance of informing and transmitting the benefits of university education to new incomes and their families; the results obtained by the students in the various years of studies are indicated, and the professional perspectives by careers and their role in society are presented to them.

Within the model an element is added that is very necessary above all to minimize the effects of academic achievement on student desertion and is the course of levelling prior to entry into the study of the subjects assigned in the curriculum of each career. In many cases, students who do not pass the leveling courses have a high probability of repetition in the basic subjects of the initial levels. At the UTE University in Santo Domingo, in 2015 and 2016, a study was carried out where more than 80% of the students who passed the leveling course reached the specialty subjects of the intermediate levels with a low repetition rate. In addition, these leveling courses serve to define the vocation of the students of the initial levels.

After the training course comes the system of continuous, permanent and systemic training of student coaches and mentors. This system deepens some of the actions currently being carried out in the institution such as comprehensive accompaniment, academic guidance, psychological guidance, vocational guidance, scholarship program and medical care with medical insurance. All these actions are carried out intermittently and not interrelated, where the main responsible and actors are the career teachers who must interact in most cases with significant numbers of students, losing attention in the process to the individual differences of students at the initial levels.

At this stage of the process, student coaches and student mentors are incorporated with their plan of activities and their preparation. These intermediate and terminal students, respectively, after receiving their intensive training become part of the day-to-day life of the students of the initial levels recently admitted to the university. It should be noted that the career tutor, as well as the student welfare department must make a permanent control of the coach and mentoring, because it was perceived in the research that in some cases the coaches and mentoring did not care about their own academic situation to help students at initial levels and in other cases the opposite occurred, is care more about their own academic situation and neglect their mentors (Fig. 3).

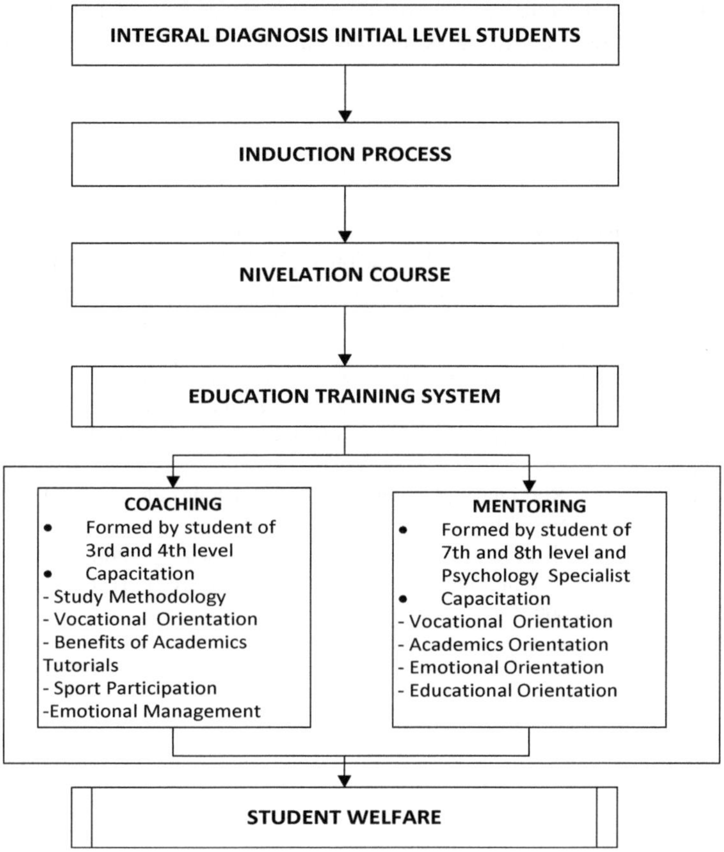

Fig. 3. Model of education training system

Coaching can be said to be a results-oriented accompaniment tool that has proven its usefulness in the professional, educational and academic spheres. Among its aims are to establish the leadership of students to clarify a problem or difficulty, help them to question their beliefs and specific academic practices, enable them to identify obstacles that hinder the development of their skills, discover alternative resources on which they can rely, implement effective personal solutions transferred from other environments or spheres of their lives that allow them to effectively achieve their objectives in Higher Education.

Generally speaking, the data collected correspond to a pilot test with a reduced sample, in which 18 students from the initial levels participated and 8 students who took part in the training and performed the function of coach. In this test the benefits that arise with the proposal were perceived, it is important to highlight that with the accompaniment of the coaches motivational manifestations were observed on the part of the students, of the 18 students of the study only one deserted for 94% of retention, and it was for economic reasons. The results of the experiences detailed in this article suggest that educational coaching is an efficient and effective way to help students succeed at university.

References

1. Sinchi, E.R., Goméz, P.G.: Acceso y deserción en las universidades. Alternativas de financiamiento. Alteridad Revista de Educación **13**(2), 274–287 Ecuador (2018)
2. SENESCYT: Informe de rendición de cuentas. Ecuador (2015). https://goo.gl/TSFPSK
3. Cabrera, L., Tomás, J., Álvarez, P., y Gonzalez, M.: El problema del abandono de los estudios universitarios. RELIEVE **12**(2), 171–203 España (2006)
4. Passailaigue Baquerizo, R., Amechazurra Tam, O., Galarza López, J.: La deserción y la repitencia en las instituciones de Educación Superior: algunas experiencias investigativas en el Ecuador. Universidad y Sociedad [seriada en línea] **6**(1), 102–107 Cuba (2014). http://rus.ucf.edu.cu/
5. Huesa, F., Reyes, A., Rodríguez, R., Narváez, R.: Coaching educativo. Optimización de recursos académicos. Sevilla, España (2013)
6. ASESCO: El coaching, España (2018). https://www.asescoaching.org/el-coaching/
7. Colomo, R., Casado, C.: Mentoring & coaching. IT perspective. J. Technol. Manag. Innov. **1**(3), 131–139 Chile (2006)

The Construction Manager's Mask: A More Purpose-Fit Selection for South Africa

Andries van Heerden(✉), Michelle Burger, and Benita Zulch

EBIT Faculty, Department of Construction Economics, University of Pretoria, Building 5, South Campus, Private Bag X 20, Hatfield 0028, South Africa
benita.zulc@up.ac.za

Abstract. The construction manager requires an educational background, technical skills, practical experience with specific soft skills in the competency domain and a specific personality profile – in order to ensure the success of a project. To enroll the correct profile student for construction management can reduce the probability of skills lost over time. The South African Personality Inventory (SAPI) was used to test the personality profiles of construction management industry participants sample (N = 10). The student group sample (N = 16) included students in their 3rd year of construction management studies enrolled at three different universities in South Africa. The online SAPI personality results suggested a remarkably similar profile between the industry group and the students' group. The online SAPI personality tests' main dimension results scored the highest in Intellect/Openness, followed by Conscientiousness, the lowest were Neuroticism and Negative Social Relational Disposition. Construction management students need to do practical work for a year before they start their studies. This will give them a better understanding of the work conditions, pressure and work-life balance associated with the construction industry, as this is not tested in the online SAPI personality test.

Keywords: Construction manager · Personality · South African Personality Inventory · South Africa

1 Introduction

In a perfect world, one would like to use an instrument that could test an individual's personality profile – in order to try to forecast the probability of that person being a purpose-fit candidate for the position. The real benefit would then depend on the quality of the fit that is formed between the employee, their work, the organisation and the environment [1, 2].

It is pivotal for the construction industry, and even more so for individual construction companies, to retain critical skills. To enroll the correct profile student for construction management can reduce the probability of skills being lost over time, provided that purpose-fit student enters the industry. Companies can improve their systems and processes and apply new technology, but this would all be fruitless if they do not invest their efforts in their intellectual capital. It would be even worse if a

W. Karwowski et al. (Eds.): AHFE 2019, AISC 963, pp. 229–237, 2020.
https://doi.org/10.1007/978-3-030-20135-7_23

construction company invests resources in the development of an individual, only for that individual to leave the industry and apply their skills and knowledge somewhere else [3–5].

2 Literature

Every individual who has worked on a construction project knows the impact that a key person has on the success of such a project and the opposite reaction if that key person decides to resign [6]. The expensive replacement cost, disruption of the project and involved team members typically create frustration within the organisation [7]. Some sources indicate that to replace a professional individual can cost as much as one and a half times their annual salaries. The bulk of the cost of recruiting or replacing management is in the cost of training the new employee; furthermore, there is the cost involved with the loss of work by the outgoing employee [1].

The research of Guion and Gottier [8] concluded that personality measurements should not be used when selecting personnel. This view was only contested a few years later by other researchers [9]. Previous research studies indicated that if a test is intended to specifically measure personality within a work context, the criterion-related validity escalates and this increases the value of personality constructs from an industrial and organisational perspective [2].

Schmidt and Hunter [10] were involved in a meta-study that stretched over 100 years to scrutinise the predictive validity of various instruments. The research findings concluded that different methods and combinations of methods have very different validities for predicting future job performance i.e. General Mental Ability (GMA) combined with a structured interview increased in terms of their prediction. Their research also concluded that methods, such as personal-job fit, personal-organisational fit and the amount of education have little validity and graphology has in essence no validity. During their research, they realised that the combination of two predictors is practical for use in recruiting and it has a high composite validity. GMA and above-an integrity test and GMA and above-a structured interview, both proved to be good predictors of performance in job-training program. These two combinations can also be used for recruiting both entry-level and expert-job applicants.

Thus, when the HR manager recruits by using GMA, the probability is good that they will select an employee who has a high level of performance on the job. This recruited individual will learn mostly from job-training programs on the job, as well as acquiring job knowledge more quickly from experience on the job [10]. The research also indicated that even personality has a certain predictive validity. In a local meta-study conducted in South Africa on personality, the results indicated that students with an education lower than grade 12 on the Big Five, were found to be low in conscientiousness. However, extraversion is a good predictor for overall job performance. Conscientiousness proved to be the best predictor for task performance in the South African context [11].

In an academic environment, conscientiousness also proved to be an important predictor but combined with an introverted personality. As such, an individual is more focused and not easily distracted from the task at hand. To predict overall performance, extraversion, emotional stability and openness proved to be the best predictors.

According to Holland's theory "personality is an important determinant of career choice and that career choice is an expression of personality". It was further explained that the interaction which individuals experience when they embrace their environments, which correlates with their personal orientation [1].

An instrument in South Africa based on the Five-Factor Model (FFM) considered the importance of local context. Strong psychometric properties across the ethnic groups were found through using their Basic Traits Inventory (BTI). Although the BTI instrument was not designed to assess indigenous personality concepts, it was commonly used in South Africa as an FFM instrument [12].

The South African Personality Inventory Instrument (SAPI) was constructed around the main ethno cultural groups of South Africa. Embracing a combined emic-etic approach includes measures of the Big Five traits and social-relational traits that are particularly striking in South Africa's Bantu ethnic groups. Parallel scales were concurrently developed in the eleven official South African languages [2, 11, 12]. The SAPI was based on the same principle as the Chinese Personality Assessment Inventory (CPAI) that was designed with the whole concept of the indigenous perspective [13].

With the development of this instrument, more focus was given to cultural specific indices and ideas more specific in certain cultures, since previous instruments neglected this area [12].

SAPI's blueprint did not start from a predefined model, such as the FFM, but it opted for a more all-inclusive reportage of the personality ideas relevant across the different groups. The SAPI instrument was developed in two stages: (1) "the qualitative stage of a conceptual model development; and (2) the quantitative stage of instrument development". The latter stage was broadly conceptual and it tried to disentangle the implied personality structure dimensions reflected in natural language by speakers of all eleven official languages in South Africa. The second part ultimately concentrated on the empirical validation of the preliminary personality inventory [13].

During the instrumental development stage, ethnographic data were collected from all eleven languages [12]. Characteristics of translated items and factor replicability across groups were used as the main criteria of selection. In their research they further stated that replicability across ethnic groups can increase the probability of succeeding factor duplication, and lessen the existence of distinctive elements. It was also suggested that structure replicability must be increased if cross-cultural comparisons are visualised [12].

They further suggested that during social desirability, a distinction must be retained between impression-management-focused aspects (associated with personality in different cultural groups) and lying aspects (more circumscribed effects) [12].

Although there are noticeably different ethno-cultural groups within the South African context, the model development was allowing for transcendence from narrow emic constructs. Cross-cultural personalities, where the emic-etic was combined in a balanced manner were represented in the instrument. From a local perspective, this gave rise to an analogous structure to the Big Five form, although not all of these factors looked as if they were equally replicable [12].

3 Methodology

An interpretivism philosophy was followed through a deductive approach and action research strategy. A multi-method was utilized over a cross-section period. In the data collected for the construction management industry participants the sample (N = 10) included both males and females from the general population in Gauteng, South Africa. These industry individuals were selected based on their experience in specific construction sectors, their age and the fact that they operate in the Gauteng area. The sample consisted of 10 white participants (9 males; 1 female; age 32 to 71 years).

The student group sample (N = 16) included students in their 3rd year of construction management studies enrolled at the University of Pretoria, University of the Free State and Nelson Mandela University. The students participated voluntarily and therefore, not all four of the major ethnic groups could be tested. This sample consisted of 13 white participants (12 males; 1 female; age 21 to 22 years), 2 black participants (1 male; 1 female; age 21 to 23 years) and 1 Indian participant (1 male; 0 females; age 22 years). The SAPI instrument was administered in English and it consisted of 188 questions. A 5-point Likert scale from 1 (strongly disagree) to 5 (strongly agree) was used. The SAPI version that was used contained seven main dimensions. Under each of the seven main dimensions was a variation of the sub-dimensions, which contained typical SAPI questions in the text.

4 Discussion and Findings

The main dimensions within SAPI that proved to be the most important sections were: (1) openness/intellect; (2) conscientiousness; and (3) positive social-relational disposition.

The construction manager must be someone who is well-informed and observant of external [outside the company] and internal [within the company] issues that could influence the outcome of the project. He/she must be a progressive and rational thinker to find practical solutions for difficult problems. To effectively improve a construction project, this individual must constantly seek new experiences, knowledge, skills and ideas. Every day is different and plans can typically change a few times during a day. Obstacles and challenges occur on a daily basis; consequently, it is important to be adaptable to constant change, with never a boring moment. The construction manager cannot have all the resources that he/she always needs on a project; and he/she must therefore be innovative and continuously able to find new ways to achieve the project goals within these constraints. A construction manager with more experience should have the ability to find more solutions to solve a problem, make better decisions, and resort to the best possible outcome.

In this environment of constant change, the construction manager must be able to learn and adapt to new information rapidly. The lack of necessary skills forces young construction managers to tackle the huge responsibility of managing relatively large construction sites. This can create even more stress in an already very stressful environment.

Conscientiousness was also indicated as a strong predictor for a construction manager. It is important for such an individual to persevere, to be ambitious and to be hard-working. The construction manager must be self-driven and pro-active, in order to minimise possible problems that could occur on the project. Construction projects always have deadlines, problems and difficulties that need to be resolved under tremendous pressure. He/she must keep everybody motivated, who is involved in the project.

Construction projects are most of the time very difficult to manage; and it is not always easy to motivate individuals in the robust construction site environment. Being organised, neat, precise and punctual with managing all aspects of a construction site is very important. Most construction managers fail in their paper work, especially when claims and disputes arise. This typically has a definite negative impact on the outcomes of the dispute, as well as on the progress and motivational climate of the project.

In the multi-cultural context of the South African industry, the construction manager must also respect different cultural and religious aspects. Although religion would not determine whether such an individual is successful or not on a project, it does indicate that such an individual would probably be more trustworthy.

Construction managers must understand and be sensitive towards the needs of everybody on the construction site. If the construction site is situated in an area, where it is very cold or hot, working hours should be adapted, as required. In some instances, the people involved in the project go through difficult family issues, where a more sensitive approach is required. Construction managers must be approachable at all times during the project, to give advice and instructions, as and where needed. If the project is running into trouble and it falls behind schedule, people might have to work over weekends for long periods. In some cases, the construction site is lit-up and construction work continues until late at night.

The stress levels increase; as quality becomes an issue people get tired and demoralised. During these difficult times, the construction manager must have the ability to give the necessary guidance, upliftment and motivation, as and when needed. He/she must be able to forgive people if mistakes occur, and accommodate sub-contractors in tight schedule situations. Construction managers must create a site culture of being dependable, loyal, honest and fair. He/she must be fair towards everybody, and create loyalty between teams. With the responsibility-assigned matrix (RAM) the construction manager can empower dependable individuals to take responsibility and be accountable for their actions (Table 1).

Extraversion and social desirability did not feature as important as the last main dimensions, but they did prove to be positive additional personality traits. Without a tendency to be sociable and talkative, much-needed interaction might be hampered. Personality traits for a construction manager are of paramount importance.

Construction managers must also have the ability to see the comical side of life, to survive every day. If he/she is not approachable and does not enjoy having people around all the time, it would be difficult to operate in the construction industry. Again, although being playful and sociable has their place on a construction site, they are not critical personality traits that one would need on a daily basis.

It is necessary for a construction manager to have a positive self-image as this will reflect well for the company and to an extent give security to everybody involved on

Table 1. SAPI: Most important personality traits for a construction manager

SAPI [Main dimensions]	SAPI [Sub-dimensions]	SAPI [Characteristics]
Openness [1] *(The quality of being well-informed and observant of external and internal things, being a rational and progressive thinker and acquiring new experiences, knowledge, skills and ideas)*	***Broad-Mindedness*** *(Being imaginative and seeking new experiences and ideas)* ***Epistemic Curiosity*** *(Being inquisitive, investigative and eager to acquire new information)* ***Intellect*** *(Being knowledgeable, a quick learner, adaptable, articulate, innovative and perceptive)*	Imaginative, seeking new experiences, seeking new ideas, inquisitive, investigative, eager to acquire new information, knowledgeable, quick learner, adaptable, articulate, innovative, perceptive.
Conscientiousness [2] *(Orientation toward achievement, order and traditionalism)*	***Achievement Orientation*** *(Being motivated, perseverant, ambitious and hard-working towards achieving things in life)* ***Orderliness*** *(Being organised, neat, punctual, precise and thorough in everything one does)* ***Traditionalism-Religiosity*** *(Being traditional by respecting one's own culture and being religious)*	Being motivated, perseverant, ambitious, hard-working, focus to achieve things, organised, neat, punctual, precise, thorough, being religious, respect own culture.
Positive Social-Relational Disposition [3] *(Positively managing relations with others)*	***Empathy*** *(Valuing and showing compassion to others by showing sensitivity towards their needs and emotions)* ***Facilitating*** *(Guiding, uplifting and motivating others through their lives by giving them advice, instructions and encouragement)* ***Integrity*** *(Being consistently dependable, loyal, honest and fair towards others)*	Value compassion, sensitivity towards needs & emotions, guiding, uplifting, motivating, give advice, instructions, encouragements, dependable, loyal, honest, fair, accommodating, forgive, helpfulness, peace, understand feelings, considerate, protective, supportive, approachable, attentive.

(*continued*)

Table 1. *(continued)*

SAPI [Main dimensions]	SAPI [Sub-dimensions]	SAPI [Characteristics]
	Interpersonal Relatedness	
	(Being accommodating in one's relationships and actively maintaining relationships through forgiveness, helpfulness and by preserving peace)	
	Social Intelligence	
	(Relating to others by being understanding of them and their feelings)	
	Warm-Heartedness	
	(Being considerate, protective, supportive of others as well as being approachable and attentive to others' needs)	

the project. The construction manager would also have to be able to absorb vast critique that flows from various decisions that have been made. One should, however, be careful that this does not spill over to becoming an arrogant person.

Extraversion and social desirability are additional personality traits for a construction manager to possess. Extraversion is the tendency to be sociable and playful. The characteristics are the ability to see the positive side of life, being lively and being able to make people laugh. With social desirability, the tendency to give a negative or positive self-description is emphasised. These characteristics consist of being easy-going, talkative and enjoying being around people.

There are certain personality traits that are not favourable for a construction manager to possess; these are neuroticism and negative social-relational disposition. Neuroticism indicates the tendency for a person to be impulsive and to fluctuate between emotions. The construction manager must not have the tendency to be impulsive. He/she must not always have all the answers, but within a reasonable timeframe to resolve the issue. This individual should reflect back before they reply to claims and issues. Construction manager's emotions should not fluctuate and easily be aggravated during negotiations, meetings and general labour issues.

Because the construction industry is so complex, the construction manager must be able to handle stress well. If he/she is a very nervous individual, worried all the time and afraid to make important decisions, he/she would fail as a construction manager. The construction manager must rather show respect, knowledge and acceptance, where needed, and be composed in difficult situations, to be able to make rational decisions and conduct educated calculations.

A negative social-relational disposition is also a main dimension that should not be part of the personality of a construction manager. The construction manager needs to work with people all the time for the duration of the project and build relationships for possible future projects.

There is, consequently, no place for an individual that thinks he/she is better or more important than another person or team member, neither would it be good if the construction manager is actively deceiving others or cheating them. Integrity in the industry is highly valued and a deceitful individual could do irreparable damage to the project and the construction company's image. Being disruptive, intrusive and indiscreet about the private affairs of others would break the trust, and be detrimental to the trustful relationships cultivated between teams and individuals.

5 Conclusions

The construction management students need to do practical work for a year before they start their studies. This will give them a better understanding of the work conditions, pressure and the work-life balance associated with the construction industry, as this is not tested in the online SAPI personality test. The difficulty of working in the construction industry can only be experienced during practical work. The difficulty when starting a family, managing stress and to leave work-related problems at work and prevent them from spilling over to your personal life, can create further conflict. For female construction managers, there is the issue of cultures that do not accept female authority and that makes it even more difficult to be successful. It is obvious that the construction manager has a special character with a specific educational background, knowledge and the experience needed to ensure the success of a project. This person would also need a personality trait or traits that will assist him/her on the road to the successful completion of projects.

During the selection process, personality traits associated with the profile of a successful construction manager would assist in selecting a purpose-fit candidate. This could not only assist in successful projects, but also increase the probability that the construction manager would be satisfied on a construction site for a longer sustainable period.

Fewer construction managers leaving the industry (taking their experience and expertise to their new workplace) should slowly create an elite pool with the appropriate levels of critical skill sets.

This retained expertise would be vital to the survival and even the expansion of the South African construction industry. After the relevant literature study was scrutinised and the SAPI results were received, a possible profile for a construction manager could be formulated.

References

1. Swanepoel, B., Erasmus, B., van Wyk, M., Schenk, H.: South African Human Resource Management: Theory and Practice, 2nd edn. Juta & Co (Ltd), Cape Town (2000)
2. Mouton, S.: An investigation into the first- and second-order factor structure of the South African Personality Inventory (SAPI) on a relatively large South African sample. Masters of Commerce (Industrial Psychology), Stellenbosch, University of Stellenbosch (2017)
3. DeNisi, A.S., Griffin, R.W.: Human Resource Management. Houghton Mifflin Company, New York (2005)
4. Walker, A.: Organisational Behaviour in Construction, 1st edn. UK, Wiley-Blackell (2011)
5. Jantan, H., Hamdan, A.R., Othman, Z.A.: Knowledge discovery techniques for talent forecasting in human-resource application. Int. J. Hum. Soc. Sci., 5–11 (2010)
6. Burma, Z.A.: Human-resource management and its importance for today's organisations. Int. J. Educ. Soc. Sci. **1**(2), 85–94 (2014)
7. Sanders, B.A.: Using personality traits to predict police officer performance. Polic.: Int. J. Police Strat. Manag. **31**, 129–147 (2008)
8. Guion, R.M., Gottier, R.F.: Validity of personality measures in personnel selection. Pers. Psychol. **18**, 135–164 (1965)
9. Morgeson, F.P., Campion, M.A., Dipoye, R.L., Hollenbeck, J.R., Murphy, K., Schmitt, N.: Reconsidering the use of personality tests in personnel selection contexts. Pers. Psychol. **60**, 683–729 (2007)
10. Schmidt, F.L., Hunter, J.E.: The validity and utility of selection methods in personnel psychology: Practical and theoretical implications of 85 years of research findings. Psychol. Bull. **124**, 262–274 (1998)
11. Van Aarde, N., Meiring, D., Wiernik, B.M.: The validity of the big five personality traits for job performance: Meta-analysis of South African studies. Int. J. Select. Assess. **25**, 223–239 (2017)
12. Fetvadjiev, V.H., Meiring, D., Nel, J.A., van der Vijver, F.J.R., Hill, C.: The South African Personality Inventory (SAPI): A culture-informed instrument for the country's main ethnocultural groups. Am. Psychol. Assoc. **27**(3), 827–837 (2015)
13. Cilliers, C., Meiring, D.: (2014). Determining the best factorial fit for the South African personality inventory: Comparison of block-and random-item formations. Master's thesis. http://hdl.handle.net/2263/45929. Accessed 26 Aug 2016

Education in Medicine and Rehabilitation

Training Caregivers to Reduce Spine Flexion Using Biofeedback

Megan Kamachi[1,2(✉)], Mohammadhasan Owlia[1], and Tilak Dutta[1,2]

[1] Toronto Rehabilitation Institute, University Health Network, 500 University Avenue, Toronto, ON M5R 2A2, Canada
megan.kamachi@mail.utoronto.ca
[2] Institute of Biomaterials and Biomedical Engineering, University of Toronto, Rosebrugh Building, 1604 Room 407, Toronto, ON M5S 3G9, Canada

Abstract. Back injuries caused by poor postures are very common among home caregivers. Real-time biofeedback may be helpful for training caregivers to use safer patterns of movement. Twenty novice caregivers repeated a series of simulated care tasks to investigate the effectiveness of biofeedback in decreasing time spent in end-of-range spine flexion. Participants in the intervention group (n = 10) received real-time auditory biofeedback when their forward spine flexion exceeded a threshold during training trials, while participants in the control group (n = 10) did not. All participants returned after two weeks to repeat the care tasks and perform a new skill transfer task (sling insertion). The intervention group maintained decreased end-of-range (80th and 95th percentile) spine flexion after two weeks during care tasks compared to controls, but there was no difference between groups with the new skill transfer task. Biofeedback has good potential for reducing spine flexion but training does not transfer to new tasks.

Keywords: Biofeedback · Posture · Home caregivers · Back injury prevention · Wearable technology · Spine flexion

1 Introduction

Home caregivers suffer from low back injuries at a high rate [1–4]. Caregivers working in the home have similar responsibilities to their counterparts who work in institutions such as a hospital or nursing home, but have the added challenge of rarely having access to mechanical lifts, and having to work in tight spaces with less access to help from others. Data on back injury rates specific to unpaid home caregivers is scarce but the US Bureau of Labour Statistics report that the injury rate for home care workers is approximately four times higher than the average for all occupations and 25% higher than hospital-based workers who perform similar tasks [5]. Low back injuries have a significant negative impact on several factors including economic costs and quality of life [6].

Lifting has been shown to be one of the biggest risk factors for low back injuries [7]. The risk of injury is high due to the awkward postures required to assist and adjust patients. Heavy loads and awkward postures, mainly high amounts of forward spine

W. Karwowski et al. (Eds.): AHFE 2019, AISC 963, pp. 241–251, 2020.
https://doi.org/10.1007/978-3-030-20135-7_24

flexion, cause extreme forces within the spine. These forces are withstood by the intervertebral discs of the spine when in poor low back postures [8]. Over time the intervertebral discs become damaged, which can lead to bulging or herniated discs [9]. The risk of injury is only expected to increase as the workforce ages and the patient population gains weight [10]. It has been shown that risk of injury can be reduced by reducing end-of-range spine flexion [11]. Stress concentrations on the intervertebral discs can be reduced by hinging through the hips instead of forward flexion in the low back.

Over the past few decades, several interventions have been investigated with the goal of decreasing the prevalence of low back pain amongst healthcare providers [3]. Research shows that despite these efforts, 70% of workers believe their training is inadequate [12]. Most of these interventions have focused on education and awareness training, manual handling training, and lift equipment [10, 13]. Dawson et al.'s systematic review showed that most interventions that focused on a single intervention were ineffective [14]. It is evident that new strategies are needed to solve this problem.

In particular, multicomponent interventions have shown promise for retention of back injury prevention interventions. Multicomponent interventions can be defined as a back-injury prevention intervention that involves more than one strategy for injury prevention, for example, an educational component, equipment procurement, exercise program, stress management sessions, or cognitive-behavioural approach [3].

Another method that has shown promise in injury prevention is the use of technology and wearable devices to provide biofeedback. These wearables have been proven to assist with knowledge transfer and motor learning within the context of sports movements, rehabilitation activities, and postural control [15–18].

Motor learning is a widely researched area with many studies on structuring biofeedback to maximize retention and learning of new motor behaviours [19]. The literature includes studies of different feedback schedules using various feedback modalities and frequencies. A common finding of many of these studies on body movement is an increase in long-term retention of skills with a decrease in the relative frequency of feedback over time (also called a faded feedback schedule). Relative frequency can be defined as the number of trials where feedback was provided divided by the total number of trials for which feedback could have been provided. This information is valuable when considering a back-injury prevention intervention for caregivers.

PostureCoach is a wearable device that is designed to help caregivers learn to avoid end-of-range spine flexion by providing real-time biofeedback (Fig. 1). This device detects lumbar spine flexion and notifies the user when the amount of flexion exceeds the preset threshold. This notification can be given through vibrotactile and/or auditory feedback; only auditory feedback was used in this study. PostureCoach has been in development since 2015 and has been shown to both accurately measure flexion and reliably provide feedback [20].

The objective of this study was to assess the effectiveness of a multicomponent intervention, which included PostureCoach and a video on back injury prevention strategies. Success was measured by the retention of safer lifting postures (reduced end-of-range spine flexion) over a two-week period. Skill transfer was also evaluated by introducing a new sling insertion task at the two-week post intervention time point.

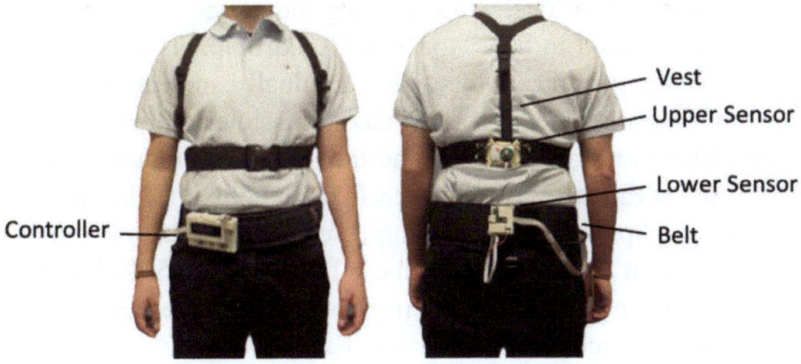

Fig. 1. Setup of PostureCoach.

2 Methods

2.1 Participants

Twenty young, healthy adults with no formal training in caregiving or patient handling were recruited to participate. Subjects were assigned to either the intervention group (n = 10) or control group (n = 10) using a coin toss for the first participant and alternating groups for each subsequent participant, ensuring groups were balanced for height and gender. Participants were recruited from Toronto Rehabilitation Institute (TRI) and the University of Toronto. This population was selected because there is a high prevalence of healthy individuals among novice caregivers. Participants were healthy, English-speaking individuals between 18 and 40 years of age. Participants were excluded if they had previous caregiving/healthcare provider training, history of back pain in the past six months or if they had any musculoskeletal issues related to the spine.

2.2 Instrumentation

Setting. This study took place in HomeLab, located in the 12th floor of Toronto Rehabilitation Institute. This laboratory consisted of a furnished bedroom, living room, bathroom and kitchen and resembled a typical single story house with functioning wiring and plumbing, similar to homes caregivers work in. A member of the research team performed the role of a patient actor in all trials for consistency. A category 3, 18 × 18 wheelchair (NRG+ Maple Leaf Wheelchair, Mississauga ON) with adjustable armrests, working push brakes and swing out standard footrests was used in this study.

PostureCoach – Wearable Device to Provide Biofeedback. PostureCoach is a wearable device that estimates the amount of sagittal lumbar spine flexion by measuring the relative angle between a sensor placed on the T10 vertebrae and a sensor on the sacrum at S2. This system was able to provide auditory feedback to the participant when the pre-set threshold angle for spine flexion was exceeded. PostureCoach stored all flexion data to an SD card. The threshold angle for providing biofeedback was customized for each participant by calculating 70% of their maximum forward flexion measurement. Maximum flexion was determined by asking the participant to reach as far as they could towards their toes, while keeping their knees straight and bending with their back. This was referred to as a “toe touch”. Participants began to hear an intermittent tone 20° before the threshold, which gradually increased in frequency until reaching a constant beep at the threshold mark. This intermittent tone was applied to warn participants they were approaching the threshold and reduced the abruptness of the constant beep when the threshold was reached. Benefits of a looming risk approach have been thoroughly explored in car driving research [21]. Also PostureCoach is able to provide biofeedback at a reduced frequency, for example, 50% feedback means the user receives feedback only every other time the threshold is reached. This function was used in this study to achieve a faded feedback schedule.

2.3 Procedure

A patient actor (male, age 25, 175 cm, 87 kg) and a research coordinator were present to guide all participants through each data collection session. An overview of this study’s procedure is shown in Fig. 2.

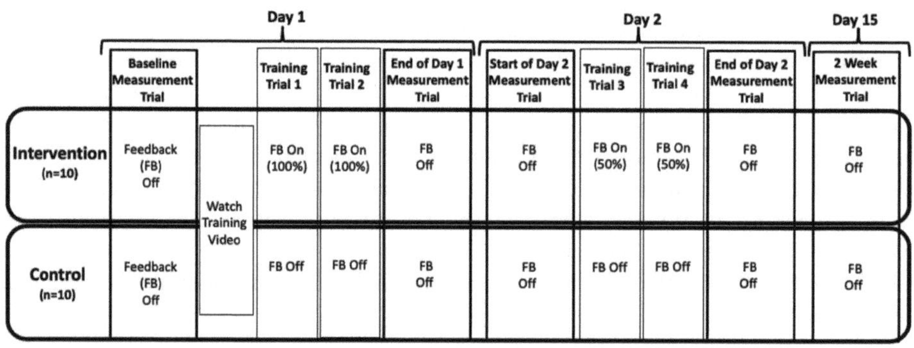

Fig. 2. Experimental protocol

Experimental Protocol. On day 1, participants were introduced to HomeLab, the patient actor and the series of caregiving tasks (described below) after providing informed consent to be part of the study. The research coordinator then helped the participant don PostureCoach. The participant performed the *Baseline Measurement Trial.* All trials (measurement trials and training trails) consisted of three box lifts and one repetition of each caregiving activity. Next, they watched a seven minute video on back injury prevention strategies. The video was developed in collaboration with the Occupational Health & Wellness team of SE Health (a home healthcare provider organization). The points discussed in the video were adapted from [11, 22, 23].

After the video, participants completed the first two training trials. The intervention group received 100% feedback in each of these trials and the control group did not receive any feedback. In other words, participants in the intervention group heard a beep when their spine flexion exceeded the threshold value described above. The *End of Day 1 Measurement Trial* was completed with neither group receiving feedback.

On the next day, participants performed the *Start of Day 2 Measurement Trial* consisting of the same box lifts and simulated care activities without feedback, which measured baseline performance for this session. Participants then repeated the tasks two more times in training trials three and four; the feedback for the intervention group was set to only provide an audible prompt for every other time the participant exceeded the preset threshold (50% feedback) and feedback remained off for the control group. In the final trial of the two-day intervention, *End of Day 2 Measurement Trial*, no participants received biofeedback.

Participants all returned two weeks (15 days ± 2) post-intervention for the *2 Week Measurement Trial* in which no feedback was provided to participants in either group. This consisted of one trial of the original box lifts and caregiving tasks and one trial of a skill transfer task, a patient sling insertion activity (described below). Prior to the skill transfer task, the researcher gave a demonstration on how the task was to be done and answered any questions the participant had.

Caregiving Activities. Participants were asked to complete the following simulated care activities in each trial, the patient actor always ended where he started and the order of the three tasks (bedroom, couch, washroom) was counterbalanced (Table 1):

Table 1. Caregiving activity requirements

	Activity requirements
	Toe touch
Bedroom	• Wheel empty chair to patient lying in bed, engage breaks and move footrests
	• Transfer patient from bed to wheelchair, assist with putting feet onto footrests and disengage breaks
Couch	• Wheel patient to couch, engage breaks and move footrests
	• Transfer patient from wheelchair to couch
	Toe touch
	• Transfer patient from couch to wheelchair, put feet on footrests and disengage breaks
Washroom	• Wheel patient to bathroom, engage breaks, move footrests
	• Assist patient to stand, doff pants, and sit on toilet
	Toe touch
	• Assist patient to stand, don pants, transfer to wheelchair, put feet on footrests, disengage breaks
Bedroom	• Wheel patient back to bedroom, engage breaks, move footrests
	• Transfer patient to lying position in bed
	Toe touch

Box Lifts. Prior to each trial of caregiving tasks, participants performed three box lifts. This task required the participant to lift a weighted, 6 kg milk crate with dimensions 33 × 33 × 28 cm (length x width x height) from the ground (28 cm) to knuckle height when standing. The load of 6 kg was chosen because according to the Snook tables, this load is acceptable for 90% of males and females using the carry distance of 0 m, lifting frequency of 35 lifts/min and lifting height mentioned above [24].

Skill Transfer Task. At the 2 week session, in addition to the patient care and box lifting tasks participants were asked to perform a new sling insertion task to test skill transfer. This transfer task required participants to insert a sling under the patient who was lying in the bed as shown in Fig. 3. The skill was demonstrated by the researcher beforehand to ensure the participant understood the activity. In the real world, a trained individual teaches the caregiver how to perform this task and it is recommended to have two caregivers complete it together, although it is assumed that it is often completed by a lone caregiver in the home [25].

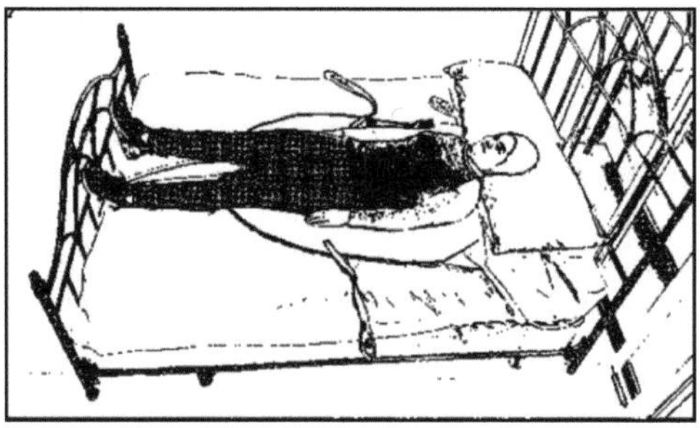

Fig. 3. Final position of patient and sling in skill transfer task

To perform this task, the participant was required to roll the patient onto his side, line up the top of the sling to the top of his shoulders and the bottom to the tailbone. The participant then had to push the far edge of the sling under the patient and ensure the midline was aligned with the midline of the patient. Next, the participant rolled the patient back into a supine position, went around to the other side of the bed and rolled him onto his other side. They could then pull the sling towards them, out from under the patient. Lastly, the patient was rolled onto his back again and the leg straps were placed in position. A toe touch was performed at the beginning and end of the task.

2.4 Data Analysis

Data was analyzed using MATLAB 2017 (MathWorks Inc., Natick, MA). Maximum forward flexion was calculated by obtaining the maximum angle achieved on the toe touches for each trial. This value was used to normalize all of the spine flexion data collected by PostureCoach. Normalization helps eliminate differences in forward flexion angles measured that were due to variations in sensor placement, as well as individuals' height and flexibility. Toe-touching portions of trials were then manually trimmed from the data.

3 Results

3.1 Caregiving Activities

The 80th and 95th percentile values for each trial were calculated. The participant used this amount of forward flexion or less for 80% of the time they were completing the caregiving tasks. Mean values for each group across each trial are shown in Fig. 4. An independent t-test concluded there was no significant difference between the intervention and control groups in the *Baseline Measurement Trial.* The repeated measures ANOVA on the intervention group concluded that trial number had a significant effect on percentage of spine flexion used, $\chi^2(4) = 12.48$, $p = 0.014$. Post hoc tests showed percentage of flexion was significantly lower in the *2 Week Measurement Trial* compared to baseline, $p = 0.010$. The repeated measures ANOVA on the control group revealed no main effect for trial number.

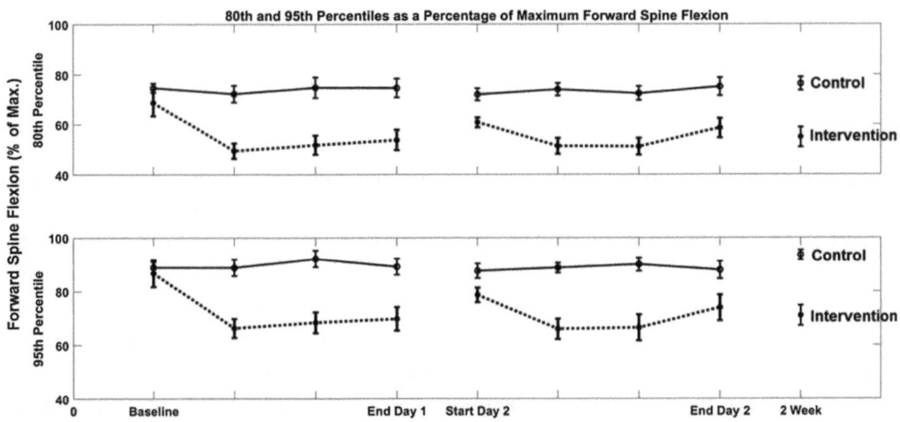

Fig. 4. 80th percentile (top) and 95th percentile (bottom) spine flexion value as a percentage of maximum forward spine flexion for control group (solid line) and intervention group (dotted line) across each trial

Box Lifts. Data collected during box lifts has not yet been analyzed.

Skill Transfer Task. Maximum forward spine flexion was determined using the toe touches of this trial. The toe touches were then trimmed from the data and 80th and 95th percentile values were calculated for each participant. Group averages were evaluated and an independent t-test showed no difference between groups on use of spine flexion (Fig. 5).

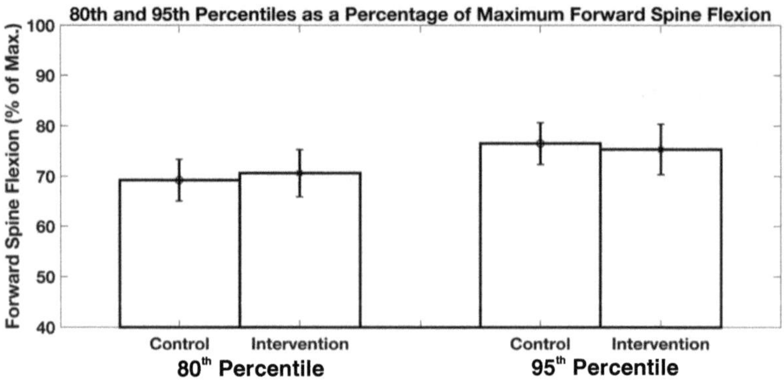

Fig. 5. 80th percentile (left) and 95th percentile (right) spine flexion value as a percentage of maximum forward spine flexion for control group (left bar) and intervention group (right bar) for the skill transfer task which involved inserting a sling under the patient

4 Discussion

4.1 Caregiving Activities

The result of the caregiving activities indicate the two-day, multi-component intervention including a back injury prevention video and auditory feedback provided by PostureCoach effectively decreased the time spent in end-of-range lumbar spine flexion. Participants in the intervention group were able to perform caregiving tasks with significantly less spine flexion in the 2 week measurement trial compared to the baseline measurement trial and the control group in the 2 week measurement trial. Future work will evaluate whether lower reliance of end-of-range spine flexion will lead to reductions in back pain and/or injury risk. This study builds upon the current body of literature on motor learning and skill retention in the area of postural control. Previous research has been conducted on training participants how to use different postures while performing a variety of tasks [26–31]. In particular, the short-term effects of biofeedback on lumbar spine mechanics have been studied and biofeedback has proven to result in improved lumbar spine mechanics [20, 32]. The significance of the data in the current study collected after 2 weeks suggests these improvements may be sustainable long-term. This study also aims to evaluate participants in one final measurement trial at 2 months post-intervention.

It is evident that biofeedback had the most influence on training novice caregivers to prevent back injuries. Both the control group and the intervention group watched the same back injury prevention video between the baseline measurement trial and training trial 1. Results of the caregiving activities show no changes to the control group between these trials, which is similar to other occupational task training literature that conclude a purely didactic approach to teaching lifting tasks is not effective [33].

The faded feedback schedule used in this study was effective at reducing end-of-range spine flexion and future work will study whether different relative feedback frequencies may offer further benefits.

4.2 Skill Transfer Task

Contrary to results of the caregiving tasks, the intervention group was not able to perform the sling insertion task with improved posture compared to the control group. This finding suggests that feedback received on postures during activities does not transfer to postures used while completing other activities.

This finding highlights the challenging nature of training caregivers. There is a large variety of tasks that caregivers are required to undertake in varying environments [34]. The results of this study suggest that training with feedback may be needed with each type of task that will be performed. Future work will determine whether certain types of tasks are similar enough that skill transfer does occur so that training for some tasks can be grouped together.

5 Conclusion

This study evaluated the effectiveness of real-time biofeedback to reduce the time novice caregivers spend in end-of-range spine flexion. Participants in the intervention group, who received biofeedback from PostureCoach, used lower amounts of end-of-range spine flexion while performing simulated caregiving activities after a two-day training period compared to their baseline measurement. Participants in the control group, who did not receive biofeedback in the training period, saw no change. This finding demonstrates that the use of biofeedback in training may be helpful for reducing the risk of back injury for caregivers as part of a multicomponent intervention. The lack of skill transfer to the new sling insertion task suggests that training with PostureCoach may need to take place for each care task a worker may be asked to perform.

References

1. Jensen, R.C.: Disabling back injuries among nursing personnel: research needs and justification. Res. Nurs. Health **10**, 29–38 (1987)
2. Yassi, A., Lockhart, K.: Work-relatedness of low back pain in nursing personnel: a systematic review. J. Occup. Environ. Health **19**, 223–244 (2013)
3. Mock, C., Cherian, M.: The global burden of musculoskeletal injuries: challenges and solutions. Clin. Orthop. Relat. Res. **466**, 2306–2316 (2008)
4. Nelson, A., Baptiste, A.S.: Update on evidence-based practices for safe patient handling and movement. Orthop. Nurs./Natl. Assoc. Orthop. Nurses **25**, 367–368 (2004)
5. US Bureau of Labor Statistics: Injuries to Caregivers Working in Patients' Homes (1997). https://www.bls.gov/opub/btn/archive/injuries-to-caregivers-working-in-patients-homes.pdf
6. Van Hoof, W., O'Sullivan, K., O'Keeffe, M., Verschueren, S., O'Sullivan, P., Dankaerts, W.: The efficacy of interventions for low back pain in nurses: a systematic review. Int. J. Nurs. Stud. **22**, 222–231 (2018)
7. Urquhart, D.M., Kelsall, H.L., Hoe, V.C.W., Cicuttini, F.M., Forbes, A.B., Sim, M.R.: Are psychosocial factors associated with low back pain and work absence for low back pain in an occupational cohort. Clin. J. Pain **29**, 1015–1020 (2013)

8. Marras, W.S., Lavender, S.A., Leurgans, S.E., Fathallah, F.A., Ferguson, S.A., Allread, W.G., et al.: Biomechanical risk factors for occupationally related low back disorders. Ergon. **38**, 377–410 (1995)
9. Callaghan, J.P., Mcgill, S.M.: Intervertebral disc herniation: studies on a porcine model exposed to highly repetitive flexion/extension motion with compressive force. Clin. Biomech. **16**, 28–37 (2000)
10. Collins, J.W., Wolf, L., Bell, J., Evanoff, B.: An evaluation of a "best practices" musculoskeletal injury prevention program in nursing homes. Inj. Prev. **10**, 206–211 (2004)
11. McGill, S.M.: Evolving ergonomics? Ergonomics **52**, 80–86 (2009)
12. Richardson, A., McNoe, B., Derrett, S., Harcombe, H.: Interventions to prevent and reduce the impact of musculoskeletal injuries among nurses: a systematic review. Int. J. Nurs. Stud. **82**, 58–67 (2018)
13. Cheung, K.M.R.: The influence of organizational factors on occupational low back injuries. Home Healthc Nurse. **18**, 463–469 (2000)
14. Dawson, A.P., McLennan, S.N., Schiller, S.D., Jull, G.A., Hodges, P.W., Stewart, S., et al.: Interventions to prevent back pain and back injury in nurses: a systematic review. Occup. Environ. Med. **64**, 642–650 (2007)
15. Fragala, G., Bailey, L.: Addressing occupational strains and sprains: musculoskeletal injuries in hospitals. AAOHN J. **51**, 252–259 (2003)
16. van Breda, E., Verwulgen, S., Saeys, W., Wuyts, K., Peeters, T., Truijen, S.: Vibrotactile feedback as a tool to improve motor learning and sports performance: a systematic review. BMJ Open Sport Exerc Med. **3**, 1–12 (2017)
17. Santos, O.: Toward personalized vibrotactile support when learning motor skills. Algorithms **10**, 1–11 (2017)
18. Lieberman, J., Breazeal, C.: TIKL: development of a wearable vibrotactile feedback suit for improved human motor learning. IEEE Trans. Robot. **23**, 919–926 (2007)
19. Papi, E., Senn Koh, W., Mcgregor, A.H.: Wearable technology for spine movement assessment: a systematic review. J. Biomech. **64**, 186–197 (2017)
20. Owlia, M., Kamachi, M., Dutta, T.: Preventing back pain among caregivers using real-time movement-centered feedback. In: 20th IEA 2018, pp. 750–758. AISC (2018)
21. Gray, R.: Looming auditory collision warnings for driving. Hum. Factors **53**, 63–74 (2011)
22. Burgess-Limerick, R.: Squat, stoop, or something in between? Int. J. of Ind. Erg. **31**, 143–148 (1999)
23. McGill, S.M.: The biomechanics of low back injury: implications on current practice in industry and the clinic. J. Biomech. **30**, 465–475 (1997)
24. Snook, S.H., Ciriello, V.M.: The design of manual handling tasks: revised tables of maximum acceptable weights and forces. Ergonomics **34**, 1197–1213 (1991)
25. Dutta, T., Holliday, P.J., Gorski, S.M., Baharvandy, S.M., Fernie, G.R.: A biomechanical assessment of floor and overhead lifts using one or two caregivers for patient transfers. Appl. Ergon. **43**, 521–531 (2012)
26. Ribeiro, D.C., Sole, G., Abbott, J.H., Milosavljevic, S.: The effectiveness of a lumbopelvic monitor and feedback device to change postural behavior: a feasibility randomized controlled trial. J. Orthop Sport Phys. Ther. **44**, 702–711 (2014)
27. Vignais, N., Miezal, M., Bleser, G., Mura, K., Gorecky, D., Marin, F.: Innovative system for real-time ergonomic feedback in industrial manufacturing. Appl. Ergon. **44**, 566–574 (2013)
28. Çelenay, Ş.T., Kaya, D.Ö., Özüdogru, A.: Spinal postural training: comparison of the postural and mobility effects of electrotherapy, exercise, biofeedback trainer in addition to postural education in university students. J. Back Musculoskelet Rehabil. **28**, 135–144 (2015)

29. Breen, P.P., Nisar, A., OLaighin, G.: Evaluation of a single accelerometer based biofeedback system for real-time correction of neck posture in computer users. In: 2009 Annual International Conference of the IEEE Engineering in Medicine and Biology Society, pp. 7269–7272. IEEE (2009)
30. Lou, E., Lam, G.C., Hill, D.L., Wong, M.S.: Development of a smart garment to reduce kyphosis during daily living. Med. Biol. Eng. Comput. **50**, 1147–1154 (2012)
31. Magnusson, M.L., Chow, D.H., Diamandopoulos, Z., Pope, M.H.: Motor control learning in chronic low back pain. Spine **33**, E532–E538 (2008)
32. Naudé, Y.: The effect of real-time biofeedback on lumbar spine and lower limb kinematics and kinetics during repetitive lifting. Thesis report (2018)
33. Chan, V.C., Tyson, B.: Comparing augmented feedback and didactic approaches to reduce spine motion during box and paramedic lifting tasks: a laboratory-based motor learning study. Thesis report (2018)
34. Darragh, A.R., Sommerich, C.M., Lavender, S.A., Tanner, K.J., Vogel, K., Campo, M.: Musculoskeletal discomfort, physical demand, and caregiving activities in informal caregivers. J. Appl. Gerontol. **34**, 734–760 (2015)

The Erasmus+ Strategic Partnership as a Way to Attract Young People to Engineering

Mariusz Stępień[1(✉)], Kévin Berger[2], Anton Rassõlkin[3], and Toomas Vaimann[3]

[1] Silesian University of Technology, Gliwice, Poland
mariusz.stepien@polsl.pl
[2] University of Lorraine, Nancy, France
kevin.berger@univ-lorraine.fr
[3] Tallinn University of Technology, Tallinn, Estonia
{anton.rassolkin, toomas.vaimann}@taltech.ee

Abstract. The paper is focused on new educational activities focused on attracting young people for technical education. The idea is based on realization of the project under Erasmus+ Strategic Partnership (Key Action 2). The project based on exchange of good practices involves seven partners. Three of them represents sector of higher education and four are schools located in the neighbor of the technical university. Partners exchange ideas how to increase number of students who starts education in technical area, which seems a bit difficult, but engineers are very expected by the industry. The second project goal is to examine possibilities how the university can contribute in increasing quality of technical education at schools. Main project ideas and results of realized project activities are reported in the paper.

Keywords: Technical education · Cooperation school to academia · Exchange of good practices · Internationalization of young people

1 Introduction

Technological market and industrial development is still growing. It requires number of highly qualified engineering personnel to be employed. Unfortunately, recent years number of candidates entering to technical universities, particularly in Europe, decreases. It causes that unemployment in area of technology and engineering almost vanished [1], but number of graduated engineers is still too low. Companies are asking university to support them with qualified personnel, while students at technical university starts their work as part time employment at mid stage of higher education.

For example, according to official statistic [2] shown in Fig. 1, nowadays, in Estonia a total number of admitted students is decreasing, what can be seen from the line Total. Admission in technical fields decreasing as well, in last decade drop was ca. 25%. What is interesting about the data in Fig. 1, is that due to global financial crisis in 2008 a percentage of students admitted to business (BUS) dropped from ca. 30% in 2006/07 to 20% in 2013/14, at the same time admission on technical specialties (TECH) grows a little between 2008/09 and 2012/13. Despite the negative

W. Karwowski et al. (Eds.): AHFE 2019, AISC 963, pp. 252–259, 2020.
https://doi.org/10.1007/978-3-030-20135-7_25

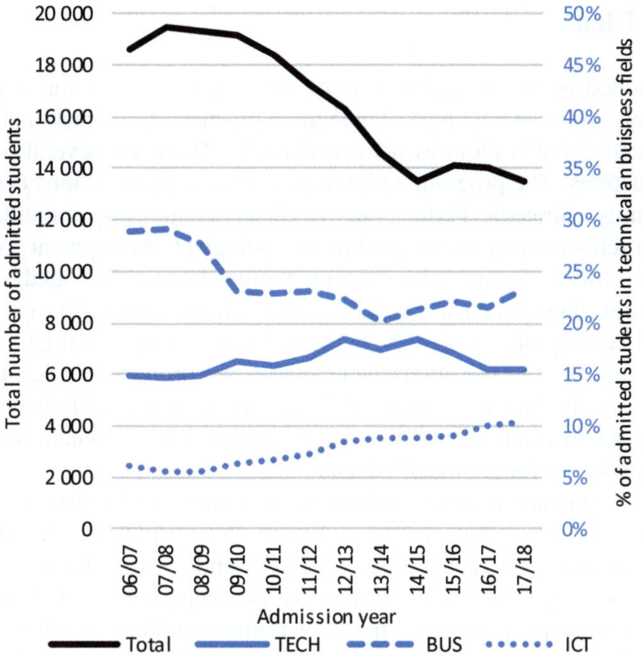

Fig. 1. Total number of admitted students in Estonia vs. percentage in technical (TECH), business (BUS) and information and communications technologies (ICT) fields.

trend of total admission, the number of students applying to information and communications technologies (ICT) in Estonia is increasing.

Such a situation results with decreasing quality of student's education because of not enough time for self-learning and smaller involvement of students in university live (they spend at the university as small amount of time as possible). It means, that in order to saturate the market with well qualified engineers the situation requires to undertake actions which increase number of students studying technical sciences instead of fields with higher unemployment (recently observed at some sectors) [3].

In order to increase number of young people entering to technical universities a better cooperation between sector of school education (lower level) and university sector of technical sciences (higher education) is needed. It relates mainly to cooperation between educational units located in relative small distance one to each other (neighbor institutions).

Useful method to increase quality of such a cooperation is exchange of good practices between institutions located in different European countries and at different educational levels. Such an activity, reported here, has been realized and supported by the European project funded in frames of Erasmus+ Strategic Partnership program.

2 Project Ideas

As was mentioned in above section, one of the best method to improve cooperation between schools and academia is exchange good practices at international level. Such a method is effective but it requires dedicated funds. There are several possibilities to support such actions. The program which was used to support actions described in the paper is Erasmus+ Strategic Partnership (KA2) under the project entitled: "Efficient cooperation in school education focused on technological development", with acronym EDUTECH [4]. Mostly, projects for strategic partnership are dedicated to given sector (school education, higher education, VET, etc.). An invention of proposed project is cross-sectorial activity of two sectors: higher education and school education.

It is important to note that direct beneficiaries are primary and secondary schools who participate with students (pupils) and teachers in project activities and trainings. The university sector is an indirect beneficiary group who will benefit from the project with future candidates for technical education.

The project structure is composed of seven educational units. It contains three academic partners (technical universities or general universities technically oriented): the Silesian University of Technology (SUT), Poland (project leader), University of Lorraine (UL), France and Tallinn University of Technology (TTU), Estonia. Each university partner cooperates with schools of primary and/or secondary level. So, the project structure contains also four school partners: two primary schools and two vocational education schools (secondary technical level). The primary schools are Primary School no. 5 from Pyskowice (SP5), Poland and St. Leon IX School Ensemble (SL IX) from Nancy, France. The secondary technical schools are Technical School Ensemble (ZST) from Wodzisław Slaski, Poland and Tallinna Polutehnikum (TPT) from Tallinn, Estonia. The age of primary schools participants was between 13 and 15 and the age of secondary school students was between 17 and 19. The structure of consortium is presented in Fig. 2. Such a composition of consortium creates different links (in respect of type of cooperation, level of education and distance between partners) for cooperation which ensure exchange of good practices of cooperation between universities and both primary and secondary schools.

Each of project partners had their own methods and tools in order to advertise advantages of technical education and attract young people to the university. The most popular activities organized by universities are festivals of science, open days, academies of young researcher, etc. Part of this activity is also cooperation with schools that educates future candidates for study. University staff invite schools pupils with teachers for activities at the university or visit schools for lectures or presentations. So, different experience of each of partners can be exchanged in range of good practices between international partners. It allows to obtain the cooperation the most efficient and profitable for everybody.

The project idea is to spread widely project results to introduce project results (good practices) in a number of schools located in the environment of the university. The most important channels for public dissemination to spread the knowledge and share experience are project webpage [5], web pages of project partners and social networks portals as Facebook, Instagram and Twitter.

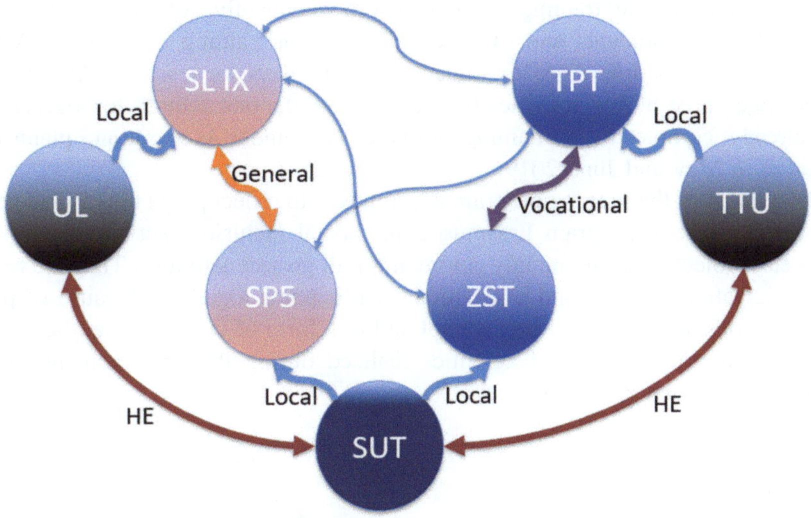

Fig. 2. Structure of project consortium and different links of mutual cooperation: local cooperation between universities and schools, general cooperation between primary schools, vocational cooperation between secondary schools and cooperation at higher education level.

The structure of the partnership presented in Fig. 2 is related to the project management structure. The structure is a hierarchical model. The general management unit is a project leader – SUT. The project coordinator affiliated at SUT animates all project activities and measures results. He also coordinates realization of the project activities at local SUT team. Each project partner has own project coordinator who manage the project activities at given partner. All project coordinators create Project Management Board which is the most important body for the project realization. Administrative and financial administration is supported by proper academic and school units responsible for related areas. Described management structure guarantee smooth decision making process and internal evaluation of the project progress.

3 Project Activities

The project is based on two types of activities – administrative type and training type, both focused on exchange of good practices between participants: teachers and students. The administrative type activities are dedicated mostly to project coordinators and school authorities, and the most important are Transnational Project Meetings (TPM). During such meetings people responsible for project management at different levels meet one to each other, visit partner facility and discuss project objectives and results. Each of project meeting is organized by different project partner. EDUTECH project includes in total five international project meetings. TPMs are a core activities in respect of project management process.

Activities based on trainings include seven events, five of them dedicated for training of school students and two are dedicated for training of teachers. Average length of training is nine days for students and five days for teachers. At the current project stage, three of five activities for students and the one activity for teachers have been already completed. Remaining activities are under preparation, planned for realization in May and June 2019.

The content of the training program is composed to reflect project goals and general (horizontal) goals of program Erasmus+, like social inclusion, networking, internalization etc. Project goals are mainly common for all project activities. They are realized with participation of local and international participants together. Because of project goals, activities are held at hosting school and in a form of workshops at the neighbor university. Part of schedule of activities realized during the first exchange held in Poland are presented in Fig. 3.

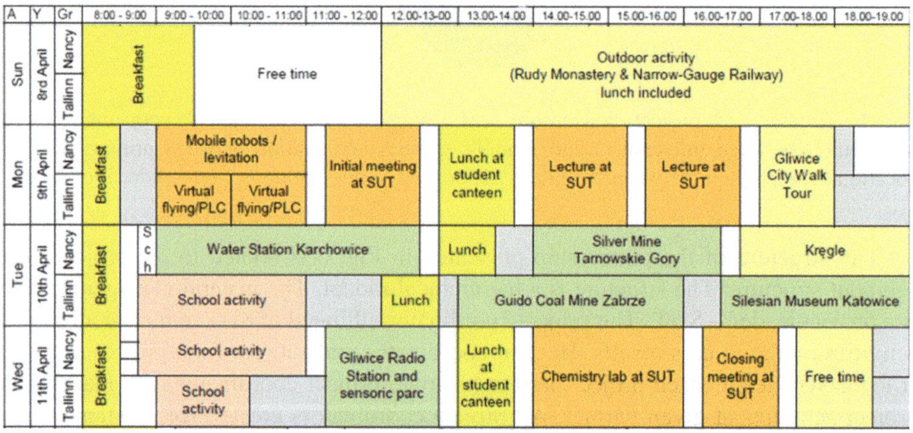

Fig. 3. Part of a schedule for first training activity realized at Silesian University of Technology and at Silesia region (part of activities divided for primary and secondary level of schools).

Poland participates in the project with three partners – one university, one primary school and one secondary school, so the training activity was created as cross-sectional activity at the university and split for two different levels at schools (Tallinn–secondary level; Nancy–primary level). Additionally cross-sectional type were activities realized outside of school and university, mostly at the "Route of industrial monuments" at Upper Silesia region (the route is dedicated for educational purposes). It included narrow gauge railway, radio tower, silver mine, coal mine, water station and Silesian museum. One of photos done during visit at the "Route of industrial monuments" in the coal mine "Guido" in Zabrze is shown in Fig. 4.

Workshop at the university included: lecture "Touch, understand, remember" of physics and material science, workshop on mobile robots and superconductivity, chemical lab and lecture about DNA and its presence in the environment. At schools students attended chemical games and 3D printing workshop. Part of free time was arranged also as an optional outdoor/social activity having cultural and touristic character, like trip to Krakow (historic capital of Poland) and evening disco party.

Fig. 4. Participants of the EDUTECH training at "Guido" Coal Mine in Zabrze – a monument/museum at the "Route of Industrial Monuments" of Silesia Region.

The second training mobility was held in Nancy, France and was dedicated for exchange of good practices at the level of primary education. Two schools participated in the activity: Primary School of Pyskowice (Poland) and St. Leon IX School from Nancy, France (as a hosting school). Supporting partner was University of Lorraine. The structure of training was similar to first, above described activity in Poland. The training was focused on digitalization of the world. Students attended dedicated workshops at the university and organized students games at school. Supporting activities contained e.g. visit at Nancy aquarium and nature museum and history games in the city. It is important to note that Nancy has a history with interesting links to Poland – one of the most powerful governors of Nancy in eighteen century was Stanislaw Leszczynski, former king of Poland. So, Polish participants of the exchange explored history of "King Stanislas" (which is his common name in Lorraine) in Nancy.

The third activity took place in Tallinn in Estonia and was dedicated for secondary level of education. The leading topic were media and communication. Students visited laboratory of printing at Tallinna Polutehnikum and participated in workshops at Mektory – technology transfer center at Tallinn University of Technology. In the part of culture and history participants visited Museum of Estonian History in Tartu and TV Tower – a historical building constructed for the Olympic Games in Tallinn (organized by Soviet Union in 1980 in Moscow with water games in Tallinn). Student mobility in Estonia was a joint activity with short term exchange training for teachers. A group of teachers from two Polish schools participated in five days trainings organized in parallel to students trainings at the Tallinn University of Technology.

All above described project activities were focused on attracting young people (attending primary and secondary education) to their interests with technical education. Trainings at schools, universities and neighbor institutions technologically oriented was organized in order to show a variety of possibilities and future profits.

4 Assessment Methods and Dissemination

During the project activities different assessment method are applied to measure project success and improve upcoming activities. It included general assessment like visual inspection of project activities and discussion with participants and teachers. Very important was quantitative analysis of project results. It has been done by satisfaction surveys. Analysis of surveys enables to find strong parts and weakness of the project. An example survey results obtained for first project activity in Poland and fulfilled by students from St. Leon IX school is presented in Fig. 5.

How you assess the project meeting in Poland (mark "x" in a proper box):	Excellent	Very good	Good	Moderate	Poor	Very poor	Fail
a) Overall assessment?	8	8					
b) Training program at the university?	3	9	3		1		
c) Technical visits at Industrial Monuments Route?	3	6	5	1	1		
d) Touristic program (e.g. Kraków trip)?	5	8	2	1			
e) Accommodation in the hotel?	9	4	3				
f) Meals?	1	4	4	2		1	

Answer the following questions:	Strongly yes	Yes	Rather yes	neither yes or no	Rather no	No	Strongly no
Have you participated in this type of project for the first time?	===	14	===	===	===	2	===
Was this your first foreign trip?	===	2	===	===	===	14	===
Was it your first trip by plane?	===	3	===	===	===	12	===
Was this your first visit to a university?	===	12	===	===	===	2	===
Have you established any new contacts / connections?	10	4	2				
Have you acquired any new knowledge (something you have learned)?	5	7	4				
Have you learned anything new about local culture / history?	7	7	1			1	
Are you interested in technical sciences?	2	7	2	2	2	1	
Would you like to study at a technical university?	1	3		7	1	4	
Did participation in the project encourage you to technical issues?	1	2	2	3	3	4	1
Would you like to take part in such an activity once again in the future?	13	3					
Did you have problems getting to know other participants?				2	3	7	4
Have you had problems with communication / language?		4	1	2	4	4	1
Did you have any problems understanding the knowledge that was presented?		3	1	5	3	4	

Fig. 5. Results of satisfaction surveys reported by 16 students from St. Leon School (Nancy) after their participation in the first exchange activity organized in Poland.

General conclusions which can be drawn after survey analysis are related to a few different aspects. The first one is general satisfaction. All students assessed the training as very good or excellent. Also all answered that would like participate in such an event once again. Very important is also that most of them established new contacts and acquired new knowledge. The second group of answers are related to question about technical education and communication. Students answers are spread at the whole range of scale. The answer that 5 participants consider the activity as attracting to technical education and 4 participants who are going to study technical sciences can be considered as the positive result of the training and a success of the project.

The part of the project which is planned for realization at the final stage, but is very important for project's results durability is the project dissemination. Dissemination of the project ideas is planned at partner institutions by dedicated technical science events and also as a seminar dedicated for different educational units located in local environment of project partners. Participants (both teachers and students) who took part in project activities are planned to be involved in dissemination actions.

5 Conclusions

The paper relates to the ongoing project under Erasmus+ programme focused on exchange of good practices (Key Action 2). Current advances of project realization, namely completed three coordinator meetings, three students trainings and one teachers training, allow to drawn the most important conclusions. The most important are:

- Attracting young people for technical education requires dedicated actions – most of young people is interested with trainings at technical labs.
- All project partners (as pairs school-academia) established cooperation before the project stated, but exchange of international experiences allowed to increase quality of this cooperation. Additionally new alliances were created under the project.
- Analysis of project surveys shows that students are interested to participate such events, create easy new friendships, get new knowledge, both in technical and cultural range. All of participants want to participate again.
- The project seems to be good example how to attract young people for technical education.

Acknowledgments. The paper describes activities realized under the project within Erasmus+ KA2 scheme. Authors thanks all contributing partners, particularly teachers from partner schools.

References

1. Engineering Labour Force in Germany, Report (2012). http://www.albertacanadacom/files/albertacanada/Ger_EngCFS_w.pdf
2. Haridus Silm (Education Eye) Webpage. https://www.haridussilm.ee/
3. Young, qualified and jobless: plight of Europe's best-educated generation, The Guardian (2013). https://www.theguardian.com/world/2013/jul/01/jobless-europe-young-qualified
4. EDUTECH Project page at European Commission website. https://ec.europa.eu/programmes/erasmus-plus/projects/eplus-project-details/#project/2017-1-PL01-KA201-038758
5. Project home page. http://we-edutech.eu/

Methods and Tools for Acquiring High-Quality Skills in Digital Era - Innovative Practices and Results from 3DSPEC and e-MOTIVE Projects

Dariusz Michalak(✉) and Magdalena Rozmus

KOMAG Institute of Mining Technology, Gliwice, Poland
{dmichalak,mrozmus}@komag.eu

Abstract. The ultimate result of each training should be acquisition of practical skills. 'Practical' means not only to obtain abilities (to carry out an activity or use something) but also to be able to apply those abilities for particular purposes. Process of development of such a training depends on its goal, subject and nature of trainees. In the article two approaches followed during development of online training related with acquiring and/or improving vocational competences were presented – one applied in e-MOTIVE project (Methods and tools for efficient use of existing resources and for support of the development of effective training content by VET Teachers) and one applied in 3DSPEC project (Specialized training in 3D printing and practical use of acquired knowledge).

Keywords: 3D Printing · E-learning · Knowledge pills

1 Introduction

Dynamic development of ICT for years has been contributing to creation and implementation of new solutions - and related ways of conducting activities - in all areas of vocational operation, giving concrete benefits and new opportunities. To exploit those benefits and opportunities, relevant competences are necessary. In the article we focus on training methods and tools the objective of which is providing high quality skills for use of up-to-date computer-based solutions in vocational practice. Two cases are presented: (i) training in use of 3D printing in work tasks, (ii) training in use of ICT for teaching. They are a subject of 3DSPEC project and e-MOTIVE project, respectively.

A requirement stablished for development of methods and tools mentioned above was that the training has to based on modern but not high-end and/or niche solutions or tools. In case of 3D printing, use of open-source solutions and tools is limited, but possible for some activities within the whole 'from an object concept to its 3D printed materialization' process. On the other hand, for all tasks within teaching process there are open-source ICT solutions and tools available, as well as not open-source but commonly used ones.

Another requirement was to provide high-quality skills for particular the vocational applications (defined within each of the projects mentioned above). 'High-quality'

W. Karwowski et al. (Eds.): AHFE 2019, AISC 963, pp. 260–270, 2020.
https://doi.org/10.1007/978-3-030-20135-7_26

skills means here that they: (i) cover use of solutions and tools which are modern (but if possible: open-source or commonly available) and relevant for trainees' work tasks, (ii) are based on good practices, proved for given applications, (iii) are truly useful - applicable for work tasks which are actually carried out by a trainee, (iv) are possible to expand by a trainee if necessary or desired (solid grounds are given), (v) develop trainees' self-efficacy and willingness to search and use another tools and solutions [1].

Development of each training mentioned above, i.e.: (i) training in use of 3D printing in work tasks and (ii) training in use of ICT for teaching requires other approach due to complexity of subject covered – Fig. 1. Use of 3D printing is a very wide topic which consists of a number of interrelated sub-topics which as whole provide necessary knowledge and skills. On the other hand, ability ICT for teaching is based on a number of individual skills in use of particular ICT tools and solutions, therefore training materials and solutions should be in form of repository in which each skill is treated as a separate topic.

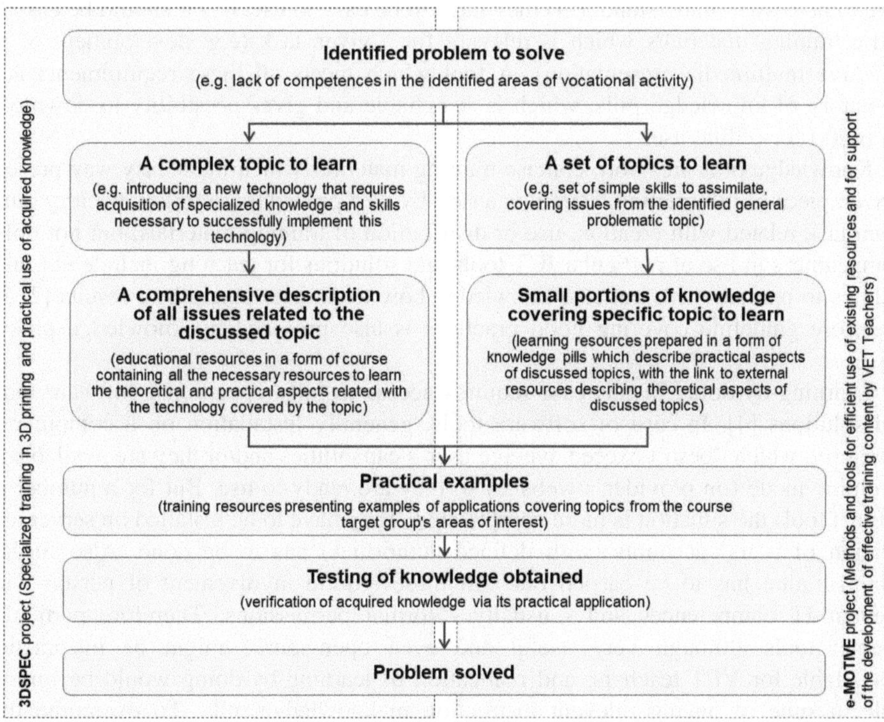

Fig. 1. Two cases of training development distinguished based on the subject complexity

Each of the abovementioned cases of training development is discussed in subsequent chapters of this article.

2 Method and Tools for Acquiring Competences Related with a Set of Individual Topics

In e-MOTIVE the problem to solve was to provide VET teachers with competences which will enable them carry out ICT enhanced training in all main modes: classroom, online and blended learning. These competences are composed of single skills in application of ICT tools and solutions. So the VET teachers have to learn how to use particular ICT tools and solutions for teaching, which in practice means - how to produce, use and distribute different kinds of ICT-based training materials and carry out ICT-enhanced teaching process with use of particular software tools and Internet tools.

Learning method which is an obvious choice to enable acquisition of the above-mentioned abilities is learning-by-doing. Therefore VET teachers have to obtain an instruction how to carry out particular activities in available software tools and Internet tools. But there are some considerations/requirements which have to be taken into account during development of training materials: (i) they have to be short, (ii) they have to be easy to understand, (iii) they have to be easy to use, (iv) it should be easy to find a training material/s which is relevant for a given task (e.g. development of an attractive multimedia presentation). A tool which meets all these requirements is a repository of knowledge pills, which is searchable and gives possibility to download the pill(s) for offline use.

Knowledge pills are short, concise training materials which in an easy way present a given piece of information. Here it is a step-by-step procedure to follow to carry out a given task related with creation, use or distribution of training materials, but not only. Competences in use of particular ICT tools and solutions for teaching include not only abilities to operate them but also knowledge how to do it to obtain best results [2, 3]. Therefore guideline covering good practices is also provided by knowledge pills – Fig. 2.

Learning-by-doing in this case requires access to fully operational software tools and solutions [4]. In case of software tools, generally installation on a computer is necessary, which doesn't exceed average user's capabilities and/or they are available in an online mode (on provider's website) so they are ready-to-use. But for a number of Internet tools the situation is more complicated. They have to be installed on server and creation of users' accounts (with defined authorities) has to be done. Also further administration has to be carried out. All these require involvement of person with relevant IT competences and - usually - formal permissions. Therefore particular Internet tools although very useful and even open-source might be in practice unavailable for VET teachers, and realization of learning-by-doing would be impossible in spite of having relevant instruction in knowledge pills. To overcome this problem software sandboxes can be used. They are software testing environments which look and operate like a ready-to-use Internet tool run e.g. on a company server. The difference is that they are provided on an external server and can be reset when needed. So a VET teacher can exercise use of a given Internet tool with no fear of errors because they wouldn't cause any permanent damage (no IT specialist's intervention would be required).

The discussion presented above is concluded by the Fig. 3.

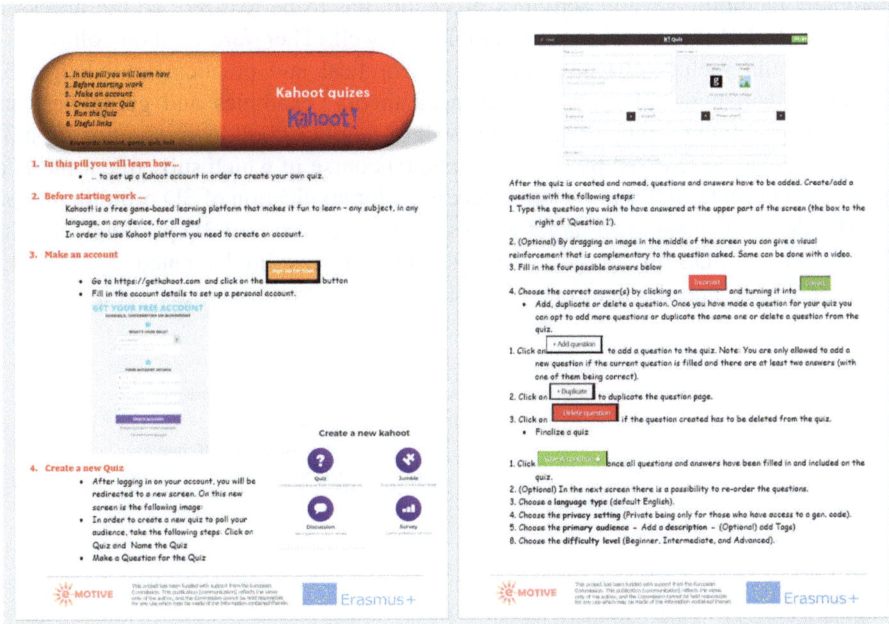

Fig. 2. Sample knowledge pill (fragment)

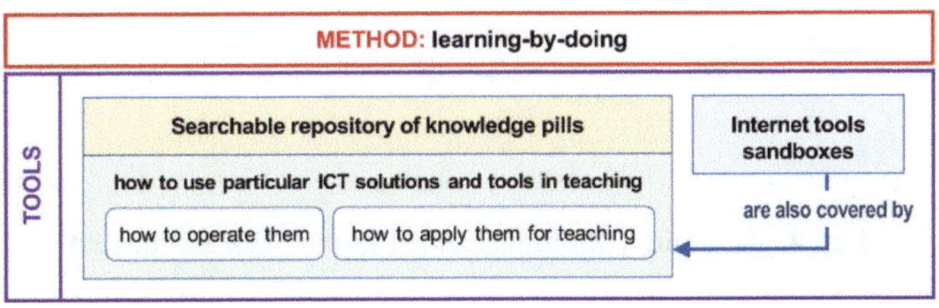

Fig. 3. Method and tools which enable to acquire competences for providing ICT-enhanced teaching

Both knowledge pills repository and sandboxes (they are described in Sect. 6.2) were developed within e-MOTIVE project.

3 Method and Tools for Acquiring Competences Related with a Wide Complex Topic

The problem to be solved within 3DSPEC was to provide competences in use of 3D printing in activities related with education, design, healthcare and SMEs' operation.

To apply 3D printing technology related fundamental knowledge and skills are indispensable, and they are common for use of 3D printing in any area of professional operation. To fully exploit possibilities given by 3D printing and apply it at work, a complex, including several activities procedure has to be realized, so the necessary knowledge and skills are wide and complex as well. Therefore method followed in learning should include: (i) acquisition of theoretical and practical knowledge: 3D printing fundamentals, 3D printing advanced theory, examples and guidelines, and (ii) learning by doing – preparing a 3D model of an object and its production on a 3D printer. This can be realized by two elements: (i) course in which subsequent modules step-by-step build trainee's knowledge necessary for application of 3D printing in work tasks (by the abovementioned target groups), (ii) practical use of 3D printing – making 3D printout. In 3DSPEC project the first element was implemented by 3DSPEC e-learning course and the later one – by 3DSPEC 3D Printing Centre – Fig. 4.

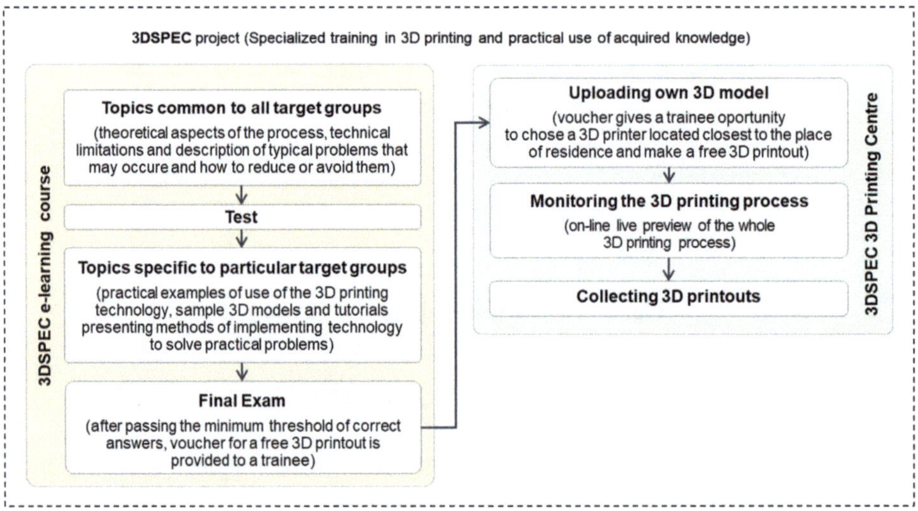

Fig. 4. Learning path implemented in the 3DSPEC project

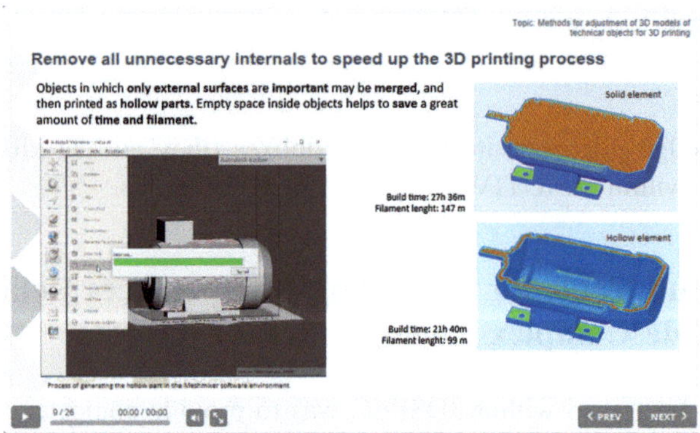

Fig. 5. Presentation available on 3DSEPC e-learning platform – sample slide

In the e-learning course, the modules are distinguished in such a way that first there are modules which provide knowledge which is necessary to be able to carry out 3D printing – Fig. 5. Therefore this knowledge has to be acquired also by representatives of all target groups defined, which is to be confirmed by passing a test. Then it is possible to enter the next module, content of which is related with use of 3D printing by particular target groups. When the whole e-learning course is completed, which is confirmed by passing of the final exam, practical application of knowledge acquired is realized in the 3D Printing Centre. This is a solution which enables to overcome the problem that a trainee has no access to a 3D printer to try and practice what they learnt – its more detailed description is given in Sect. 6.1.

4 Methodology to Provide Methods and Tools for Acquiring Competences Related with a Set of Individual Topics

A list of ICT tools and solutions use of which should be a part of VET teachers abilities is not established. ICT provides a huge number of solutions and tools which are of common use but are proved to be useful for teaching/learning purpose as well as solutions and tools which are directly addressed for this purpose (see item 1 and 2 in the Fig. 6). So the question was, what type of solutions and tools should be covered by the training materials. The main requirement was that they have to enable to produce ICT-based training materials and apply Internet solutions which are: (i) useful for particular activities within teaching (e.g. theoretical introduction, instruction, exercises, test etc.), (ii) attractive for VET students and – as a whole set – cover all main modes of teaching (item 2 and 5 in the Fig. 6). The next question was, what specific tools and solutions should be covered by the training materials. The requirements were (items 2, 3, 5 in the Fig. 6): (i) they meet the requirement established for the previous question, (ii) they are available to teachers, because they are commonly used (like e.g. MS Office) and/or they are open-source (e.g. Moodle, YouTube). Another factor which had to be taken into account is that to provide a quality ICT enhanced teaching, operation of a given ICT tool or solution is not enough. It has to be accompanied by ability to effectively apply them for teaching (item 4 in the Fig. 6).

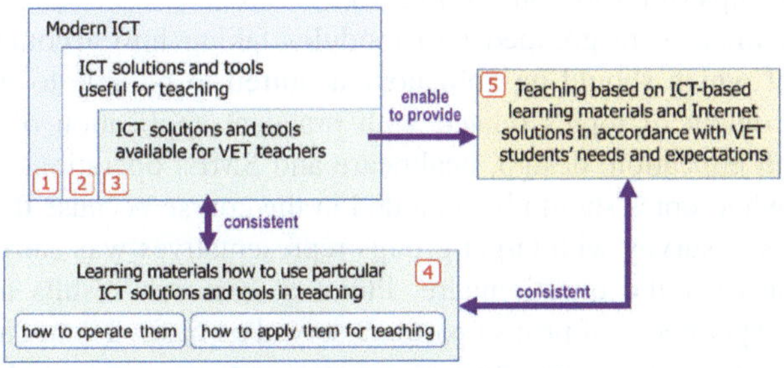

Fig. 6. Considerations to be taken into account when development materials to enable VET teachers acquire abilities to provide ICT enhanced teaching

To identify what freely available ICT tools and solutions can be applied for teaching a study was carried out (see Fig. 7). The findings served as basis for preparing of questionnaires used in a survey on VET teachers' and VET students' needs and expectations regarding types of ICT based training/learning materials and Internet tools. Conclusions drawn enabled to establish what specific ICT tools and Internet solutions should be covered by the training materials to be developed, namely knowledge pills and sandboxes.

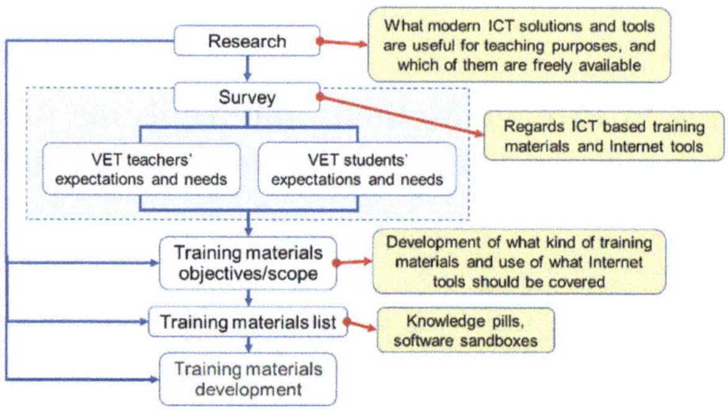

Fig. 7. Procedure followed to develop training materials within e-MOTIVE project

5 Methodology to Provide Methods and Tools for Acquiring Competences Related with a Wide Complex Topic

In case of a development of training in use of a complex technology in particular vocational applications, first it has to be established, what subtopics have to be covered and in what order they may or should be provided. In 3DSPEC project an e-learning course covering several modules composed of topics was developed [5, 6]. The content was ordered in such a way that two phases of training were carried out: (i) phase covering knowledge common and necessary for all 3D printing applications, (ii) knowledge regarding application of 3D printing in sample tasks carried out by representatives of selected target groups (refer to point 4, Fig. 6).

All topics defined were grouped into modules taking into account that there are topics content of which should be obligatory acquired to be able to fully understand and assimilate content of topics related with practical application of 3D printing in tasks related with education, design, healthcare and SMEs operations.

To identify what topics should be included in the course because they are expected by future trainees, a survey with target groups representatives was carried out (Fig. 8). To correctly construct the questionnaire, literature research results as well as own knowledge and experiences of project partners served as basis. Then, data collected via the questionnaire as well as expertise of project partners were used for the course curriculum development. In this document not only topics were listed but also their order was defined but also issues regarding self-evaluation and obligatory test were established. It served as a roadmap for the actual e-learning course creation. During the course content development, the course curriculum was modified if it occurred desired from the course quality point of view. As source for training materials, project partners used they own experiences as well as literature and Internet resources. In parallel 3D Printing Centre was created.

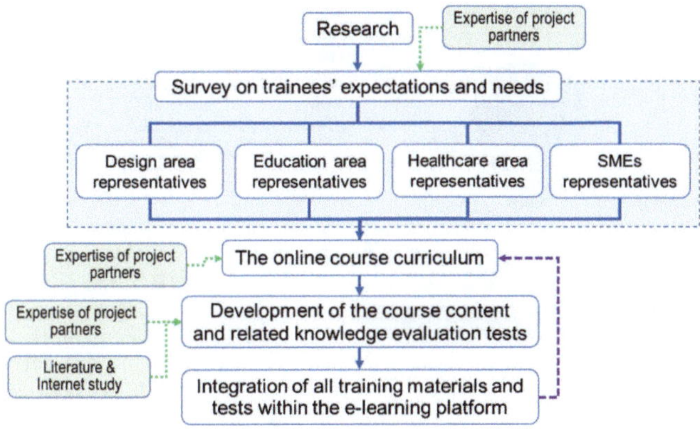

Fig. 8. Procedure followed to develop training materials within e-MOTIVE project

6 Testing of Acquired Knowledge

6.1 3D Printing Centre as a Tool for Verification of Acquired Knowledge on 3D Printing

Practical verification of acquired knowledge on a given technology is crucial especially when its improper use may generate lost rather than profit for organizations. Acquiring new competencies connected with hardware based topics might be hard to verify in situations when a trainee has no access to the real object (machine, device, system of objects). Such situation is characteristic for 3D printing technology.

3D printing technology evolves very quickly and nowadays is more affordable and accessible than ever. But there are still a number of practical traps and problems not yet solved by providers of 3D printers, which may have a big influence on the interest in and use of their final product [7]. Proper use may positively influence on commercial success and speed up the new product development process [8]. Improper use of technology may be frustrating and discouraging, which may cause that this use will be abandoned. To avoid such situation within 3DSPEC project: (i) 3D Printing Centre was developed, (ii) prior completion of the 3DSPEC online course (e-learning) is required.

The 3D Printing Centre is a free service offered by the project consortium members for 3DSPEC e-learning course attendees. The goal is to enable the trainees to test their knowledge on real examples. The 3D Printing Centre consists of: (i) 3D printers located at the project partners' sites (Poland – 2 partners, Slovene – 1 partner, Spain – 2 partners), and (ii) Internet platform (Fig. 9) which makes these 3D printers accessible to the trainees.

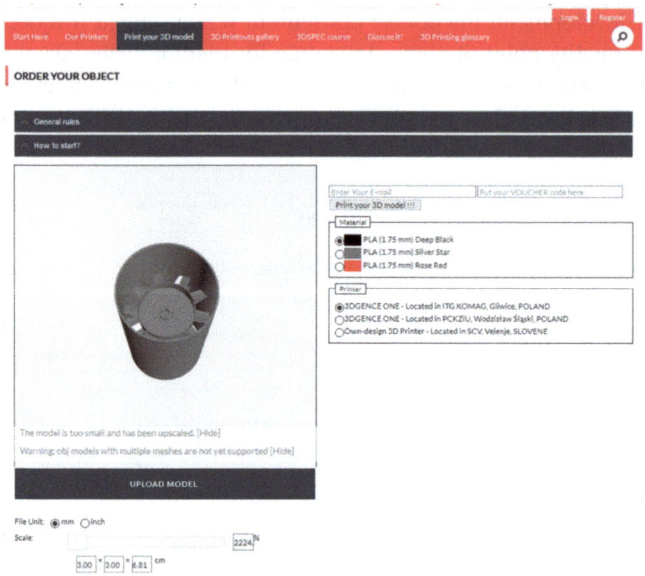

Fig. 9. Internet platform of the 3D Printing Centre.

Completion of the 3DSPEC course is a prerequisite for use of the 3D Printing Centre because it gives the trainees theoretical background as well as guideline for possible problems solving and presentations of practical examples. Practical examples bring trainees closer to real use but without the opportunity to make their own 3D printouts this process would incomplete. Another reason is that prior completion of the online course ensures that 3D printing of own objects will be carried out by appropriately motivated and prepared people.

6.2 Software Sandboxes as a Tool for Verification of Acquired Knowledge on Use of Internet Tools

Verification of obtained knowledge about use of new software can be difficult, especially for a newbie. Typically software can be downloaded and installed on PC or is available online as a ready-to-use tool. But how to proceed if we want to learn e.g. Moodle system, LimeSurvey or other Internet tool, but we have no access to them at all or have no permission to use available ones for learning/exercising purposes. Such solutions are offered within a new observable trend to provide software as an Internet service, not as a standalone application. Such approach has a number of advantages, e.g. software is available via almost every web browser, you can get access to your resources from every computer, your data are safely stored on the Internet servers. But from operational point of view, even open-source Internet software environment needs at the beginning advanced IT skills - for installation, user accounts creation, customization etc. After this step, the majority of the functions is available with no coding skills, but the installation must be done by professionals. Such problems might be solved by software sandboxes. The idea of software sandboxes is similar to the real sandboxes in which we 'play with our toys', having still opportunities to construct and destroy over and over again. Software sandboxes are prepared by professionals as ready-to-use systems which can be used with no worries that we can damage something. Their 'look and feel' imitates the fully operational tool (used by companies, institutions etc.), the sandbox is created for. Even if something goes wrong, system might be restored to the initial state, often by trainees themselves. So you can learn how to use advanced Internet tool by hands-on exercises in application of its particular functions with no fear about consequences of your actions.

In the e-MOTIVE project, a number of software sandboxes were prepared, e.g. Moodle, LimeSurvey, BigBlueButton and Wiki system. The objective is to enable VET teachers to acquire skills in use of particular Internet tools. Necessary instructions are provided to them in related knowledge pills. Giving the VET teachers solid grounds for expanding their skills in use of these tools is also a desired effect. Another objective is to enable them making proper decisions regarding relevance of particular Internet tools for intended teaching tasks or the whole process.

Software sandboxes are typically used during testing new code or during development process of a new application [9]. Implementation of the idea in the field of training of VET teachers is unique and innovative, and brings valuable benefits – effective mastering of particular tools with no stress related with fear about consequences of improper use.

7 Conclusions

During development of training the aim of which is acquisition of skills based on use of computer-related technology, the methodology followed as well as methods and tools provided depend on how complex and interrelated are those skills. It also has to be taken into account that hands-on activities must be included in the learning process, and relevant guidance (instructions, good practices, problem-solving tips etc.) has to be provided as well.

When skills to obtain are strongly interrelated – as for 3D printing - it is desired to apply e-learning to maintain structured learning in which subsequent topics are properly ordered. Additionally, in e-learning the learning process can be divided into stages in such a way that completion of one stage (passing test) is necessary to enter the next topic(s). This way, a situation when a topic is poorly understood due to lack of some knowledge/skills (covered by previous topics) is avoided. It can also be assured that some practical activities with hardware and software won't be carried out if necessary level of knowledge and/or skills hasn't been obtained.

Other situation is in case of ICT enhanced training - a set of individual skills is needed for practical use of a given technology, and there is no obligatory list of them. Acquisition of skills can be done in a flexible way – in accordance with current needs regarding development of training materials and/or application of Internet tools. Training materials provided to teachers should be short and easy-to-understand, and cover a single activity to be carried in a given software tool or Internet tool. Therefore a searchable repository of knowledge pills is best choice to provide relevant instruction.

Regarding Internet tools, there might be a problem that a teacher has no opportunity to use them (unavailable at workplace, or there is no permission to use them as a newbie). A solution desired to overcome this barrier are software sandboxes, and related knowledge pills giving easy to comprehend instruction.

Acknowledgments. The paper was written as part of the work carried out under the European projects 3DSPEC "Specialist in 3D printing - specialized training in application of 3D printing and practical use of acquired knowledge", 2016-1-PL01-KA202-026335 and e-MOTIVE "Methods and tools for efficient use of existing resources and for support of development of effective training content by VET Teachers", 2017-1-PL01-KA202-038850. These project has been funded with support from the European Commission. This communication reflects the views only of the author, and the Commission cannot be held responsible for any use which may be made of the information contained therein.

References

1. Tokarczyk, J., Dudek, M., Jorda, O., Martinez, E., Penuelas-Herraiz, A., Primo-Capella, V.-J.: OVOMAX online course as w way to improve competencies and qualifications for designing and manufacturing of custom-made orthopaedic implants. In: AHFE 2018, International Conference on Human Factors in Training, Education, and Learning Sciences, Orlando, USA, 21–25 July 2018, pp. 362–374
2. Rozmus, M., Michalak, D.: Computer aided shaping of safe behavior at work place. Mach. Dyn. Res. **39**(1), 93–102 (2016)
3. Michalak, D., Rozmus, M., Winkler, T.: Shaping of working conditions using ICT technology. In: Handbook of Loss Prevention Engineering, vol. 1, pp. 423–443 (2013)
4. Latos, M., Stankiewicz, K.: A method for improving the competences of personnel of electrical divisions in mines. Eduk. Ustawicz. Doros. nr 2 s, pp. 157–163 (2015). ISSN 1507-6563
5. Bartnicka, J., et al.: Educational needs in neurological rehabilitation–"Brain4Train" survey outcomes from European countries. In: International Conference on Applied Human Factors and Ergonomics, pp. 385–396. Springer, Cham (2018)
6. Michalak, D., et al.: Specialized training in 3D printing and practical use of acquired knowledge–3DSPEC online course. In: International Conference on Applied Human Factors and Ergonomics, pp. 339–350. Springer, Cham (2018)
7. Bałaga, D., Kalita, M., Siegmund, M.: Zastosowanie technologii druku 3D do szybkiego prototypowania dysz zraszających. Masz. Gór. nr 3 s, 3–13 (2017)
8. Prostański, D.: Development of research work in the air-water spraying area for reduction of methane and coal dust explosion hazard as well as for dust control in the Polish mining industry. In: MEC 2018, Mineral Engineering Conference, Zawiercie, Poland, 26–29 September 2018, pp. 1–13 (2018)
9. Tokarczyk, J.: Dispersed software environment in virtual prototyping of underground mining mechanical systems. In: Kim, H.K., Ao, S.-I, Amouzegar, M.A., Rieger, B.B (eds.) IAENG Transactions on Engineering Technologies: Special Issue of the World Congress on Engineering and Computer Science 2012. Lecture Notes in Electrical Engineering 247, pp. 611–620. Springer, Dordrecht (2014)

Educational Strategies

Entrepreneurship Education and Digitization

Martin Kröll(✉)

Institute for Work Science, Ruhr-University Bochum,
Universitätsstraße 150, 47780 Bochum, Germany
martin.kroell@rub.de

Abstract. The present article focuses on an entrepreneurship education concept which enables experienced professionals to further develop their knowledge as mentors in order to provide vocational orientation to young people searching for a job against the background of digitization, thus supporting them, for example, on their way to independence. The corresponding three-year EU project "Job Developer" was funded by the European Commission. Potential aspects to improve the quality of the concept itself but also in relation to the implementation of this concept were worked out. The following article focuses on the question which quality criteria can be used to improve the quality of the implementation of the Job Developer concept with the help of specific existing or future IT-solutions. On this basis, a potential and resistance analysis was carried out during the project.

Keywords: Youth unemployment · Entrepreneurship education · Digitization · Service quality

1 Introduction

1.1 Problem Context

Digital change in the world of work and industry 4.0 has a profound impact on the processes of vocational orientation and transition between school and work [1]. Together with international partners from six EU countries (Bulgaria, Germany, Greece, Lithuania, Spain, Hungary), new measures to assure employability were examined within the EU-funded "Job Developer" project. From 2015 to 2018, the project team developed a programme within this framework with which experienced professionals can develop their knowledge as mentors in order to support young people seeking employment on their way to independence (http://www.jobdeveloper.eu/).

A particular focus of the project was the megatrend "Digitization of Work" (http://www.jobdeveloper.eu/). In this regard, the fusion of production (Cyber-Physical Systems; Industry 4.0) and services (Smart Services) to Product-Service Systems (Big Data, Crowdwork, Internet of Things and Augmented Reality) takes centre stage. The corresponding developments create the prerequisites for generating new value-added opportunities. These changes result in new demands regarding the competencies of the actors, the associated personnel and organisational development as well as the use and further development of digital media. The aim is to face the challenge of the increasing gap between the world of education and the world of work. Many young people leave

W. Karwowski et al. (Eds.): AHFE 2019, AISC 963, pp. 273–283, 2020.
https://doi.org/10.1007/978-3-030-20135-7_27

education and training systems without sufficient preparation to enter the labour market. This does not just affect the learning of professional skills but also extends to the acquisition of valuable key competences.

1.2 The Job Developer-Tools

The Job Developer concept comprises the three tools Talent Diagnosis, Employment Radar and Expert Hearing [2–4] which serve to record the potentials and talents of young adults in a broad spectrum and to draw up a personal development plan against this background [5]. The focus of the talent diagnosis is on the personality of the young people. The (professional) biography of the young people as well as their inclinations, interests and strengths are examined by means of various activities [6]. The deficits are only taken into account to the extent that they (can) prevent young people from using and building on their strengths. The employment radar, on the other hand, analyses the immediate geographical surroundings of young people with regard to the labour market and economic developments. It comprises the following steps: (a) Analysis of economic framework data, (b) Detection of social megatrends, (c) Preparation of information from the sinus meta milieus, (d) Use of geodata to localise customers (regional/IT-based), (e) Interview technique to determine market potential and (f) Creation of services. The new services will be created in the sense of product service systems. In the final expert hearing, the young people can benefit from the advice and network of external experts. A personal development plan is drawn up to record the individual development and qualification steps of the young people so that they can work towards a professional future that combines both their talents as well as the general conditions of the labour market.

1.3 Theoretical Approach as Frame of Reference

In order to evaluate the success of the implementation of the three tools and to identify success or failure factors, the present analysis identified potentials and resistances in the five EU countries in order to answer the following central question of this work: In what way is it possible to constructively deal with potentials and resistances that can arise during the implementation of the Job Developer programme?

In connection with the implementation of IT-supported systems, scientific metastudies [7] come to the conclusion that their acceptance depends on the perceived usefulness of the tools and ease of use. The evaluation of the quality of the services offered or rendered by the job developers plays a central role in the implementation of the three tools and provides indications of their perceived usefulness.

The initial goal of the resistance and potential analysis is to ensure a certain degree of acceptance of the developed and tested tools. According to the results of the research group around Gondo and Amis [8], securing acceptance alone is sufficient, and a minimum amount of reflection is required. For this reason, the quality of the mentor's or job developer's services was measured, the extent of resistance and potential was assessed, and ways of dealing with them were worked out.

In order to classify, interpret and reflect on the results of the empirical studies of the Job Developer project from the perspective of economic education, the Cologne structural model of economic didactics [9, 10] is used, which serves as a structuring and orientation approach. The central components of the model are: (a) Determination of the basic value judgments, (b) Elaboration of the prerequisites of the addressee system, (c) Decision field: Topic, (d) Decision field: Methodology and (e) Teaching-learning control and impulses for curriculum development.

2 Measurement of Service Quality

2.1 Methodology

The project considers the activities of the trained job developers as a service to the young participants. To measure service quality, a special SERVQUAL questionnaire was developed [11, 12] consisting of a total of 15 items relating to the following five quality dimensions: environmental characteristics, reliability, obligingness, sovereignty and empathy. A total of over 65 participants took part in the survey.

The first step was a qualitative study and the second a quantitative study. In the context of the qualitative study, the criteria for assessing the services were developed in several workshops. The assessment of the quality of the services by the young people took place in two steps: (I) In the first step, they assess the worked-out quality criteria on the basis of a scale from insignificant = 1 to very important = 5. (II) Subsequently, they assess the performance of the job developers on the basis of the corresponding criteria on a scale from one to five (extent to which the criteria were fulfilled: 1 = very low and 5 = very high).

2.2 Results

In the following, the results from Hungary and Greece will be discussed as examples. In Hungary, 24 young participants evaluated eight job developers (see Table 1). It was of most importance for the respondents that the job developers listened to them, answered their questions and responded to their statements (M = 4.71). Following that, it was important for them that the job developers had empathy and that they were friendly and respectful towards them (M = 4.67). The two points that the job developers had sufficient time for them and their project (M = 4.58) and that the cooperation is characterized by mutual trust (M = 4.58) were named third in terms of significance. In terms of job developer performance, the respondents from Hungary gave the best rating to their ability to introduce themselves (M = 4.67). For this criterion, the performance corresponds to the perceived importance. In addition to the criterion of empathy, the accessibility of job developers is also highly rated. Here is an actual "overfulfilment" apparent (M target = 4.04; M actual = 4.17).

Of the 15 respondents from Greece, the two criteria (1) trust (M = 4.88) and (2) empathy (M = 4.75) were rated as particularly important. Third place (M = 4.63) was given to the criteria of quick response, listening, leading discussions and experience. The job developers in Greece succeeded in achieving a better performance than expected by the participants in eight of the 15 examined criteria.

Table 1. Evaluation of service quality of job developers in Hungary, Bulgaria and Greece and overall (plus Lithuania and Spain)

No.	Criteria	Bulgaria	N = 17	Hungary	N = 24	Greece	N = 15	Total	N = 65	Fulfilment
		Target	*Actual*	*Target*	*Actual*	*Target*	*Actual*	*Target*	*Actual*	
1	Experience	3,76	4,35	4,08	4,04	4,63	4,25	4,16	3,93	95%
2	Relevance of industry	2,41	3,29	3,75	3,45	4,25	3,88	3,53	3,38	96%
3	Interest	4,11	4,64	4,5	4,29	4,38	4,88	4,33	4,42	102%
4	Empathy	4,11	4,35	4,67	4,67	4,75	4,88	4,57	4,57	100%
5	Trust	4,23	4,7	4,58	4,33	4,88	4,75	4,55	4,45	98%
6	Accessibility	4,29	4,7	4,04	4,17	4	4,63	4,15	4,36	105%
7	Quick response	4,35	4,7	4,5	4,29	4,63	4,5	4,51	4,35	96%
8	Time	4,41	4,76	4,58	4,46	4	4,5	4,37	4,39	101%
9	Listening	4,47	4,75	4,71	4,46	4,63	4,75	4,62	4,51	98%
10	Own mistakes	2,35	2,47	3,74	2,88	3,63	3,25	3,31	2,90	88%
11	Motivate	4,35	4,64	4,54	4,04	4,25	4,44	4,38	4,21	96%
12	Problem identification	4,35	4,82	4,46	4,17	4,5	4,5	4,44	4,34	98%
13	Leading discussions	4,35	4,64	4,24	4,21	4,63	4,5	4,25	4,18	98%
14	Feedback	4,35	4,76	4,33	3,92	4,5	4,75	4,39	4,25	97%
15	Interpreting all ideas	4,41	4,94	4,29	4,17	4	4,88	4,24	4,42	104%

Overall, the assessment is positive across all countries, despite some negative differences in target performance. The ability to respond to ideas and the ability to empathize as well as trust were consistently rated as good in all countries. The unimportant criteria, on the other hand, were the reporting of own mistakes and relevance of industry. In the overall evaluation it becomes apparent that the criteria of interpreting all ideas with 104%, interest with 102%, time with 101% and accessibility with 105% exceeded the expectations of the participants. On average, the empathy, listening skills and trust of job developers were rated as the most important. This shows that social skills and interpersonal interaction are most important for the young people. At the same time, however, it becomes clear that the points of quick response, time, availability and feedback play an important role in the assessment of service quality.

2.3 Discussion and First Conclusion

Against the background of these results, the possibilities on the one hand and the limits of IT-supported solutions for quality improvement on the other hand become clear. For example, IT-supported solutions can be based on sufficient time, fast response, availability and feedback. On the other hand, it proves more challenging to increase empathy and mutual trust with the help of IT-supported solutions.

The implementation of the Job Developer concept within the framework of the EU project can be interpreted as a pilot application. Even with this application, the search for persons for the job developer's and expert's tasks proved to be difficult. It is assumed that the problems with regard to the above results will become even more acute if a larger number of people are to be found to take on these tasks. The importance of developing appropriate IT-solutions becomes apparent at this point.

3 Analysis and Design of Potentials and Resistances

3.1 Methodological Approach

In order to adequately reflect on the project's activities, a study was conducted to analyse and shape the resistance and potential of the job developers in the context of a qualitative study with experts in several workshops, in addition to measuring the service quality of the job developer. Based on the results of the qualitative study, a total of 52 items were worked out. These form the basis for the questionnaire used in the quantitative study. On the one hand, questions were asked about the importance (0 = not at all important to 5 = very important) and on the other hand about the ability to expand potentials or change resistances ("not at all high" = 0% to "very high" = 100%). In addition, the questionnaire consisted of 15 open questions. A total of 93 completed questionnaires were evaluated.

3.2 Results

In terms of talent diagnosis, the most potential for improvement is seen in the fact that it leads to an increase in self-confidence and promotes the willingness to show initiative (M = 75.05; SD = 22.81). In second place is the recognition of the young people's strengths (M = 74.54; SD = 19.35). The possibility of saving time for further support by using their tools is rated as less advantageous (M = 69.88; SD = 24.56). From the perspective of the interviewees, their most important potential is that the young people can use them to reveal their strengths (M = 4.39; SD = 0.82). The fact that companies receive employees who are better acquainted with their skills comes in second place (M = 4.28, SD = 0.94).

The fact that job developers can acquire the necessary skills is classified as the most changeable (M = 62.65; SD = 27.72). The point that the young people do not receive an answer as to what they should do is also classified as changeable (M = 46.20; SD = 30.69).

The most significant resistance to talent diagnosis is that job developers do not have the necessary skills (M = 3.58, 1.41). At the same time, however, this resistance is considered to be the most variable (M = 62.65; SD 27.72). A second important problem is that the results of using different tools for talent diagnosis are contradictory (M = 3.48; SD = 1.34). This barrier is also regarded as the least changeable (M = 37.64; SD = 28.35). The third largest barrier is that the results of the talent diagnosis do not correspond to the expectations of the young people and thus lead to confusion among them (M = 3.33; SD = 1.34). At the same time, however, this point is also classified as changeable (M = 54.87; SD = 29.11), there ranking second.

These results can be supplemented by the answers to the open questions. The answers were suggestions for solutions, strengths and weaknesses of the tools and further needs. It was regarded as a strength that the tools for talent diagnosis help the participants become aware of their potential and at least partially aware of the possibility of future talent development. As a weakness the occurrence of discrepancies between the discovered and perceived talents of the young people was mentioned, connected with the danger that the results demotivate the participants, beyond that the required expenditure of time and the width. In addition, it was noted that the use of the tools was of no use if the results were not subsequently worked on and if no evaluation discussions were held with the young participants.

In the following, the results with regard to the potentials and resistances of the employment radar are explained: The fact that young people feel inspired by sharing positive experiences is assessed by the respondents as the most expandable potential of the employment radar (M = 70.24; SD = 27.44). At the same time, this potential is rated as the most important (M = 4.06; SD = 1.01). In addition, the possibility of promoting creativity through the use of the employment radar is rated as particularly capable of development (M = 68.53; SD = 25.34). In terms of importance, this point is mentioned in second place (M = 3.94; SD = 1.16). Last but not least, the easy comprehensibility of the employment radar concept is highlighted (M = 3.85; SD = 1.08). Furthermore, the fact that young people are given easy access to information on potential labour markets with the help of the employment radar (M = 66.91; SD = 24.73) is considered to be particularly expandable.

The strict rules regarding the establishment of new services (M = 46.75; SD = 31.62) are regarded as rather changeable in the context of the employment radar. The lack of data on the social milieus in each oft he respective EU countries is considered feasible (M = 45.50; SD = 30.35). The fact that, despite the heterogeneity of the group of young people, new services are jointly worked out is also classified as feasible (M = 43.64; SD = 29.37). In contrast, the insufficient motivation of young people to participate actively in the development of new service ideas is classified as the least changeable (M = 34.07; SD = 29.05). At the same time, this resistance is regarded as the most significant of all resistances (M = 3.71; SD = 1.36). The absence of the time and financial resources to organise interviews with potential clients is considered to be the second biggest resistance (M = 3.65; SD = 1.41). The third biggest challenge is that there is a high need for country-specific adjustments in terms of market research questionnaires and market segmentation (M = 3.44; SD = 1.23).

The answers to the open questions could complement the previous findings: The use of these employment radar tools could encourage young people to develop their potential and strengths. They could also be motivated to shape their possible personal development. In addition, the employment radar tool is considered to be advantageous in the way that young people seeking employment receive the necessary support from third parties (job developers and/or other experts). The sinus meta milieus were found to be a simple and beneficial tool to connect employees with new business ideas with their target groups. An additional strength of the employment radar is that it helps young people get to know their customers and identify market trends comparatively quickly and easily. Moreover, new employment opportunities are uncovered. However, it was unclear to the participants where they could obtain up to date information about the market., nor are there any suitable methods for processing the results of the interviews with potential customers. In regard to the last two points, IT-solutions could help.

The following two potentials of expert hearings are in first and second place with regard to their expandability: the young people gain first-hand access to work experience (M = 73.06; SD = 24.27) and there are positive synergy effects due to the fact that the young people receive feedback from various experts (M = 72.80; SD = 23.39).

The last point is classified as the most significant potential of the expert hearing (M = 4.26; SD = 0.88). In addition, it is regarded as important and advantageous that the experiences of the experts are passed on to the young people through the expert hearing (M = 4.24; SD = 1.01).

The biggest problem with expert hearings is the lack of time for experts (M = 3.86; SD = 1.32). The chance of chaning this is classified as rather low (M = 35.94; SD = 26.28). The second largest and third largest resistance are: the success of the expert hearing depends decisively on the competence of the experts (M = 3.73; SD = 1.34) and there is no network available to fall back on experts (M = 3.71; SD = 1.36). The transformability of this barrier is rated rather cautiously. The following resistances are classified as most likely to be changed: the high expenditure of resources to attract experts (M = 43.03; SD = 28.48) and the misinterpretation of the expert-youth relationship in the sense of a teacher-student relationship (M = 40.77; SD = 29.63). However, the last point is assigned a subordinate role (M = 3.37; SD = 1.53) when it comes to importance. In addition, it is regarded as changeable that the experts receive more knowledge about which forms of feedback prove to be particularly suitable (M = 40.00; SD = 26.75).

These results can be supplemented by the answers to the open questions: The expertise of the experts was experienced as extremely enriching. It is also particularly advantageous that the same service ideas as well as the respective personal development plans are examined by the experts from different perspectives. The fact that an interaction takes place between the job-seeking young people and experts is a strength of the motivating and inspiring process based on the personal contact between the experts and the job-seeking young people.

4 Discussion

4.1 Recommendations for Action

On the basis of the following measures, the teaching and learning objectives, the contents, the methodology and the possibilities for monitoring teaching and learning in qualification programmes for job developers, the young people concerned and the experts can be worked out in line with the Cologne structural model.

The greatest challenge for job developers is the skills required to use the appropriate tools. In addition, more intensive training of job developers is considered necessary. The selection criteria for job developers should be defined in advance. Possible selection criteria could be the extent to which they have skills in mentoring and coaching and what experience they already have in career orientation and job placement. These selection criteria and the type of selection procedure should be further specified. On the basis of this result, the qualification of job developers is necessary in order to deal with the tools in a professional manner.

The greatest challenge for job developers is the skills required to use the appropriate tools. In addition, more intensive training of job developers is considered necessary. The selection criteria for job developers should be defined in advance. Possible selection criteria could be the extent to which they have skills in mentoring and coaching and what experience they already have in career orientation and job placement. These selection criteria and the type of selection procedure should be further specified. On the basis of this result, the qualification of job developers is necessary in order to deal with the tools in a professional manner.

The starting point for developing a training programme for young jobseekers might be the realisation that they can utilize the results of the talent diagnosis. The aim of such programmes would be for them to acquire the ability to constructively use feedback that does not meet their expectations for their further development [13]. In connection with the identification of the strengths of young people, the problem that they increasingly do not perceive their strengths as such and overlook them must be taken into account. Against the background of these results, it needs to be clarified which methods are particularly suitable for uncovering the strengths and promoting them individually.

The use of targeted practical exercises within the framework of the employment radar concept is proving to be a way of encouraging young people's creativity and inspiration. In addition, the teamwork that takes place in this context also has a positive effect on finding new service ideas. On the basis of this knowledge, further exercises could be developed in order to make even more intensive use of the corresponding advantageous group-dynamic effects. The prerequisite, however, is that the respective responsible actors, e.g. the job developers, have group-dynamic skills. This aspect should therefore be included in the qualification programme for job developers. The training documents for job developers could also be further developed in such a way that concrete examples and exercises, e.g. related to micro trends, sinus milieus and possible new business ideas, are worked out and included.

In the empirical study it turned out to be a central point that young people seeking employment need support in finding "information for the existing milieus". This task could be taken on by the job developers. If they are not in a position to do so, this point could be included in their qualification programmes. The question is to what extent the employment radar will make it even easier for young people to obtain information about the potential labour market. This point was considered by the respondents to be particularly capable of improvement.

To process the extensive data resulting from the interviews with potential clients appropriately was considered a particular challenge. Data processing tools should be provided to process the information obtained in connection with the employment radar. IT solutions can help. In view of the lack of time and other resources to conduct sufficient interviews within the framework of the employment radar and, for example, to adapt the market research questionnaires to country-specific conditions, IT-solutions could also be implemented.

In order to make even better use of the experts' experiences for the actions of young people in the future, it should be noted that the experiences are primarily available in the form of implicit knowledge. At this point, competence development activities in the sense of the "community of practice" concept are of particular interest [14]. For young people to receive feedback from different experts, a sufficient number of experts must be available. This was classified as the central and most expandable potential of expert hearings. In addition, the experts should be trained in giving feedback. Constructive feedback is important in order to motivate the young people, but the experts should also convey a realistic picture, e.g. of the respective economic and labour market situation, through their point of view.

If the forms of expert hearings used so far are only suitable to a limited extent for structuring the implementation of the personal development plan, then the question arises as to what measures are necessary for this to succeed. It should be kept in mind that the central goal of the Job Developer programme is to jointly develop a personal development plan that will be implemented as far as possible later on - especially as the development of this development plan is seen as a very helpful starting point for every young person looking for a job.

If the effort required to find experts is estimated to be comparatively high, suitable measures would have to be examined to reduce this effort. Here it is useful to establish a social network in which experts are involved in an appropriate way. The possibilities of online consulting should be explored in terms of advantages and disadvantages within the framework of expert hearings. All in all, an IT-supported platform for the exchange of knowledge between experts, job developers and young job seekers is necessary. In this sense, either an existing platform should be used or a new one created.

4.2 Quality Criteria

Against the background of the findings of the empirical studies, the following quality criteria are worked out in order to improve the utilization of the three tools. The tools for talent diagnosis should offer young people the opportunity to identify their own strengths. Consequently, it cannot be the primary goal of the talent diagnosis that third parties work out a profile of the young people's strengths. Consequently, the requirement is placed on the talent diagnosis that it should serve to increase the motivation of the young people to take their life into their own hands or to turn themselves into a project. On the basis of this quality criterion, it can be examined which IT-supported tools are helpful in talent diagnosis. The strengthening of young people's self-confidence, the disclosure of previously unknown strengths as well as the use and realistic assessment of these strengths by the tools can also serve as further quality criteria for talent diagnosis. A further criterion would be the extent to which the results of the use of talent diagnosis help to concretize the goals and steps of the implementation of the personal development plan from the perspective of the participants. Based on the listed quality criteria for talent diagnosis, various procedures could be examined more closely for suitability.

The previous explanations have shown that it proves logical to develop qualification programmes with different objectives, contents and methods for the three target groups (job developers, young job seekers and experts). For both the job developers and the experts, it should be clarified to what extent they are in a position to provide appropriate forms of constructive feedback adapted to the respective situation (e.g. development of new service ideas or a personal development plan) after the corresponding training.

With regard to the question of which criteria are relevant for measuring quality improvements connected to the employment radar, the following points can be made: To what extent have young people been reoriented? How many service ideas were developed? How high is their degree of novelty and as how promising is their implementation assessed? Another criterion could be the number of young people who have improved their employment opportunities and/or taken up new employment chances within the framework of the employment radar.

The use of the employment radar could also be examined in terms of the extent to which it is possible to find specific IT-solutions tailored to the situation. These IT-solutions could, among other things, offer the possibility of dealing appropriately with the multitude of information in the sense of finding new service ideas.

The absence of networks is a particular barrier to expert hearings. The quality of an expert network can be characterised, for example, by the number of experts or the fields of experience they cover. They can come from different branches of industry, commerce or trade or from different areas such as marketing, finance, production or services. As quality criteria for the expert hearing, the willingness and competence of the experts to reflect on their experiences and to give constructive feedback to the young people can be recorded. It should also be examined to what extent it has been possible to draw up a personal plan for each young person looking for a job, tailored to his or her life situation and talents, and to reflect this plan in a sufficient way with the appropriate experts. With regard to improving feedback, it is advantageous to differentiate between the following three feedback situations: (1) feedback in relation to the talents or strengths of the individual youths, (2) in relation to the new service ideas and (3) in relation to the development of a personal development plan.

A further step in quality improvement is the development of criteria for selecting job developers and experts. This involves clarifying the minimum requirements to be met by these groups of people. When selecting experts, attention could be paid to the combination of different experts. It could also be assessed to what extent the experts have mentoring or coaching experience. With regard to the selection of experts, special IT-solutions to be developed could help. An additional quality criterion for assessing the expert hearing could be the extent to which it is possible to better structure the implementation of the personal development plan and to create opportunities for a more promising implementation.

In view of the fact that the quality of the expert hearing depends on the respective competencies of the experts, it would be useful to clarify which competencies they should possess. In order to get an overview of the competences, it would be advantageous to set up a corresponding database. This database could also contain information on the interests and accessibility of the experts. Since they have little time, it should also be analized how social media can be used to reduce their time expenditure.

5 Outlook

The above remarks have highlighted the need to ensure the quality standards and professionalism of activities in order to win and maintain the confidence of jobseekers. Starting from the goal of improving the employability of young people, the various megatrends pose a further challenge. In addition to digitalisation, the following megatrends (https://www.zukunftsinstitut.de/dossier/megatrends/) [15] should also be mentioned at this point: Demographic change, globalisation, increasing mobility, progress in the health sector, individualisation, development towards a knowledge society and a growing logistics sector. With regard to the megatrend "digitization of work", the focus is shifting to the fusion of production (Cyber-Physical Systems; Industry 4.0) and services (Smart Services) to Product-Service Systems. Although the

Job Developer programme is primarily focused on the creation of new services, the Job Developer concept is also of interest for the production sector against the background of the merge of production and services. The next step would be to examine how this link can be achieved in the context of digitization.

Acknowledgements. A part of the present explanations can be found in German-language essays by the author: Kröll, M. (2019a): Entrepreneurship Education und Digitalisierung: GFA (Hrsg.): Frühjahrskongress 2019, Dresden; Kröll, M. (2019b): Strategien zur Bekämpfung der Jugendarbeitslosigkeit in der EU. In: Zeitschrift für ökonomische Bildung.

References

1. Hirsch-Kreinsen, H., Ittermann, P., Niehaus, J. (Hrsg.): Digitalisierung industrieller Arbeit. Die Vision Industrie 4.0 und ihre sozialen Herausforderungen. 2. Aufl., Nomos Verlagsgesellschaft, Baden-Baden (2018)
2. Keicher, I.: Beschäftigungsradar - das Marktpotential für innovative Dienstleistungen darstellen. In: Hartz, P., Petzold, H.G. (Hrsg.) Gegen Jugendarbeitslosigkeit – Innovative Ideen, Modelle, Strategien, pp. 242–254. Aisthesis, Bielefeld (2015)
3. Kröll, M.: Job Developer: Kreierung von Dienstleistungen im Kontext der Digitalisierung. Bochum: Institut für Arbeitswissenschaft der Ruhr-Universität Bochum. Unveröffentlichtes Manuskript (2016)
4. Mühlhausen, C.: Neue Jobs für Arbeitssuchende in 7 Trendmärkten. In: Hartz, P., Petzold, H.G. (Hrsg.) Gegen Jugendarbeitslosigkeit – Innovative Ideen, Modelle, Strategien, pp. 285–297. Aisthesis-Verlag, Bielefeld (2015)
5. Kröll, M., Vos, B.: Selbstlernmaterial zur Kompetenzmessung und –beurteilung. In: Kröll, M. (Hrsg.) Studienkonzepte zur Qualifikation von Führungskräften, Mitarbeitern von Personalabteilungen und Kompetenzexperten, pp. 1–77. Bochum (2013)
6. Erpenbeck, J., Rosenstiel, L.V., Grote, S., Sauter, W.: Handbuch Kompetenzmessung. Erkennen, verstehen und bewerten von Kompetenzen in der betrieblichen, pädagogischen und psychologischen Praxis. 3. überarbeitete und erweiterte Auflage 2017, Schäffer-Poeschel Verlag, Stuttgart (2017)
7. Kohnke, O.: Anwenderakzeptanz unternehmensweiter Standardsoftware. Springer, Wiesbaden (2015)
8. Gondo, M.B., Amis, J.M.: Variations in practice adaption: the roles of conscious reflection and discourse. Acad. Manage. Rev. **38**(2), 229–247 (2013)
9. Bank, V.: Von der Organisationsentwicklung zum "systemischen Change Management". Kiel (2004)
10. Jongebloed, H.-C., Twardy, M.: Strukturmodell Fachdidaktik Wirtschaftswissenschaften (SMFW). In: Twardy, M. (Hrsg.) Kompendium Fachdidaktik Wirtschaftswissenschaften, pp. 163–203. Bd. 3, Teil I, VH-Verlag, Düsseldorf (1983)
11. Bruhn, M.: Qualitätsmanagement für Dienstleistungen. 9. Aufl., Berlin-Heidelberg (2013)
12. Zeithaml, V.A., Parasuraman, A., Berry, L.L.: Qualitätsservice. Was Ihre Kunden erwarten - was Sie leisten müssen. Campus-Verlag, Frankfurt a. Main - New York (1992)
13. Hattie, J.: Visible Learning for Teachers: Maximizing Impact on Learning. Routledge, London (2012)
14. Webber, E.: Building Successful Communities of Practice. Drew London Ltd., London (2016)
15. Zukunftsinstitut (Hrsg.): Megatrend-Dokumentation, Frankfurt a. M. (2018)

Entrepreneurship Education for Fashion Design

Graziela Sousa[1(✉)], Ines Simoes[1], and Paulo Maldonado[1,2,3]

[1] CIAUD - Centro de Investigação em Arquitetura, Urbanismo e Design, FA-ULisboa, Lisbon School of Architecture, University of Lisbon, Lisbon, Portugal
grazielasousa@fa.ulisboa.pt, ines.sa.simoes@gmail.com, paulomaldonado@inspaedia.com

[2] CITAD – Centro de Investigação em Território, Arquitetura e Design, Universidades Lusíada, Lisbon, Portugal

[3] CHAIA – Centro de História e Investigação Artística, Universidade de Évora, Évora, Portugal

Abstract. This paper imparts part of the results of an ongoing PhD research, which focuses on the promotion of entrepreneurship in Fashion Design education. The paper addresses the development of relevant skills that might prepare fashion design students for their upcoming careers in the fashion industry. The results, which were withdrawn from literature review and semi-structured questionnaires administered to alumni from our institution, suggest that there is a need for an immersive and hands-on approach to education so as to foster the development of entrepreneurial skills among the next generations of fashion designers, while giving them further preparation for the challenges of a career in this particular field.

Keywords: Fashion design education · Entrepreneurship education · Fashion industry

1 Introduction

This paper develops from an undergoing PhD research project entitled 'Fashion Design (and) Entrepreneurship: a Strategic Vision for Higher Education', which aims to advance a conceptual model for the development of entrepreneurial skills in Fashion Design Higher Education in Portugal. This paper imparts part of the developed research and its resulting insights, which connect the review of relevant literature on the various implicated issues with the responses of many fashion design alumni to a semi-structured questionnaire.

The research question consists of: How can we foster the development of entrepreneurial competencies among Fashion Design students? In order to address this question, the research is being developed with the aim of empowering the next generations of fashion designers with the skills that would enable them to perform successfully as professionals and to empower them to behave entrepreneurially (either employed in companies, self-employed or starting and owning their businesses).

W. Karwowski et al. (Eds.): AHFE 2019, AISC 963, pp. 284–291, 2020.
https://doi.org/10.1007/978-3-030-20135-7_28

The collected insights showed that:

1. Entrepreneurship education in a creative field is, although controversial, something fundamental and must foster the development of behavioral skills;
2. Entrepreneurship is one of the most relevant skills in today's economies and its definition goes beyond its business acceptation – an entrepreneur is an agent of change, someone who is able to find, exploit and develop opportunities;
3. Fashion is a very particular creative industry which has a massive economical impact and raises several environmental and ethical issues. Therefore, education in this field must empower future professionals with contents, experiences and opportunities for learning and developing several types of abilities/skills;
4. The Portuguese fashion industry is rising in reputation and expertise, and the country is highly recognized for being genuinely promoting entrepreneurship in the last decade;
5. The alumni's perspectives pointed out that the Bologna process impoverished the curricula in Higher Education and that more practical as well as individualized approaches are needed for fashion students to be prepared for (and to succeed in) the professional field of their choice.

We conclude, therefore, that students are more likely to acquire knowledge, drive, ability and self-confidence to act entrepreneurially in the future if they are stimulated through practical and immersive experiences while in school. We believe that those experiences can show them the features, challenges and opportunities of the professional field they are about to enter, as well as provide knowledge about the specific fields they wish to specialize in and the skills they want to develop.

2 Methodology

The adopted research methodology for the development of this paper includes secondary and primary sources, such as: (a) the review of relevant literature concerning the fields of Fashion Design Education, Entrepreneurship Education especially for the Creative Industries, Fashion Business and Higher Education; and (b) semi-structured questionnaires, implemented through Google Forms and sent to Fashion Design alumni of our institution, FA-ULisboa – the older Higher Education institution in Portugal that offers under-graduate and graduate studies in the field of Fashion Design –; the questionnaires were sent between November and December 2015, through email and social media messaging (using Facebook Messenger app and Linked'In). Seventy-four alumni, who attended any of the three BA in Fashion Design programs that were launched from 1992 to the present day, responded to the questionnaire.

3 Background

3.1 Entrepreneurship Education in the Creative Industries

The idea of connecting Entrepreneurship with the creative disciplines (generally speaking, be it design, fashion, music or the visual arts) is controversial, in a certain sense, from the academic and artistic perspectives, since there is prejudice against the business prosecution of one's projects/products, a certain aversion to discuss 'money', let alone to find ways to pursue it, make it or even build a business [1]. It is the University's role to foster the students' economic independence and professional success after graduation [2]. Yet, creative individuals tend to lack skills related to selling and profitability, therefore, it is necessary to address and develop those skills through approaches that are rewarding and fulfilling for them.

Learning through professional experiences, internships and practical projects, using an experiential-learning approach [3] in a setting that fosters apprenticeship through improvisation and practice [4], might be more effective and even more rewarding than having formal lectures about business or trade issues [2, 5]. The identified traits that should be developed in order to foster an entrepreneurial behavior in the creative industries are opportunity recognition, designing, risk managing, resilience, and effectuating skills [6], to which Pollard and Wilson [7] add self-confidence and collaboration. These seven traits seem to be fundamental in the field of fashion as well. We intuit that these might enable future professionals to be agents of change, capable of developing and implementing novel and pertinent ideas in our field.

3.2 Entrepreneur(ship) and Higher Education

It is the capacity of creating, bringing ideas to life, developing and making new products that impel economies, countries and societies to move forward. From a business definition, being entrepreneurial means building new enterprises and taking risks [8, 9]. However, from the etymological acceptation of the word, which derives from the French 'entreprendre', it means 'to do something' or 'to undertake' [10] – the latter is much more enriching for a discussion in the context of education towards entrepreneurship in the field of fashion design.

Entrepreneurship is one of the most relevant skills in today's economies and its definition goes beyond its business acceptation – an entrepreneur is an agent of change, someone who is able to find, exploit and develop opportunities [11]. Thus, being entrepreneurial today is absolutely fundamental across all disciplines and should not be regarded merely in the economic/business sense of creating new enterprises or start-ups. Entrepreneurship is the ability to recognize and exploit opportunities, which fill up existing gaps or solve issues/problems. In today's society an entrepreneur is an individual who is empowered to innovate and make an impact (adding value) through his/her actions [5, 8, 9, 12–14]. Entrepreneurship is associated with a capacity for innovation, initiative [15, 16], creativity [13, 17] and it stimulates employment and growth [12].

There is an urge for Higher Education institutions to qualify their students to become entrepreneurs and/or behave in an entrepreneurial manner, but also to be themselves entrepreneurial institutions. In current knowledge based-economies, universities must be open to society and create conditions for open 'dialogues' with it: either by solving problems/issues or capitalizing on innovations/advances (e.g. converting them into businesses and putting them into practice – processes of knowledge transfer [18–20]).

3.3 Fashion and Education

The contemporary fashion industry scenario has many challenges as well as opportunities: the need to solve/stop over-consumption and overproduction and to offer sustainable and ethical alternatives can be regarded as opportunities for fashion innovators/entrepreneurs – who might be able to solve prominent societal and environmental issues provoked by the fashion industry [21, 22]. The character of the fashion discipline is somewhat controversial: it is both an artistic and stylistic celebration, yet being a trillion-dollar industry, one of the largest in turnover and employment in the world – which makes business and profitability mandatory to be addressed, because in the end, fashion products have to be sold and worn [5, 23].

Several perspectives of educators were analyzed regarding Fashion Design education. It is possible to state that schools offering this program must: (1) deliver up-to-date content about the fashion industry [24, 25], (2) show the tasks and professions that a fashion professional might perform [26], besides 'designing' garments/products [21], (3) address current issues like the (urgent matters of) sustainability, ethics and technology [21, 23], and (4) prepare the students with fashion business knowledge while developing relevant skills that might enable/empower them to behave as innovators, entrepreneurs and industry game-changers [23, 25, 26].

Schools also must provide facilities, partnerships and access to opportunities for their students, supporting them during and after their attendance: alumni networks, incubators, mentoring and job banks are some of the key features which not only allow young entrepreneurs to exploit and develop opportunities as well as helping/supporting them throughout their professional paths [25, 27].

Students attending Higher Education nowadays are being more recently described as Generation Z – they are digital natives, their attention span is smaller then the Millennial's and they show several features which indicate that hands-on, collaborative and learning-by-doing experiences might work better for their knowledge apprehension [28].

3.4 Fashion and Entrepreneurship in Portugal

Portugal is now recovering from the severe economical crisis of 2007/2008 and the fashion industry has been showing a steady recovery in the last years. With proven reputation for manufacturing excellent textiles and leather goods, the country is affirming itself as a production country for many high-end brands [14, 29]. The country's location also adds up to the quality it delivers, since premium brands are gradually bringing their overseas' production to Europe – which cuts transportation costs, makes quality control easier and guarantees a sustainable and ethical attitude from the perspective of manufacturing [30].

From the perspective of design, Portugal offers Fashion Design programs (Bachelor and Master's degrees) in seven Higher Education institutions and launches two major contests twice a year for young/emerging designers (Sangue Novo and Bloom) in two fashion weeks (ModaLisboa and Portugal Fashion, in Lisbon and Oporto respectively). Both events promote senior and young fashion designers in Portugal and abroad.

Portugal (particularly Lisbon) also provides many opportunities for entrepreneurs – as the host country for the Web Summit[1] –, making efforts to foster entrepreneurship through incubators, initiatives and funding in the last decade [31]. Lisbon, the Portuguese capital, has been in the forefront of such efforts, while trying to affirm itself as a 'StartUp city' [32], having been building conditions for the development of new businesses and fostering creative entrepreneurship, namely with the creation of the 'Criativas' network – which gathers the city's infrastructures like Fab-Labs, incubators and innovation hubs, providing mentorship and making facilities to young talents and their projects [33].

3.5 Alumni's Questionaires

FAUL alumni's insights reinforce the fact that learning must combine different approaches and that the transmission of knowledge represents just a portion of the role that Higher Education should play. Internships, direct experiences with the industry/professionals, networking opportunities, business skills, employment opportunities and career advice are seen by alumni as important factors for Fashion Design education to be effective.

Semi-structured questionnaires were carried out, implemented at Google forms and sent via email and Facebook and LinkedIn social networks. The seventy-four alumni who filled in the forms attended any of the three BA in Fashion Design programs that were launched from 1992 to the present day. The BA in Fashion Design program offered at FA-ULisboa has gone through three redesigns and the most drastic one occurred in 2004–2005, forced by the Bologna Agreement for Higher Education, which shortened the BA program from 6 to 3 years. The institution also offers a Master in Fashion Design program since 2008.

According to the collected responses it is possible to infer that the attendance at FA-ULisboa has, in some way, lost quality, because of the curricular redesigns provoked by the Bologna Agreement for Higher Education.

The internship (which was part of the original 6-years BA in Fashion Design) is considered useful and essential for a large majority of the alumni (23 out of the 25 alumni who took internships) and was recognized as fundamental to acquire knowledge about the profession and the industry. From the responses it became clear that an internship could provide a link to the professional world, besides networking and employment opportunities, which are seen by the respondents that graduated after 2005 as the biggest flaws of their experience at FA-ULisboa.

Among the alumni's suggestions for the improvement of the Fashion Design programs, we highlight the contact with professionals and the industry's reality (38), internships (16), learning fashion business/entrepreneurship tools (13), the recognition of other jobs/roles besides designing garments and accessories (10), learning fashion business/entrepreneurship tools as well as counseling/mentorship about their academic/professional careers (6).

[1] One of the main events for entrepreneurs in the world, which lasts four days and gathers thousands of entrepreneurs from several fields, to show their innovative businesses (specially in the technology field), meet investors, build partnerships and listen to experts' keynotes about contemporary matters [34].

4 Discussion

It is possible to conclude that there is a need for a strong and directed input to promote Fashion Design entrepreneurship in Portuguese Higher Education. It is also relevant to state that such an approach would be more likely to succeed if it was focused, in the beginning, on the development of soft skills such as communication, effectuation, opportunity recognition, analytical/critical thinking skills and an adaptation/actualization of the lectured contents [5, 6, 35].

Internships or apprenticeships, immersive experiences, problem-solving approaches/ challenges and the direct contact with professionals would be the most successful and appreciated teaching/learning strategies in preparing the next fashion designers for the industry and the possibilities (and problems) they will come across when entering the professional world.

Regarding the ongoing doctoral research, the collected inputs and information communicated in the present paper will be used as the basis of a Model for Fashion Design Entrepreneurship Education.

Acknowledgments. CIAUD – Centro de Investigação em Arquitetura, Urbanismo e Design; FA-ULisboa, Lisbon School of Architecture, University of Lisbon, Portugal.

References

1. Bridgstock, R.: Not a dirty word: arts entrepreneurship and higher education. Arts Hum. High. Educ. **12**, 122–137 (2013)
2. Penaluna, A., Penaluna, K.: Business paradigms in Einstellung: entrepreneurship education, a creative industries perspective. In: Internationalizing Entrepreneurship Education and Training Conference, IntEnt (2006)
3. Kolb, D.A.: Experiential Learning: Experience as the Source of Learning and Development. Prentice-Hall, Englewood Cliffs (1984)
4. Lave, J.: Situating learning in communities of practice. Perspect. Socially Shared Cogn. **2**, 63–82 (1991)
5. Sousa, G.: Pop-Up store DEMO: an entrepreneurial experience for fashion design graduates. Int. J. Des. Educ. **12**, 41–53 (2018)
6. Duening, T.N.: Five minds for the entrepreneurial future: cognitive skills as the intellectual foundation for next generation entrepreneurship curricula. In: USASBE 2008 Proceedings-Page (2008)
7. Pollard, V., Wilson, E.: The "Entrepreneurial Mindset" in creative and performing arts higher education in Australia. Artivate J. Entrep. Arts. **3**, 3–22 (2014)
8. Schumpeter, J.A.: The Theory of Economic Development: An Inquiry Into Profits, Capital, Credit, Interest, and the Business Cycle. Transaction Publishers, New Brunswick (1934)

9. Drucker, P.: Innovation and Entrepreneurship. Routledge, New York (1985)
10. Kreft, S.F., Sobel, R.S.: Public Policy, Entrepreneurship, and Economic Freedom. Cato J. **25**, 595 (2005)
11. Redford, D.T.: The State of Entrepreneurship Education in Portugal - An Empirical Study on a Nascent System in the European Union Policy Framework (2008)
12. Naia, A.M.P.: Entrepreneurship Education in Sport Sciences: Implications for Curriculum Development (2013). http://www.repository.utl.pt/handle/10400.5/6203
13. Baron, R., Shane, S.: Entrepreneurship: A Process Perspective. Cengage Learning, Mason (2007)
14. Sousa, G., Simoes, I.: Fashion design and entrepreneurship: a strategic model for higher education in Portugal. In: The Virtuous Circle. McGraw-Hill Education, Milan (2015)
15. Drucker, P.F.: Management Challenges for the 21st Century. HarperCollins e-books, New York (2009)
16. Maldonado, P.: Inovação, Design et cetera (Innovation, Design and so on). Ph.D dissertation, Universidade Técnica de Lisboa (2012)
17. Shane, S., Locke, E.A., Collins, C.J.: Entrepreneurial motivation. Hum. Resour. Manag. Rev. **13**, 257–279 (2003)
18. Etzkowitz, H., Webster, A., Gebhardt, C., Terra, B.R.C.: The future of the university and the university of the future: evolution of ivory tower to entrepreneurial paradigm. Res. Policy **29**, 313–330 (2000)
19. Hannon, P.D.: Why is the entrepreneurial university important? J. Innov. Manag. **2**, 10–17 (2013)
20. Sá, E., Dias, D., Sá, M.J.: Towards the university entrepreneurial mission: Portuguese academics' self-perspective of their role in knowledge transfer. J. Furth. High. Educ. **42**, 784–796 (2018)
21. Edelkoort, L.: Anti Fashion Manifesto. Trend Union, Paris (2015)
22. Fletcher, K., Grose, L., Hawken, P.: Fashion and Sustainability: Design for Change. Laurence King Publishing, London (2012)
23. Corner, F.: Op-Ed | Schools Must Respond to the Needs of the Industry. https://www.businessoffashion.com/community/voices/discussions/how-should-fashion-schools-integrate-business-learning-into-a-creative-education/op-ed-schools-must-respond-to-the-needs-of-the-industry
24. Op-Ed | The Gap Between Design and Business Education Has Been Filled. https://www.businessoffashion.com/community/voices/discussions/how-should-fashion-schools-integrate-business-learning-into-a-creative-education/op-ed-the-gap-between-design-and-business-education-has-been-filled
25. Corner, F.: Why Fashion Matters. Thames and Hudson Ltd, London (2014)
26. Rissanen, T.: Possibility in fashion design education—a manifesto. Utopian Stud. **28**, 528 (2018)
27. Meadows, T.: How to Set up & Run a Fashion Label. Laurence King, London (2012)
28. Simoes, I., Ribeiro, M.M., Nogueira, N.: Sense and sensibility in fashion design education. In: Nazir, S., Teperi, A.-M., Polak-Sopińska, A. (eds.) Advances in Human Factors in Training, Education, and Learning Sciences, pp. 151–160. Springer, Cham (2019)
29. Khalip, A.: Analysis - Portugal Textile Firms Weave Way Out of Crisis. http://uk.reuters.com/article/2012/02/08/uk-portugal-textile-idUKTRE8170YF20120208
30. ModaLisboa: PORTUGAL DESTACOU-SE NA PREMIÈRE VISION. https://modalisboa.pt/noticias/portugal-featured-at-the-premiere-vision_860

31. Startup Lisboa | About Us. http://startuplisboa.com/aboutus/
32. Sítio da Câmara Municipal de Lisboa: Empreendedorismo. http://www.cm-lisboa.pt/investir/empreendedorismo
33. Lançamento das Criativas de Lisboa | FabLab Lisboa. http://fablablisboa.pt/?p=14090
34. Web Summit | Lisbon 2019 | Where the Tech World Meets. https://websummit.com/
35. Gardner, H.: Five Minds for the Future. Harvard Business Review Press, Boston (2009)

Manguinhos' Shantytown: Investigating Low-Income Students' Education Problems and Their Environmental and Social Context

Nathália Pompeu[1(✉)], Luiza Helena Boueri Rebello[1], and Arlene Pompeu[2]

[1] GDI Graduação em Desenho Industrial, Universidade Federal Fluminense - UFF, Rua Passo da Pátria, 156, Niterói, RJ 22471-210, Brazil
nathapompeu@gmail.com, luizarebello@id.uff.br
[2] PROFELETRAS Mestrado Profissional, Universidade do Estado do Rio de Janeiro - UERJ, R. Francisco Portela, 1470 – Patronato, São Gonçalo, RJ 24435-005, Brazil
arlenepompeu@gmail.com

Abstract. This paper has the goal of investigating the Brazilian low-income student's education problems and their relation with environmental and social context. It was observed that many low-income teenager's students have difficulties to properly read and write. Thus, the students find a large gap in their literacy level and often cannot even read and write although they are at a middle level school classroom. This investigation was conducted in a public school from Manguinhos' shantytown in Rio de Janeiro city, Brazil. The investigation's development consisted of systematic observation, interview, meeting and gamification's activities. Through an ergonomics approach with Participatory Design and Action Research, it was possible to analyze the relations of social critical literacy and gamification approach toward inclusive and effective education.

Keywords: Participatory Design · Action Research · Education · Social critical literacy · Gamification

1 Introduction

Since 2014 the Brazilian Government has increasing source investments in search of develop a better public education system [1]. The Organization for Economic Co-operation and Development reported that Brazil spends more than 5, 4% off the Gross Domestic Product in education, above the Latin America's Countries investments average. Although the Brazilian education's level was identified as lower than the Latin America's Countries level average (OECD 2015f). Thus, according the [2] United Nations Children's Fund (UNICEF 2018) the shantytown[1] areas education in Brazil has the classification of emergency. More than 28% of the students did not finish the school.

[1] According the Cambridge Dictionary: an area in or near a city in which poor people live in small, badly built houses.

W. Karwowski et al. (Eds.): AHFE 2019, AISC 963, pp. 292–300, 2020.
https://doi.org/10.1007/978-3-030-20135-7_29

The Manguinhos's Shantytown is in Rio de Janeiro City, Brasil [3]. It has 44.051 inhabitants (Cavallieri and Lopes Peres 2008) and most of them live below the poverty line. It was observed that many students in this area cannot read and write, although they are at middle level school classroom. Moreover, the students (14–16 years old) face problems as violence and segregation. They do not enjoy been at classroom and have high numbers of early school-leavers (Fig. 1).

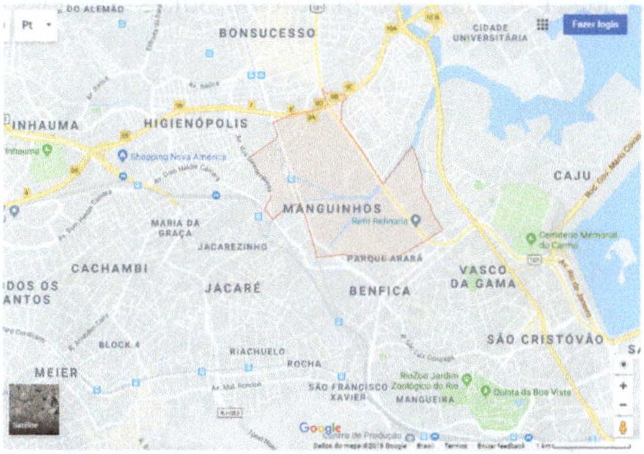

Fig. 1. Google Maps. Manguinhos' Shantytown.

The students' families' houses were built under the Manguinhos' railway line. The houses have no infrastructure and no basic sanitation. On the streets, there is trash everywhere, and also smells badly. The houses are heaped together, with no recreation space. The school is situated in front of a prison and is often confused with the prison, because the architecture (Fig. 2).

Fig. 2. Students families' houses.

The classroom has 40 students delayed (they should be in the high school but they are in middle school level). In addition, the students live in gangs' domain areas and reported that they wake up because listened shooting and police invasion. Most of them were forsaken by the parents or they do not receive the familiar attention.

The students cannot read and write and they are resistant to do classroom and homework tasks. Besides that, the students' learning level is much lower in relation to the materials provided by the government. Therefore, most of schoolteachers do not use the available school materials and need to develop others activities (Fig. 3).

Fig. 3. Classroom.

2 Methods

This study is about investigating Manguinhos' low-income education problems and their relation with environmental and social context in search of effective solutions. It was developed in the main public school in Manguinhos' Shantytown which receives a large quantity of students.

The relation between student and classroom is an important part of learning and developing knowledge. Thus, to understand this relation it was necessary conducted this investigation using not only Participatory Design and Action Research methods but also gamification technics. The ergonomic investigation not only was engaged in fully understanding the classroom, focusing on how the students interacted with the environmental but also was engaged in make the data collection enjoyable and humanized to the students.

Through a series of consistent and frequent visits to the classroom and game activities approach we could analyze the social environmental relation with the students and the impact of those in their education development.

2.1 Discussion About Participatory Design and Action Research on Education Environment

Participatory Design refers including users on design process [4]. According to Kesing and Blomberg (1992) users' involvement increase the systems' quality because it possible to make better need's and problem's analysis, better reflection and evaluation about systems' components and design useful solutions. The students face problems as violence, deep poverty and segregation daily, which are often not considered in the classroom environment and learning materials.

Thus, the Participatory Design and Action Research approach enable the debate about the students' perspective in search of understand their real need and give them voice about their education [5]. Preece, Rogers and Sharp (2005) argue that engaging the users in participatory research is not easy. For that reason, planning and making activities with students not only act as an ergonomic intervention but also as an effective approach to communicate with students and develop efficient solutions (Fig. 4).

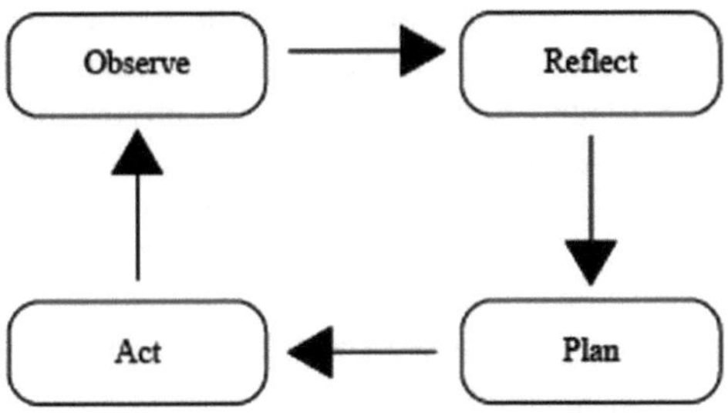

Fig. 4. Action participatory model.

2.2 Discussion About Gamification Approach on Education Environment

According to [6] Stott and Neustaedter (2013) gamification refers the application of game dynamics into non-games settings. And this approach has been increasing student motivation and achievement in the classroom. On the other point, [7] Werbach and Hunter (2012) argue that one challenge in gamification approach is the difficulty provided by the mixture of science and experience.

The students analyzed have a particularly undisciplined conduct. [8] Arroyo (2014) content that when students' behaviors (teenagers and children) scandalize and surprise us, they force us to understand that ethics, values, behaviors and culture have never been static. By the way, the dynamics of free human being are revealed in them. For that reason, the gamification approach facilitates problems investigation with the students because the activities help teachers and designers to understand the students' behavior without making this process oppressive.

3 Ergonomic Investigation's Overview

The ergonomic investigation was developed with three stages. Firstly, we analyzed the inputs, outputs and the organization of the system. Secondly, we looked for the system's mains problems. Lastly, we analyzed the interactions between users and system and collected data in search of creating effective solutions.

Looking at the system's structure organization provided us a fully comprehension about the classroom's components, users' functions and environmental context. Therefore, we were able to analyze the ineffective education approach results. For example, even the teacher following correctly the class' plan, the students have not fully understood the subjects' topics.

In order to understand the system's main problems (which was providing an ineffective result) we focused on how the students and teachers interact during the class. Thus, through a series of frequent visits to the classroom we were able to analyze the system's variables, tasks and problems. For instance, we observed how the classroom's proceedings changed as consequence of the discrepancy between students' literacy level and the textbooks' literacy level. The teacher always changed the activities because the students could not fully understand the textbook's questions. Moreover, the students always complained about how the activities were monotonous.

Thus, in search of diagnosing the problems we developed a series of activities and interviews with students and teachers. Through participatory design, action research and gamification we investigated how the social and environmental context influence the education quality. Furthermore, this stage was essential to give users voice and collect real and useful data in search of effective solutions (Table 1).

Table 1. Ergonomic investigation stages.

Stage	Investigation's goals
1- System's overview	Analyze the system's structure organization
2- Problems' research	Looking for the system's main problems
3- Diagnosis	Diagnosing real problems and give users voice in search of efficient solutions

4 Problem Investigation and Diagnosis

It was observed that the emotional insecurity and self-esteem (due to the school delay, deep poverty, hungry and violence) interfere in the students' education and adaptation process. [9] Oborne (1995) argue that what is specific in Ergonomics is analyzing more than human properties but also understanding how the people's personal story, their individual wishes, experiences and yearnings constantly influence their lives and work.

During the research, the students daily reported their social and cultural experiences, mentioned that they felt hungry because they do not have what to eat at home. It was observed that sometimes the students did bullying because they live in poor houses. Moreover, the students usually reported violence facts experienced in Manguinhos' shantytown. The gamification was incorporated as an investigation tool into the Participatory Design and Action Research Methods (Fig. 5).

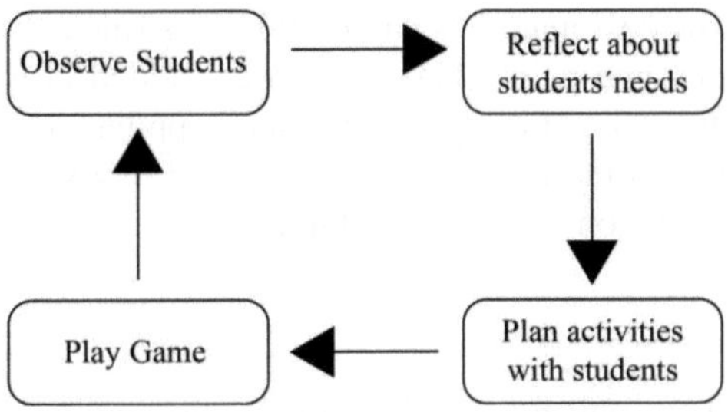

Fig. 5. Gamification approach as investigation tool.

4.1 Activities

Firstly, it is important understand that the school does not have enough resource for school supplies and the students have difficulty to understand tasks, primarily write and read even though they have 14–16 years old. That is the reason why the game activities

Table 2. Game activities and goals.

Game activity	Game dynamic	Specific Goals Data collected
1- Building my school	The students developed a comic showing how they see theirs school and how they would build the school	Identifying main problems and solutions find by the students. Identifying how much enjoyable is the activity and students' behavior at classroom
2- Building my future	The students chose a random magazine and they should make a collage of how they saw each other today and how they dream to be in the future	Identifying how much school is important in the students' lives., their skills and self-esteem. Identifying how much social and environmental context influence them
3- Song Challenge	Firstly, the students should suggest songs for one to be chosen by the school teacher. Secondly, the students should form groups and go through four stages (the first one, they should complete the words that were missing in the lyrics text while they were listening the song; the second, they should write what they understood about the song; in the third they should debate about the song context and their live reality. In the end, they should draw the song lyrics. In each stage the students receive a score, the group with the highest score win the game	Identifying reading and writing level, level of interpreting rules and do tasks. Identifying students' engagement. Stimulate reading and writing, debate about their social, text interpretation and environmental context and critical literacy

were developed with low cost materials and easy level's rules. As it was noted that the students enjoyed to draw and paint, these tasks were added in the activities.

Thus, three activities in game dynamics were applied and developed with the students to understand what motivate them to study and to give them voice about what they think about the education's system (Table 2).

4.2 Results

After each activity, it has analyzed the data information collected in search of identifying how the social and environmental context influenced the students on their education and understand which problems were identified by the students (Fig. 6 and Table 3).

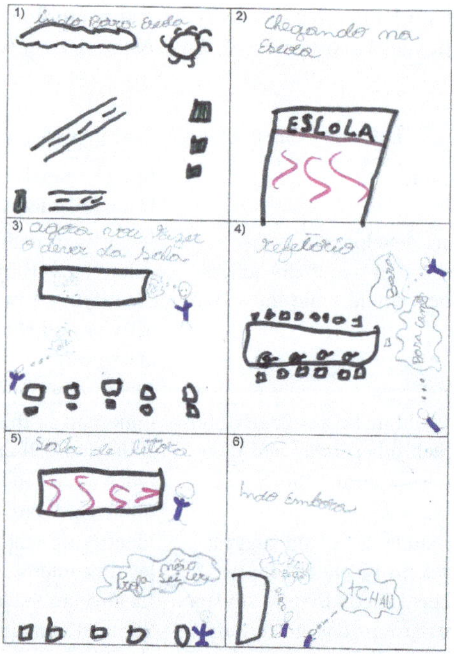

Fig. 6. First Game activity applied. Text: (1) "Going to school" (2) "Arriving at school" (3) "Now I will do my homework" (4) "Canteen" (5) "Read classroom – teacher, I do not read" (6) "Going home - bye".

Table 3. Game activities results and data collected.

Game activity	Results and data collected
1- Building my school	The students draw the school with big walls and the classroom as a boring activity. They complained about the room temperature, the video classroom and about the lack of school supplies. Another point observed was the fact that many students draw themselves in the reading room and write "teacher, I do not read". On the other side, many students draw if they built the school, they design open areas, game rooms and would do inclusive facilities implementation. The students were engaged with the draw activities
2- Building my future	Many students did not know how they see themselves in the future. Most of them do not understand the activity. The students draw their family and social problems as hungry, poverty, violence and death of relatives. Many personal problems faced by students were identified
3- Song challenge	The students enjoyed the activity and asked to play several times. They fully understood the rules and the tasks, because the support materials were designed. Was identify the need of easy level writing lyrics. The students debated about their social context and were feeling stimulated to study literacy

5 Discussion About Critical Literacy and Inclusive Education

Critical literacy focuses on promotes reflection, transformation and encourages readers to be active participants in the education process. [10] Rosenblatt (2002) noted that no reading experience is purely aesthetic or purely efferent, but rather that readers are always making choices about their thinking.

How could the students enjoy school where their voice and perspective are ignored by the system? Introducing critical literacy means focus on ensuring inclusive education and make properly tasks for the students' efficient education. [3] Jordão (2013) argue that the critical literacy is an educational perspective and philosophy action.

Thus, critical and inclusive practices recognize the value in appreciating difference and looking for marginalized people have equal access to education. It means reflect about students' social context and include this debate on the education's system.

6 Conclusion

Through participatory design and action research we could understand the real difficulties and problems faced students from Manguinhos' Shantytown. Moreover, a solid collaboration was developed with the participants (designer, schoolteacher and students) by gamification activities. Moreover, the importance of debating the students' social environmental context in classroom was identified for creating effective approaches since this directly influences the student's willingness.

Even though increasing financial investments, ignore students' experiences, social problems and yearnings makes the lack of efficient in education even bigger. Critical and inclusive practices are imbedded in sociopolitical and cultural ways of viewing the world. Thus, listening and stimulating the students' critical voice is the first step to promotes the search for fairer and more efficient education' system.

References

1. OECD, PISA 2015 Results (Volume I): Excellence and Equity in Education, OECD Publishing, Paris (2015f)
2. UNICEF HOME PAGE. https://www.unicef.org/brazil/pt/panorama_distorcao_zidadeserie_brasil.pdf. Accessed 10 Jan 2019
3. Jordão, C.M.: Abordagem comunicativa, pedagógica crítica e letramento crítico farinha do mesmo saco? Língua estrangeira e formação cidadã; por entre diversos dircursos e práticas. Pontes, Campinas (2013)
4. Kensing, F., Madsen, K.H.: Generating visions: future workshops and metaphorical design. Design at Work. ACD Digital Library (1992)
5. Preece, J., Rogers, Y., Sharp, H.: Interaction Design: Beyond Human-Computer Interaction. Wiley, New York (2002)
6. Stoot, A., Neustaedter, C.: School of Interecctive Arts and Technology, Simon Fraser Univerity, Canada (2013)
7. Werbach, K., Hunter, D.: For the Win. Wharton Digital Press, Philadelphia (2012)
8. Arroy, M.G.: Imagens Quebradas. Rio de Janeiro. Editora Vozes (2014)
9. Oborne, D.J.: Ergonomics at Work, 3rd edn., 442p. Wiley, London (1995)
10. Rosenblatt, L.: La literatura como exploración. Fondo de Cultura Económica, México (2002)

Educational Strategies for Coping with Problems of Stay up Late Based on the Psychological Characteristics of Contemporary Chinese College Students

Yueyi Lin(✉) and Wuzhong Zhou

School of Design, Shanghai Jiao Tong University,
800 Dongchuan Road, Minhang District, Shanghai 200240, China
linyueyi123@sjtu.edu.cn, indi_go@126.com

Abstract. The damage of staying up late to our body is enormous, which can lead to physical and psychological problems such as decreased immunity, fatigue, endocrine disorders [1]. The paper takes Shanghai Jiao Tong University as an example, and is based on the psychological characteristics of contemporary Chinese college students. This paper proposes an educational strategy to cope with the problem of college students staying up late. The innovation of this paper lies in the two-dimensional perspective of students and college education, in-depth analysis of the motivation behind students staying up late, establishing a user psychological model of interview cases, and proposing strategies to ameliorate students' staying up phenomenon from the psychological acceptance level and actual results of students.

Keywords: Stay up late · Psychological characteristics · College education · User mental mode

1 Introduction

1.1 Research Background

At present, most people are living a fast-paced life under extremely high pressure, while modern entertainment activities and facilities are becoming more and more abundant. The phenomenon of staying up late is very common, especially among college students. Everybody knows that staying up late is not good for our health, but we don't know much about the specific degree of harm [2]. Therefore, many people still stay up all night for work, study, entertainment, while knowing that being a night owl is harmful. This phenomenon has been widespread and caused more and more social attention.

Driven by the employment pressure, academic pressure, and self-development of college students, the pace of university life is accelerating, the task of learning is aggravated, and the popularity of electronic devices has made the group of night owls growing larger and larger [3]. College students usually need to invest a lot of time to complete their homework from professional courses, as well as a large amount of

W. Karwowski et al. (Eds.): AHFE 2019, AISC 963, pp. 301–309, 2020.
https://doi.org/10.1007/978-3-030-20135-7_30

student affairs and community work. Therefore, sitting up late becomes the normal state of life for college students.

1.2 Research Objectives and Significance

Staying up late, long time sitting, unbalanced diet, irregular life hours, and other bad habits are very common in nowadays people, especially in college students. These unhealthy lifestyles are becoming a risk factor affecting college students' physical and mental health. According to the psychological characteristics of contemporary college students, this paper starts from the reasons of staying up of different groups of people and proposes some advice for college students by analyzing their needs.

This article will take Shanghai Jiao Tong University as an example to propose an ameliorate strategy for college students staying up late from the perspective of educational work.

1.3 Research Methods and Technical Routes

This paper mainly uses the questionnaire survey method, gathering questionnaire data from students in Shanghai Jiao Tong University. The content of the questionnaire mainly focuses on the regularity of college students' life and work, the psychological incentives of staying up late behavior and the acceptance of various advice for staying up.

Secondly, select typical cases for interviews, establish their user mental models and propose suggestions for individual students. Analyze the typical case and look for the educational strategies of the contemporary Chinese college students' psychological characteristics to cope with staying up late problems.

2 Motivation and Data Analysis of College Group Staying up Late at Night

2.1 Survey Data Analysis

Staying up late have a huge impact on the human internal clock. The phenomenon of staying up of college students at Shanghai Jiao Tong University is very common. To investigate whether their staying up behavior will affect their study and life, and endanger their physical and mental health, an electronic questionnaire was distributed to Shanghai Jiao Tong University, and 435 valid questionnaires were collected. The data were analyzed regarding the current situation, causes and psychological hazards of college students staying up late. Among them, 412 respondents indicated that they generally fell asleep later than 23 o'clock, which can be defined as staying up, accounting for 94.71%. There are various reasons for staying up late. The following data will explain the differences in sleep quality and staying up behavior among people of different ages in colleges and universities.

The Relationship Between Staying up late Behaviors and Age Distribution of College Students. It can be seen from the Fig. 1 that the proportion of staying up late of different age groups in colleges and universities has a certain regularity. As the grades of undergraduates increase, the proportion of staying up late increases, especially when the freshman is in the second year of the sophomore year. The proportion of people staying up late has increased significantly, indicating that after one year of university life, the life and work situation has changed dramatically. The same is true for the master's degree. It is particularly important to note that students who fall asleep at three o'clock in the third year of the undergraduate class are much higher than those in other grades. Later, this phenomenon will be analyzed as a particular case.

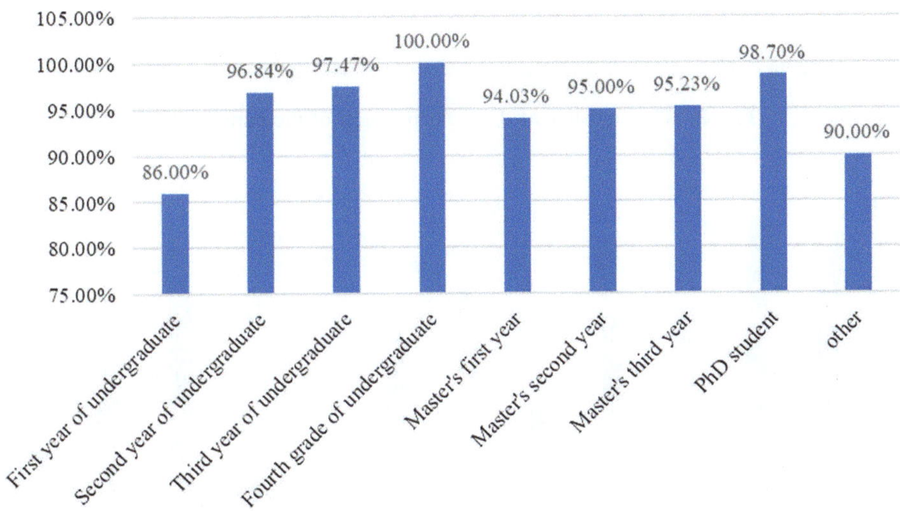

Fig. 1. The proportion of effective participants in the total number of participants.

The Relationship Between Staying up Late Behaviors and Sleep Duration of College Students. Considering that contemporary college students may have the habit of sleeping late and getting up late, the actual sleep duration of college students was investigated. As can be seen from Fig. 2, more than 65% of the students in the survey data only have 6–8 h of sleep, and only 12.86% of the population can have more than 8 h of adequate sleep. Such a phenomenon may cause many physical and mental problems in the group of college students.

Behavioral Needs Analysis. As can be seen from Fig. 3, more than half of the participants pointed out that the reason for staying up late was to learn and complete the necessary homework, and most of the crowds went to bed late because of social platforms such as Weibo. In particular, 175 college students have admitted that they grow a habit of staying up late, which will be analyzed later as a special case.

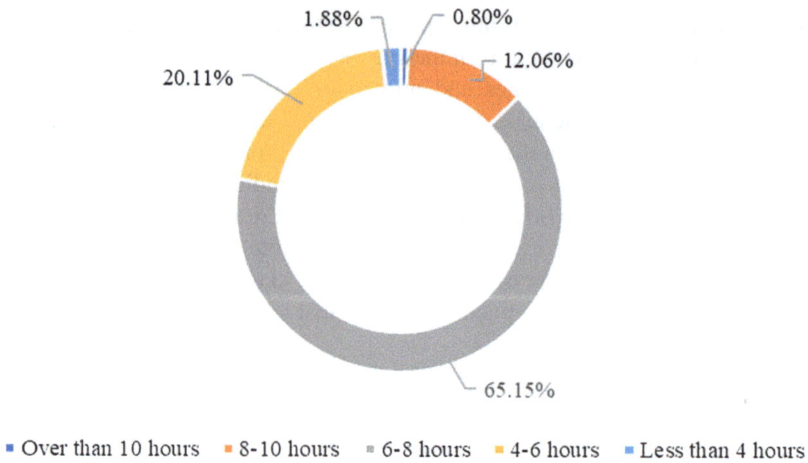

Fig. 2. The statistics of daily sleep length of the participants.

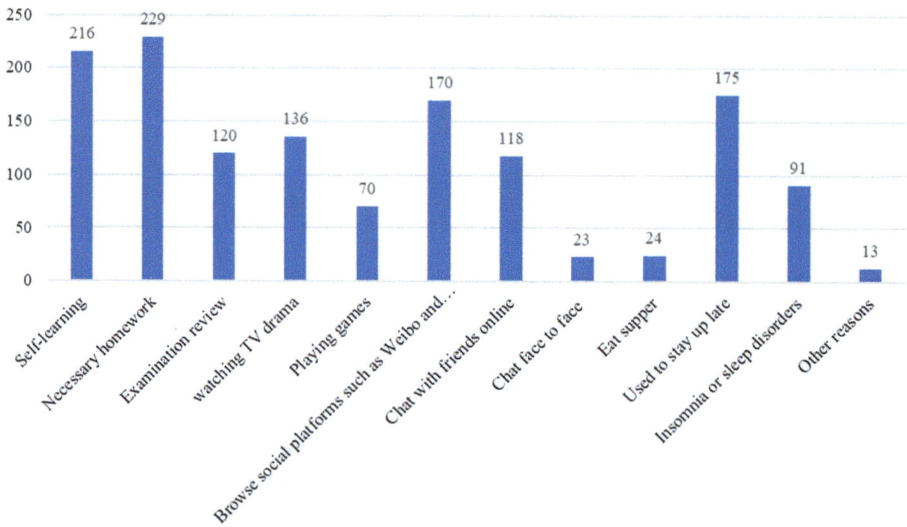

Fig. 3. Analysis of behavioral needs of college students staying up late

2.2 Summary

The scope of this questionnaire survey involves students from Shanghai Jiao Tong University. The main questions include basic information, staying up late time, length of sleep, reasons for staying up late and understanding of the hazards of staying up late. According to the survey, 77.21% of college students have a basic understanding of the dangers of staying up late, but there are still more than 90% of college students who choose to stay up late. The reason for staying up late is not as simple as the results of the survey.

3 Analysis of the Psychological Causes Behind Staying up Behavior of College Groups

3.1 Psychological Cause Analysis

The reasons behind the group's staying up late behavior are complex, the influence of students' staying up late personal factors, as well as the influence of social and university environmental factors. Only looking at the questionnaire survey cannot make us fully understand the specific factors behind the phenomenon, so the following analysis is conducted via the interview records obtained by the interview method.

Personal Factor Analysis
Procrastination caused staying up late. Procrastination refers to "knowing that it will have negative consequences if late, but one is still habitually or deliberately delaying the start or completion of a task." Many students mentioned in the interview that "I feel that my learning efficiency is not high enough. I know I have procrastination, but I just don't have the power to change it." "The procrastination" is almost a word frequently mentioned by the students in the interview. Everyone will jokingly say that they shouldn't be procrastinating, but they just can't get rid of this habit.

Procrastination has become an important reason that everyone generally believes that it causes staying up late. The homework or work that should have been done in the morning, but it has been delayed until the evening due to low working efficiency or other issues and will eventually lead to staying up late.

Inefficiency and Lack of Self-Control Ability. Some participants also mentioned that part of the reason for procrastination is that because of their weak self-control ability, they are always disturbed by such things and cannot concentrate on one thing. Many students who lack self-control ability generally work inefficiently. In the end, they can only rely on staying up late to compensate for working hours.

A Quiet Environment is Required to Complete Homework, Work, Etc. College students inevitably need to complete homework and self-enhanced learning. Focusing on learning requires a quiet environment. For example, a quiet library during the exam week is very popular.

3.2 Environmental Factor Analysis

New Media and Internet Development Lead to Colorful Night-Life. The continuous development of new media and the Internet has led many people to indulge in social platforms such as Weibo and WeChat. With the rapid development of modern science and technology, the use of advanced tools such as mobile phones and computers has led to changes in people's communication methods, and students' attention to learning has also shifted. Most students said that they have developed a late night habit because of playing mobile phones. In addition, most of the work needs to be done by computer. In the process of using the computer to collect the information, one will be attracted by other contents of the Internet, and it is easy to spend time on web pages that are not related to study and work. Participants have said that they will still play games, brush microblogs, and listen to music online in the middle of the night. In order to complete the day's learning tasks on time, they can only continue to complete the work by delaying the time to fall asleep.

Staying up Late Caused by Social Competition and Anxiety. College students are generally anxious because of their future development and the competitive pressure brought by society. The anxiety here does not mean that anxiety caused the symptom of insomnia. Instead, it was found in interviews that many participants felt that the pressure of social competition was extreme, and the students around them were working hard, so they throw themselves into tons of internships and student unions to get improved in ability, but at the same time, the pressures become higher, leading to staying up late.

3.3 Analysis of User's Psychological Model of Interview Cases

Name: Ms. yu
Age: 20
Education: Undergraduate
Interest: Reading, Travel, Photography

Introduction:
Strong personality, hard work, treat every friend around sincerely
She is the youngest co-publisher of international A-level conference papers in communication major of her university.
She is a powerful journalist in practice.
She is the president of the Student Union, a good debater...

Fig. 4. A typical case of passive staying up late crowd

Interview time: 2:50 in the night of November 23, 2018

Interview method: WeChat voice

Interview record: At present, I'm in the third year of an undergraduate degree. I think my academic performance is very good, and the academic pressure is not too much. As the president of the student union, I need to have a regular meeting every week. In terms of the festival, I have to prepare for the celebration of the college about one month in advance. I think work efficiency is very important. There are often unexpected tasks that cause me to stay up late. For example, the recent merit student model campaign requires special preparation for about two days. I sit up to 3 o'clock only finished preparing the speech. The main reason for staying up late is the unexpected work from the student union, or chatting with friends (Table 1).

The main characteristics of this group:

They don't have a strong desire to stay up late, but they have to stay up late because of unexpected events or inevitable tasks, such as the second day of the course pre, or the final exam. This group has fewer staying up late, but the irregular staying up time is very late (Fig. 4).

Table 1. Schedules of Ms. Yu in her third year as an undergraduate student.

Number of courses	Course time	Monday	Tuesday	Wednesday	Thursday	Friday	Saturday	Sunday
1	8:00-8:45	Complete homework and study by oneself	Sleep	Ambition or illusion:Europe as a global actor	Complete homework and study by oneself	Sleep	Sleep	Sleep
2	8:55-9:40							
3	10:00-10:45		French Theory			French Theory	Complete homework and study by oneself	Complete homework and study by oneself
4	10:55-11:40							
5	12:00-12:45	Lunch	Lunch	Lunch	Lunch	Lunch	Lunch	Lunch
6	12:55-13:40	Browse Weibo, Wechat					Complete homework and study by oneself	Complete homework and study by oneself
7	14:00-14:45	FEANCAIS CEBUTANT	Law and Society	FEANCAIS CEBUTANT	Ambition or illusion:Europe as a global actor	Sociology of Organisations		
8	14:55-15:40							
9	16:00-16:45							
10	16:55-17:40							
11	18:00-18:45	Dinner	Dinner	Dinner	Dinner	Dinner	Dinner	Dinner
12	18:55-19:40	Work of associations and student organizations	Complete homework and study by oneself	Work of associations and student organizations	Complete homework and study by oneself	Work of associations and student organizations	Rest time	Rest time
13	19:50-20:30							
Night	20:40-23:00	Social or TV dramas		Social or TV dramas		Social or TV dramas		
	After 23:00	Rest time		Rest time		Rest time		

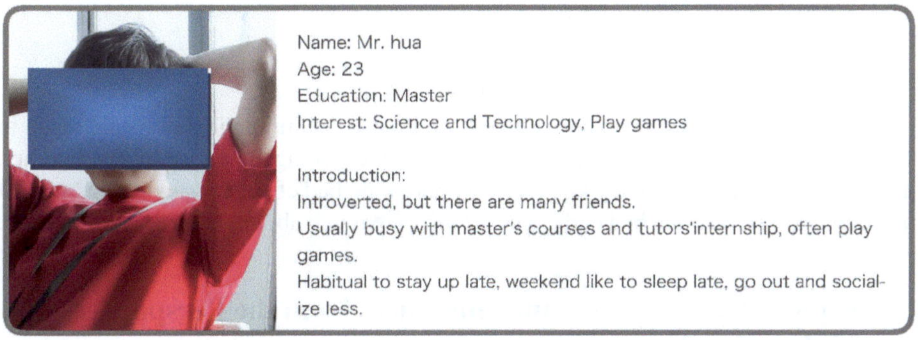

Fig. 5. A typical case of active staying up late crowd

Interview time: December 16, 2018, 3 pm

Interview method: interview

Interviews record: I am a first-year graduate student, just entering school in September. Currently, I need to adapt to the curriculum and life of the new university, there is always a lot of things to do every day. At present, I am mainly busy with the major courses and the innovative design center doing my own tasks. In addition to doing homework, it takes 1–2 h to play the game in order to relax or relieve stress. Though I am not very busy, I am always used to stay up late. In the evening, I will play games with several roommates in the dormitory. I will be willing to calm down and do my homework until midnight. The first reason for staying up late is that I need to have a quiet night environment to finish my homework and work when everyone will sleep and won't bother me. The second reason is that I am doing some work during the day, there is no free time to do what I like, and I want to stay up late to have leisure time (Table 2).

Table 2. Schedules of Mr. Hua in his first year as a graduate student.

Number of courses	Course time	Monday	Tuesday	Wednesday	Thursday	Friday	Saturday	Sunday
1	8:00-8:45	Jewelry Design	Regional revitalization design	Design Psychology	Sleep	Design Art Theory	Sleep	Sleep
2	8:55-9:40							
3	10:00-10:45				Academic English			
4	10:55-11:40							
5	12:00-12:45	Lunch	Lunch	Lunch	Lunch	Lunch	Lunch	Lunch
6	12:55-13:40	Practice in Innovation Design Center	Research and Design of Brand Culture	Practice in Innovation Design Center	Cognition and Information Design	Practice in Innovation Design Center	Complete homework and study by oneself	Complete homework and study by oneself
7	14:00-14:45							
8	14:55-15:40							
9	16:00-16:45							
10	16:55-17:40		Complete homework and study by oneself		Complete homework and study by oneself			
11	18:00-18:45							
12	18:55-19:40							
13	19:50-20:30							
Night	20:40-23:00	Social or game		Social or game		Social or game		
	After 23:00	Rest time		Rest time		Rest time	Rest time	Rest time

The main characteristics of this group:

They are willing to stay up late. There is no time-consuming job that would result in staying up late, but they will always do some entertaining like surfing the Internet, playing games and so on. They know clearly about the bad effects of staying up late, but they don't choose to go to bed early. They don't care much about bad life habits (Fig. 5).

4 Propose Reasonable Countermeasures from the Perspective of Education Management

4.1 Feasibility Analysis of Staying up Recommendations Based on Different Groups

The current viable staying up late solutions include: strengthening education and publicity, controlling school time to turn off the lights, creating a better learning environment, and making reasonable plans and understanding of health care. More than 60% believe that "making a reasonable plan to adjust the work schedule" is an acceptable staying up late method, and other improvement methods support no more than half. Some students also said that they know how to avoid staying up late but have no will to try it themselves.

According to the psychological characteristics of different groups of people, the previous chapter divides college students' staying up late motivation into active staying up late crowds and passive staying up late crowds.

According to the interview results of the passive staying up late crowd, they are recommended to make a reasonable plan for their actual situation. It is recommended that they take time out of their busy schedule to take a rest. Every day, they can arrange some spare time for the unexpected task. If there is no incident, these time can also be used to rest.

However, for people who are active late at night, they cannot be generalized. It is necessary to analyze the psychological incentives behind students who are habitually staying up late. For example, the retaliatory night owl in the interview case can create a better learning environment during the day according to his psychological characteristics, so that he can adjust his mindset and be willing to go to bed early, instead of forcing by controlling external conditions such as the time of turning off the lights.

4.2 Management Countermeasures from the Perspective of Instructors

If starting from the perspective of instructors, we should first understand the helplessness of the students staying up late, but we can't support students to stay up all night. Instead, we need to point out the dangers of staying up late. Then, we can start from the specific case of the students, help the students increase their self-management awareness, which can also guide the students to develop a reasonable learning plan, break down the task, and let the students create a quiet environment that conducive to learning. Then, we need to encourage students to establish a positive attitude, be able to calm down to study, not easily be disturbed by outside entertainment activities, and truly improve work efficiency.

If you really stay up all night, you should also provide students with timely remedial measures, such as appropriate supplements for sleep, vitamins, green tea and so on. Advise students to properly handle their emotions, learn to relax, and don't give themselves too much burden and pressure.

References

1. Wei, W.: A review of the status quo of health education for college students' lifestyle habits. Surv. Educ. **7**, 15 (2018). (in Chinese)
2. Ziqi, L., Miaoobei, X., Ru, W., Su, Lai., Wanting, W., Xueli, S.: Study on the influence of day and night on the physiological and mental health of college students—taking the new college of medicine, guangxi university of traditional chinese medicine as an example. Ind. Technol. Forum. **17**, 156 (2018). (in Chinese)
3. Yayun, Q.: Investigation on the current situation of university students' staying up late study in design colleges and countermeasures. Popular Lit. **22**, 228–229 (2018). (in Chinese)

Instructional Design and Learning Strategies

The Application of Clustering Analysis in the Assessment of Eye Movements During Flight Training Intervention

Nima Ahmadi(✉) and Matthew R. Romoser

Western New England University, Springfield, MA, USA
nima.ahmadi@wne.edu

Abstract. In this research, the application of K-means clustering analysis on pilots' gaze points was studied in order to develop a quick evaluation approach for gazed-based training interventions. Detailed analysis of eye movements' data takes a considerable amount of time which makes it difficult to use in some stages of training intervention. As a result, a more efficient indicator of trainees' performance is needed to allow trainers the ability to provide timely feedback on trainees' performance. In this research, the potential of using clustering analysis on pilots' gaze points was assessed in order to classify bad and good gaze behaviors. One expert pilot and six novice pilots participated in the experiment. Pilots were equally divided into experimental and control groups.

Keywords: Eye movements · Eye tracking technology · Training intervention · Clustering analysis · Flight simulator

1 Introduction

Gaze-based training has been used in various domains to improve novice performance [1]. For instance, in medical domains gaze behaviors have been used to train novice medical professionals by comparing their eye gaze patterns to those of experts [2–10]. The primary reason to look at eye movements is due to the fact that the way novices scan environment is often very different from experts. Novices, who have yet to develop schemas and recognize important patterns and cues, tend to more randomly scan at objects in the environment (bottom-up processing), whereas experts purposefully and more tactically scan the environment (top-down processing).

The main apparatus used in gazed-based training interventions is eye tracking technology – either fixed or portable. These devices collect eye movements data such as fixation points and fixation duration. The sampling rate of eye tracking devices typically varies from 60 to 500 Hz. The type of study being conducted often determines the most appropriate sampling frequency. For instance, eye movements are collected at the higher 500 Hz to study concussed patients gaze behaviors. This kind of fidelity is required to capture differences between concussed and non-concussed patients [11].

Assume that in a research study, the sampling frequency is 120 Hz. This means that 7,200 eye movements records will be collected within a one-min scenario. In human factors engineering, training intervention studies usually have more than one scenario,

W. Karwowski et al. (Eds.): AHFE 2019, AISC 963, pp. 313–322, 2020.
https://doi.org/10.1007/978-3-030-20135-7_31

and each scenario is performed by several participants. To draw an effective conclusion on the effectiveness of a training, the collected data should be assessed using various eye movement metrics. The primary step in data analysis is data wrangling. In this step, various data analytics skills are applied to remove unnecessary records and estimate missing data. Once the data is clean, then the gaze behaviors of trainees are evaluated. To better understand the ways trainees scanned different parts of research environment, visual areas are divided into sections. These sections are called Areas of Interest (AOIs) which are semantic or gridded [12]. Once these AOIs are defined, trainees' eye movements are mapped to them. The eye movement mapping process is a time-consuming task and with two primary steps (Fig. 1). First, gaze points are mapped automatically to AOIs using machine learning algorithms by means of eye tracking software. Second, any incorrectly mapped gaze points are manually mapped to AOIs by a researcher (Fig. 3). The automatic mapping process could take up to 30 min for a 10-min recording. Because software can make errors, double checking the mapped gaze points for accuracy is a must. To check and manually map the incorrectly mapped gaze points, a researcher could spend more than 2 h for a 3-min scenario.

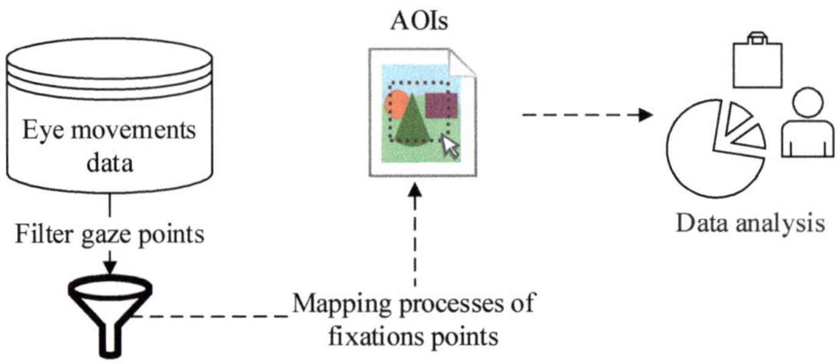

Fig. 1. Common Gazed-based eye movements' data analysis process

The software-based error of an automatic mapping process highly depends on the quality of calibration of the eye tracker to the participant. This has always been a challenge for researchers even with participants with normal or corrected to normal vision. If a researcher cannot calibrate an eye tracker, then the researcher cannot get quality data. In this case, there is a deviation between actual gaze point data and the mapped gaze point data of participants to AOIs (Figs. 2 and 3). When the research environment is rich with various sources of visual cues or AOIs are located close to each other, then a good calibration is a requirement to properly study gaze behaviors. A good example of this is when pilots' eye movement are studied.

This process of mapping gaze points is appropriate when thorough analysis is needed (Fig. 1). For instance, when a flight instructor wants to score a trainee's gaze behaviors to measure the pilots' situation awareness, a detailed analysis of eye movements should be conducted. In this case, various eye movements' metrics are measured and compared to an ideal threshold. But, sometimes a less accurate but faster method is required to have a big picture of pilots' gaze behaviors while piloting.

This also could be used as a warning system for pilots to alert them to follow recommended in-flight scanning pattern to help better keep control of the airplane. To do this, K-means clustering analysis was used to assess the effectiveness of an eye movement training intervention on the pilots' gaze behaviors. The number of clusters of gaze points of two groups of pilots were compared with each other. In this context, the research objective is to find a relationship between the number of fixations' clusters and the level of tactical scanning skill of pilots.

Fig. 2. Pilot's gaze point recorded by Tobii Pro Glasses 2

Fig. 3. Incorrect automatically mapped (left circle) & manually mapped gaze point (right circle)

2 Literature Review

Eye trackers have great potential for use as assessment and training tools [1]. They have been used in domains such as aviation, driving, medical and sports [13–18]. In medical domain, five successful training interventions were developed using eye tracking technology [8, 19–22]. In one study, a gaze-based training intervention was designed to train surgeons on experts' gaze behaviors. The outcomes of these studies showed novices could improve their scanning patterns by watching experts' gaze behavior. Gaze behaviors of trainees were analyzed using eye movement metrics such as, fixation frequency, fixation duration, and completion time. In sports, golf players, basketball players, and penalty shooters were trained to perform "quiet eye". This term, quiet eye, is defined as "a final fixation point with brief pause" [6, 9, 10]. This has shown to be effective in improving the performance of athletes.

In the domain of driving, several training interventions were developed for novice drivers to enhance novice drivers' Situation Awareness (SA). SA is defined as the operator's capability to predict the upcoming events and properly responding to them [23]. To achieve the highest level of SA, applying tactical scanning while having an active visual search is a must. In these training interventions the transfer of training, or the effectiveness of training, was measured using eye tracking technology. The eye tracker was used as an assessment tool to evaluate drivers' visual search, their abilities to apply tactical scanning, and to predict proceeding traffic situations. Two well-known driving interventions are the Risk Awareness and Perception Training (RAPT) and

Engaged Driver Training Systems (EDTS) systems [3, 18]. In both of these training interventions, driving skills were measured, in part, by watching participants' gaze points and their reaction time.

In driver training interventions, researchers deal with dynamic AOIs. Dynamic AOIs refer to moving stimuli in a dynamically changing scene – such as in an animation or movie. But in some cases, AOIs are static, which means that they do not move within the environment. A good example of this case is when pilots fly under Instrument Flight Rules (IFR). In IFR conditions, visibility is not good enough to get visual cues outside the cockpit. So, the pilots have to scan the static instrument panel to control the airplane.

Previous research in the aviation domain by application of eye tracking technology has focused on (1) understanding the way novices and experts scan the flight environment, (2) comprehending the change in pilots' gaze behaviors as a result of gaining experience, (3) developing stress management program for military pilots, and (4) the characteristics of experts and novices gaze behaviors [15, 24–28].

Generally, training interventions are designed to be used by instructors and trainees. In aviation domain, flight instructors use training interventions to train novice pilots. Providing feedback on trainees' performance and piloting assessment are an indispensable part of the learning process. In a gazed-based training intervention, a flight instructor could provide qualitative and quantitative feedback. Both of them are beneficial in improving the transfer of training. Sometimes fast quantitative feedback on a pilots' gaze behavior is needed. So, how could this be provided by a flight instructor? This study was conducted to assess the potential of clustering analysis on providing quantitative feedback on pilots' gaze behaviors.

3 Method

3.1 Participants

A total of seven pilots participated in the study: one flight instructor and six student pilots. Student pilots were recruited from nearby flight schools. As compensation for participating in the study, student pilots were paid $60. The mean total flight hours of student pilots was 20 h. On the other end of the spectrum, the total flight hours of the flight instructor was almost 5000 h. The flight instructor served more than 1600 h training student pilots as an instructor.

3.2 Experimental Procedure

Once Institutional Review Board (IRB) was approved, student pilots were invited to participate in the experiment using flyers that were posted in local flight schools. Student pilots were asked to send an email to or call the study administrator. The objective was to check the qualification of participants before the experiment. The study administrator visited the qualified participants at their flight school in order to make sure the eye tracker could be calibrated on the eyes of student pilots.

The experiment began with a 15-min practice flight on a flight simulator to (1) get used to the sensitivity of the flight simulator, and (2) learn the flight environment including flight deck and flight instruments. Once the practice flight was over, participants were divided into two groups: an experimental and control group. The experimental group trained on tactical scanning, while the control group received control treatment in the form of no formal training.

Flight scenarios were developed with the aid of expert pilots on a research committee. The research committee was composed of three experts in the field of aviation, flight training, and designing training interventions. The participants received one task, they performed one at the time in a random order. These tasks were used to define the effectiveness of a gazed-based training intervention for novice pilots.

4 Apparatus

4.1 Eye Tracker

Tobii Pro Glasses 2 was the eye tracker that was used in this experiment (Fig. 4). This device is a non-obstructive, light-weight, and head-mounted eye tracker. The used eye tracker's sampling rate was 60 Hz.

4.2 Flight Simulator

Three main parts of the flight simulator were (1) seat and frame, (2) flight control, (3) monitors (Fig. 5). Three monitors allowed the participants to have 120° field of view during the experiment. Participants used virtual and physical control to control the airplane. The main physical controls were yoke, rudder pedals, pitch trim, and throttle quadrant. Participants were able to watch the flight instrument and engine gauges on a monitor (middle) during experiment. Participants flew Cessna Skyhawk 172 in all parts of the experiment.

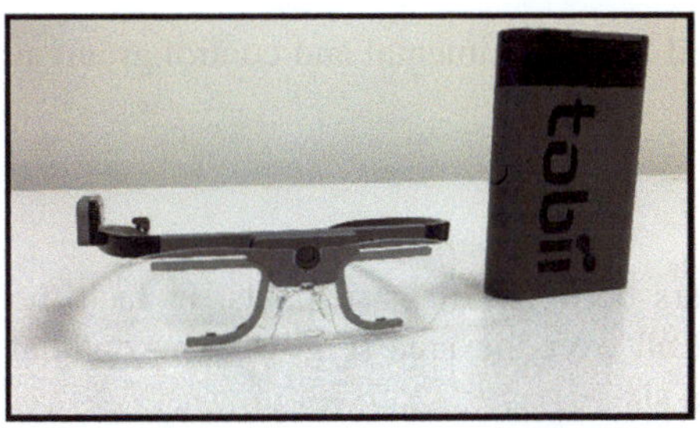

Fig. 4. Tobii Pro Glasses 2

Fig. 5. Flight simulator

5 K-Means Clustering Analysis

The goal of the study was to determine whether there is any difference between the numbers of gaze point clusters of those pilots who were in control of their airplane and those who lost the control of airplane. Analysis was conducted on those eye movements that were classified as fixation (not saccade). Saccades are quick eye movements that are made between scanned objects. Maintaining consistent visual gaze on an object (s) is called a fixation. The difference between fixating on an object and saccade is like standing on object and walking toward an object. In both cases, quick movements are made to get to an object.

The objective of the K-means analysis is to divide the observations (in this case, fixations) into clusters (groups). To implement this approach, there are main steps to take: (1) randomly choose centroid for each cluster from all points, (2) choose new centroids from the fixations to minimize the variations (distance), (3) iterate the process to find the optimal points of centroids. To find the optimum number of clusters, the within-cluster variation (distance) should be minimized. One method to find the optimal number of clusters (or clusters configurations) is the silhouette method. The outcome of this method is a number (for each number of clusters) which is called "average silhouette score"; it ranges from −1 to +1. The higher an average silhouette score, the better data points match with their own clusters. In other words, the highest average silhouette score of all clusters indicates the optimal numbers of clusters. In this experiment, the average silhouette score calculated for 2 to 6 clusters for each group of participants.

This present analysis focused on one scenario: the "Left Level Turn" scenario in IFR conditions. In this one-minute scenario, student pilots were asked to practice 180-degree standard turns while maintaining altitude and rate of turn. The trained pilots of the experimental group showed good piloting skills, while selected ones from control group lost the control of the airplane in the experiment.

6 Results

The outcome of clustering analysis is presented for: experimental and control group as well as a flight instructor.

6.1 Flight Instructor

The expert had the highest number of clusters between all participants. In fact, the fixation point of flight instructors distributed all over the area (Fig. 6). The optimal number of cluster for the flight instructor was six.

6.2 Experimental Group

These student pilots were selected from those trained pilots that showed good gaze behaviors during left level turn. The average silhouette scores were 4, 3, and 4 (Fig. 7). The mean numbers of clusters of this group is 3.6.

6.3 Control Group

The pilots of this group lost the control of their airplane while practicing left level turn. According to computed average silhouette scores, optimal number of clusters for participants of this groups are 2, 3 and 2 clusters (Fig. 8). The mean numbers of clusters of this group is 2.3. It seems that the pilots' fixation points were located pretty close to each other.

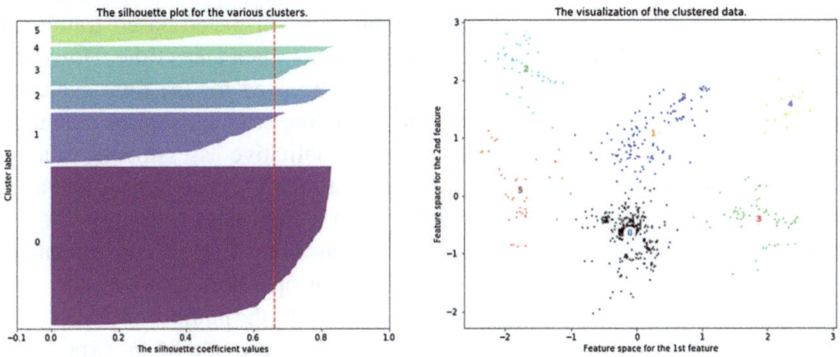

Fig. 6. Fixation clustering of the flight instructor

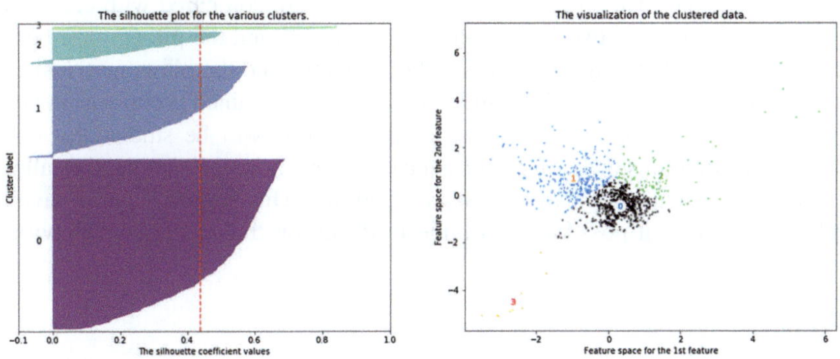

Fig. 7. Fixation clustering of the experimental group (did not lose control)

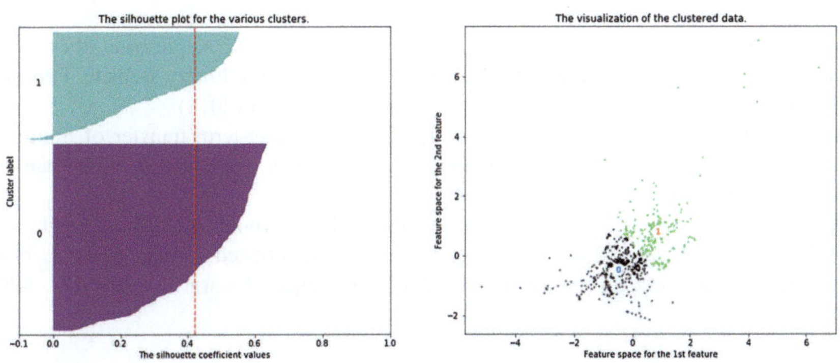

Fig. 8. Fixation clustering of a pilot of control group (lost control)

7 Discussion and Conclusion

Training intervention are designed to be used by educational institutes and instructors. The main motivations of using training interventions are reducing the time and the cost of training while improving the transfer of training. The backbone of new training intervention is providing active learning environment for trainees. In this process, trainees should receive feedback on their performance from the instructor.

Training interventions have several components, one is providing feedback. Different types of feedback are needed in different steps. In gazed-based training interventions, preparing detailed analysis of eye movements' data takes time. However, it is valuable. Gazed-based training interventions could be used for the final evaluation of pilots in order to develop a quantitative score to define whether a pilot is ready for a given maneuver or not. Used in conjunction with qualitative assessments, quantitative scores can be quite informative to instructors and students. Sometimes a less accurate, but faster metric is needed to know if a pilot is on a right path or not.

In this study, a new approach was presented for cases in which eye movements should be analyzed when (1) AOIs are statics and (2) when operators look straight ahead to perform a given task. Rather than mapping them onto AOIs, the number of clusters of pilots could be compared with the ideal numbers of clusters as provide by an expert operator baseline. A scoring method should be developed to evaluate pilots' gaze behaviors according to the comparison of the trainees' number of clusters with the experts.

In this study, the fixation points of seven pilots were compared with each other. It seems there is a pattern between the numbers of clusters and the pilots' gaze behavior. The expert pilots had the highest number of clusters (6) and the pilots who lost control of their airplanes had the least (2.3) while the mean of the trained group was in between (3.6). In future research, fixation points of more pilots will be studied for different scenarios and different levels of performance. Having a larger sample size allows to draw statistical conclusion based on research findings. This approach could save hours of analysis and make it possible to provide feedback on trainees' gaze behaviors.

References

1. Tien, T., Pucher, P.H., Sodergren, M.H., Sriskandarajah, K., Yang, G.-Z., Darzi, A.: Eye tracking for skills assessment and training: a systematic review. J. Surg. Res. **191**, 169–178 (2014)
2. Ahmadi, N., Matthew, R., Guarnieri, L., Kry, T., Porter-fyke, E.: Application of eye tracking technology in naturalistic usability assessment of an academic library website. Presented at the applied human factors and ergonomics, Orlando, Florida (2018)
3. Ahmadi, N., Katerahmani, A., Matthew, R.: Short and long-term transfer of training in a tablet-based teen driver hazard perception training program. Presented at the human factors and ergonomics society, Philadelphia Marriott (2018)
4. Jadad, A.R., Moore, R.A., Carroll, D., Jenkinson, C., Reynolds, D.J.M., Gavaghan, D.J., McQuay, H.J.: Assessing the quality of reports of randomized clinical trials: Is blinding necessary? Control. Clin. Trials **17**, 1–12 (1996). https://doi.org/10.1016/0197-2456(95)00134-4

5. Martin, J.A., Regehr, G., Reznick, R., Macrae, H., Murnaghan, J., Hutchison, C., Brown, M.: Objective structured assessment of technical skill (OSATS) for surgical residents. BJS **84**, 273–278 (1997). https://doi.org/10.1046/j.1365-2168.1997.02502.x
6. Moore, L.J., Vine, S.J., Cooke, A., Ring, C., Wilson, M.R.: Quiet eye training expedites motor learning and aids performance under heightened anxiety: the roles of response programming and external attention. Psychophysiology **49**, 1005–1015 (2012)
7. Sen, S., Kranzler, H.R., Didwania, A.K., Schwartz, A.C., Amarnath, S., Kolars, J.C., Dalack, G.W., Nichols, B., Guille, C.: Effects of the 2011 duty hour reforms on interns and their patients: a prospective longitudinal cohort study. JAMA Int. Med. **173**, 657–662 (2013). https://doi.org/10.1001/jamainternmed.2013.351
8. Sodergren, M.H., Orihuela-Espina, F., Froghi, F., Clark, J., Teare, J., Yang, G.Z., Darzi, A.: Value of orientation training in laparoscopic cholecystectomy. Br. J. Surg. **98**, 1437–1445 (2011)
9. Vine, S.J., Wilson, M.R.: The influence of quiet eye training and pressure on attention and visuo-motor control. Acta Psychol. (Amst.) **136**, 340–346 (2011)
10. Wood, G., Wilson, M.R.: Quiet-eye training for soccer penalty kicks. Cogn. Process. **12**, 257–266 (2011)
11. Samadani, U., Ritlop, R., Reyes, M., Nehrbass, E., Li, M., Lamm, E., Schneider, J., Shimunov, D., Sava, M., Kolecki, R., Burris, P., Altomare, L., Mehmood, T., Smith, T., Huang, J.H., McStay, C., Todd, S.R., Qian, M., Kondziolka, D., Wall, S., Huang, P.: Eye tracking detects disconjugate eye movements associated with structural traumatic brain injury and concussion. J. Neurotrauma **32**, 548–556 (2015). https://doi.org/10.1089/neu.2014.3687
12. Holmqvist, K., Nyström, M., Andersson, R., Dewhurst, R., Jarodzka, H., van de Weijer, J.: Eye Tracking: A Comprehensive Guide to Methods and Measures. OUP, Oxford (2011)
13. Comstock Jr., J.R., Coates, G.D., Kirby, R.H.: Eye-scan behavior in a flight simulation task as a function of level of training. In: Proceedings of the Human Factors Society Annual Meeting, pp. 391–395. SAGE Publications Sage CA, Los Angeles, CA (1985)
14. Haslbeck, A., Zhang, B.: I spy with my little eye: analysis of airline pilots' gaze patterns in a manual instrument flight scenario. Appl. Ergon. **63**, 62–71 (2017)
15. Kasarskis, P., Stehwien, J., Hickox, J., Aretz, A., Wickens, C.: Comparison of expert and novice scan behaviors during VFR flight. In: Proceedings of the 11th International Symposium on Aviation Psychology, pp. 1–6 (2001)
16. Katoh, Z.: Saccade amplitude as a discriminator of flight types. Aviat. Space Environ. Med. **68**, 205–208 (1997)
17. Katrahmani, A., Ahmadi, N., Romoser, M.: Using situation awareness as a measure of driver hazard perception ability. In: Driving Assessment Conference, pp. 256–262 (2017)
18. Younger Drivers: Risk Awareness and Perception Training, HPL, UMass-Amherst (RAPT, version 3.0). http://www.ecs.umass.edu/hpl/RAPT3.htm
19. Chetwood, A.S., Kwok, K.-W., Sun, L.-W., Mylonas, G.P., Clark, J., Darzi, A., Yang, G.-Z.: Collaborative eye tracking: a potential training tool in laparoscopic surgery. Surg. Endosc. **26**, 2003–2009 (2012)
20. Sodergren, M.H., Orihuela-Espina, F., Mountney, P., Clark, J., Teare, J., Darzi, A., Yang, G.-Z.: Orientation strategies in natural orifice translumenal endoscopic surgery. Ann. Surg. **254**, 257–266 (2011)
21. Vine, S.J., Masters, R.S., McGrath, J.S., Bright, E., Wilson, M.R.: Cheating experience: guiding novices to adopt the gaze strategies of experts expedites the learning of technical laparoscopic skills. Surgery **152**, 32–40 (2012)

22. Wilson, M.R., Vine, S.J., Bright, E., Masters, R.S., Defriend, D., McGrath, J.S.: Gaze training enhances laparoscopic technical skill acquisition and multi-tasking performance: a randomized, controlled study. Surg. Endosc. **25**, 3731–3739 (2011)
23. Endsley, M.R.: Measurement of situation awareness in dynamic systems. Hum. Factors **37**, 65–84 (1995)
24. Li, W.-C., Chiu, F.-C., Wu, K.-J.: The evaluation of pilots performance and mental workload by eye movement (2012)
25. Li, W.-C., Braithwaite, G.: The investigation of pilots' eye scan patterns on the flight deck during an air-to-surface task. In: International Conference on Engineering Psychology and Cognitive Ergonomics, pp. 325–334. Springer (2014)
26. Liu, W., Liu, C.K., Zhuang, D.M., Liu, Z.Q., Yuan, X.G.: Comparison of expert and novice eye movement behaviors during landing flight. In: Advanced Materials Research, pp. 2556–2560. Trans Tech Publ (2012)
27. Tole, J.R., Stephens, A.T., Vivaudou, M., Ephrath, A.R., Young, L.R.: Visual scanning behavior and pilot workload (1983)
28. Yang, C., Liu, Z., Zhou, Q., Xie, F., Zhou, S.: Analysis on eye movement indexes based on simulated flight task. In: International Conference on Engineering Psychology and Cognitive Ergonomics, pp. 419–427. Springer (2014)

Design of *Design*: Learning Dynamics in Design Degree

Maria João Delgado[1(✉)], Isabel Duarte de Almeida[1,2,3], Gianni Montagna[1], and J. Vilas-Boas da Silva[3]

[1] Faculty of Architecture, Center for Research in Architecture, Urbanism and Design (CIAUD), Lisbon University, Rua Sá Nogueira, 1349-063 Lisbon, Portugal
mjoaodelgado@fa.ulisboa.pt, isabel.dalmeida@edu.ulusiada.pt
[2] Lisbon Lusíada University (COMEGI), Rua da Junqueira, 188-198, 1349-001 Lisbon, Portugal
[3] IBS- Business Research Unit (BRU-IUL), Instituto Universitário de Lisboa (ISCTE-IUL), Av. das Forças Armadas, 1649-026 Lisbon, Portugal

Abstract. Though different College's curriculums in Design point to different objectives, there may be low awareness about different stakeholder's real needs in issues like design relevance, planning, production, marketing, selling and recycling of designed products. The curricular programs include the approximation of the Design students (DS) to the job market with the purpose of professional success. However, the labor-market-approach focus must also include the challenge of entrepreneurship, based on the transformation of DS projects into competitive products. Having in mind the relevance of the analysis of DS entrepreneurship predisposition, and the need to stimulate this target to transform the DS skills in order to develop business, this paper intends to: (1) profile the DS regarding their entrepreneurial competencies and capabilities, and (2) to know how to tailor Design curriculum and proposing tools (Canvas) in order to develop/align new DS skills for development/implementation of business projects.

Keywords: Design Students (DS) · Design and business · Entrepreneurship for designers · Design curriculum fitness

1 Introduction

Entrepreneurship is a complex phenomenon depending on several factors, such as: environmental, demographic, social, economic and cultural. Entrepreneurship is a way of thinking, which highlights chances over pressures, and the opportunity identification process is undoubtedly a deliberate procedure, which offers a resource to better enlighten, and foresee, entrepreneurship. In turn, the European Commission [1] defends that "Entrepreneurship refers to an individual's ability to turn ideas into action. It includes creativity, innovation and risk taking, as well as the ability to plan and manage projects in order to achieve objectives". Since the early nineties, it was possible to assist to a

W. Karwowski et al. (Eds.): AHFE 2019, AISC 963, pp. 323–334, 2020.
https://doi.org/10.1007/978-3-030-20135-7_32

growing effort on research using entrepreneurial driving models as a framework, thereby settling the utilization of the concept in this field. In literature, entrepreneurship intention is addressed in a range of issues trying to know the factors that influence entrepreneurship intention such as demographic ones (education, age, gender), personality qualities (extraversion, scrupulousness) or contextual influences (like perceived barriers and support) [2].

To Thompson [3], intent is the self-acknowledged conviction that someone has considering the establishment of new business venture and deliberately plan to do it forthcoming. Lans et al. [4], it is possible to distinguish two dimensions of entrepreneurial intentions: the intention to be an entrepreneur or corporate entrepreneur (the entrepreneurial intention) and the intention to continue operating an inherited or acquired firm (the alternative entrepreneurial intention). At the same time, the universities and other education institutions started to get aware of the relevance of teaching and learning entrepreneurship as it became evident that the Entrepreneurship Education (EE) could influence the predisposition to start-up, or at least could influence individuals to become more proactive and entrepreneurial in their lives and their jobs. This concern is transversal to all students and graduates' levels.

There is a considerable number of Design students (DS) in EU universities for whom it will be difficult to start thinking on creating their own job [5]. This target group must learn "entrepreneurship experiential elements", so a tailored entrepreneurship training program should cover the whole life-cycle of business and should be directly linked with the Design curriculum of Design M.Sc. or PhD students to whom the program is directed [6]. Given the importance of entrepreneurship to overcome the issue of joblessness that these students face, more academic research is required in order to provide useful insights developing groundbreaking entrepreneurship programs for Design students/graduates given them transversal abilities and providing them with a new professional future.

Teaching future Designers to behave as managers has become a significant and frequently essential issue in Design education [7]. In part because multiples sectors of Business are placing a larger emphasis on recruiting creative but business skilled Designers because of the impact of innovation and creativity coming out from successful business initiatives on the economic and societal dimensions [6]. If DS understand entrepreneurship context, then the designs they make can be contextual, significant, and address factual enterprises endeavors. Therefore, while training DS to develop Business knowledge and entrepreneurship skills, University pave the way to a significant element of Design education in Design degrees. Furthermore, to DS it is critical the understanding of this knowledge and is also important to recognize the significative impact in the industry, in their future profession, or in the design of new products.

Having in mind the relevance of analysis of entrepreneurship predisposition of DS and graduates, and the need to stimulate this target to transform their own skills allowing them to develop their business, this paper intends to profile these students regarding their entrepreneurial competences and capacities. Accordingly, it is also expected to know how to tailor Design curriculum in order to create and align new DS skills for development/implementation of business projects. Thus, the data was collected during the diagnosis phase of the Project "*The Value of Transdisciplinary*

Projects: A Sustainable way in Academic Research" (VTP.SAR). It is an exploratory study followed by a qualitative research approach which aims to build a sequential statistical procedure of factor, cluster and discriminant analysis allowed finding and profiling the groups of DS and graduates concerning their awareness about the requirement to identify, plan and manage all the needs, from the effective and efficient production point of view, in order to gain productive capacity and competitiveness in the market. With this approach it is intended to recognize the gaps in their formation concerning knowledge, skills and abilities towards entrepreneurship.

2 Literature Review

2.1 Entrepreneurship Education

In literature entrepreneurship education is a structured and formal transfer of knowledge about entrepreneurial capabilities, namely, the notions and mental consciousness that entrepreneurs use in the process of starting and developing their growth-oriented ventures [8]. This definition proposes that some outputs for entrepreneurship education (EE) that can be measured lead to: (i) changes in entrepreneurial values; (ii) changes in people's orientation considering entrepreneurial careers, and, (iii) changes in personal assessment of entrepreneurial knowledge and skills. Considering what has been said before, entrepreneurship education is critical in actual economy since it contributes to entrepreneurial knowledge and skills which, in turn, pave the way to job creation, and this will help meaningfully to decrease poverty [9]. Entrepreneurship education leads to knowledge enhancement, explaining how to design, launch and manage a new business project. Furthermore, empowering DS to gain necessarily practice in a real business world framework allows favorable attitudes towards entrepreneurial activities [6]. In turn, develops awareness about self-efficacy, increases the entrepreneurial skills [10], and inspires DS to choose an entrepreneurial career [6]. In several universities and other Higher Education Institutions in European Union (EU), Entrepreneurship Education (EE) is growing speedily given its potential to develop entrepreneurial skills and mindsets that benefit the society by stimulating creativity, innovation and self-employment [11]. This education and training represent an unprecedented opportunity in motivating future entrepreneurs since it develops skills for new businesses creation. If DS have undergone a proper entrepreneurship education, they had the opportunity to acquire new knowledge and entrepreneurial skills required to initiate a new business. By these reasons some authors sustain that early formal EE influences students' attitudes inducing them in the direction of a future entrepreneurship and business career. Furthermore, Collins and Moore [12] propose that an entrepreneurial role might be experimentally and culturally acquired, and therefore influenced by education and training. Likewise, Gibb [13] states that the entrepreneurial predisposition is consistently being influenced by education and training. Cooney [14] states that entrepreneurship is a skill that can be learned, i.e., some basic skills in entrepreneurship can be developed through training, and accordingly Drucker [15] entrepreneurship is not "a magical answer" and it is not related with any genetic factor. It is a subject, and like any subject, it can be learnt and encouraged. So, to accomplish different aims of EE training programs, different skills, abilities and knowledge may be required in DS formation.

2.2 Competences, Skills and Other Traits Associates to Entrepreneurs

According to Mitchelmore and Rowley [16] competencies are not simply "job tasks", but rather the "pulp" that enables individuals to accomplish a task. The competences can be defined in terms of the crucial individual skills, knowledge, qualities, and motives of the employee that leads him/her to a better managerial performance. There are many competencies related with entrepreneurship, and most of them can be acquired and developed through learning. Thus, Bird [17] suggests that entrepreneurial competencies can be defined as the fundamental characteristics (e.g. specific knowledge, purposes, self-images, behavior, social skills and roles) that will result in a project creation, application and new business growth. In turn, to Man et al. [18] entrepreneurial competences are the entrepreneur's ability to accomplish with success a job role. In their study, and making an analysis from a behavioral standpoint, the authors distinguish the following areas of entrepreneurial skills: opportunity, relationship, analytical, innovative, operational, human, strategic, commitment, learning and personal strength competencies. Additionally, Robles and Zárraga-Rodríguez [19] evidence various competencies that are considered relevant to achieving entrepreneurship effectiveness: risk assumption, autonomy/self-determination, search and information analysis, work quality, communication, self-confidence, establishment of social networks/generation, dynamism, change management, initiative, innovation, integrity, leadership, self-control, results orientation, entrepreneurship competencies, social mobility, negotiation, troubleshooting, and responsibility and teamwork. Thus, the most important competencies for entrepreneurship are linked to the perception and explanation of a worthwhile niche of market, to the development of new products and services, to product innovation market and generation of ideas, environmental scanning, recognizing and envisioning. Those competences are also related with opportunities and with the strategies' formulation to seize the present business opportunities.

A skill is thought as a behavior resultant from the knowledge and expertise of someone. Making a bridge between this description and the behavior that is needed in entrepreneurship, entrepreneurial skills imply the aptitude to design and establish a business. Thus, an entrepreneur should develop fitted skills that could be resumed in three classes: Business Management skills (such as planning, decision-making marketing and accounting), Technical skills (which include communication and organizing competences) and Personal Entrepreneurial skills (such as innovation, risk taking, and determination). According to Chell [20] skills are multidimensional concepts, namely: the cognitive (knowledge and what is learnt), the affective (emotions expression and experiences feelings), the behavior (deliberate, premeditated and personal actions), and the context (job and tasks levels, occupational, sectoral). Correspondingly, skill must be differentiated from capability and competency which seems to address to a combination of skills, abilities, knowledge, and occasionally other characteristics as well. Skills encompass an affective (the emotion of carrying out the task), cognitive (know-how), and a behavioral (the strategic, tactical or personal action) element [11].

2.3 Designing Business and the Business Model Canvas

Design gained a growing consideration in innovation management studies, however, there is a lack of understanding on how designers could start or develop their own businesses. Only recently, designers entered the founding teams of successful start-ups, but there is still a poor understanding about the specific features of design-related form of entrepreneurship. Thus, to inspire designers to be more entrepreneurial, and to push them for further new product development, starting from the idea and ending in the market, it will be necessary to give them crucial and extra awareness about the whole process of product development.

If we consider a newly formed designer (NFD), which is at the beginning of the procedures to design a business model and develop a startup, for instance, and turn into an Entrepreneur, a few issues and conditions that are still unknown, except in theory, will need to be met first. Business and management know-how are important skills for a NFD. Without them, it will be extraordinarily hard to make any kind of approach in the business world since it cannot understand and utilize the business and management tools that will allow him to create a value proposition that might be used and successfully scaled up to a gainful business. Accordingly, to Almeida and colleagues [6] designing a business model could become a real challenge for a NFD.

Understanding the importance of business models is a cornerstone to an NDF who wants to become an entrepreneur. In a world where the speed of change is accelerated, where external aspects that affect any business model or company (macroeconomics, competitors, new stakeholders, market oscillations, consumer demand) cannot be controlled, it is imperative to have the ability to understand/adapt/change/innovate in the business model to survive and grow.

This difficult task is easier when it is chosen a tool that facilitates this process of conception/creation/mapping/rethinking/reinventing the model.

The "Business Model Canvas" (BMC) [21] is a strategic map that includes the top topics that make up a business. The single biggest impact the BMC has is helping management teams and strategists to create a more future-oriented strategy and it fits perfectly in the case of NFD, that is intended to become entrepreneurial and to embark on the world of business. BMC is a visual compact presentation of nine building blocks that the authors consider underpin the description of a business model of an organization. Canvas is a summary of the key points of a business plan, but it does not exclude a business plan itself, it is a less formal tool that can be used more frequently on a day to day basis. This model encourages discussion, creativity and analysis (Table 1). This model is the most appropriate for NFD that wish to initiate management project design and enter in the world of business. According to some authors [22] Canvas has great synergy with the concepts of the Lean Startup methodology,[1] because as Canvas is something practical to do, it is possible to bring together the young entrepreneur and his team, to discuss the main components of the business model, to test the hypotheses and, gradually, to evolve the concept of the business, producing new versions of Canvas.

[1] Methodology that allows shorter product development cycles and articulates with design businesses. Both, design process and Lean Startup methodology, are procedures that indorse innovation.

For a NFD who wants to be an entrepreneur, to develop a professional activity with an innovative project and own a future business in the area of Design, he/she must have specific knowledge and skills. A Designer entrepreneur will need to know how to implement a business strategy, to be aware of the business building blocks, and as to well understand how these components will relate with resources and with the business environment. Thus, the BMC, in an early-stage work, provides a how-to visual tool for designers to think about and plan their startup or their business with industry.

Table 1. Building Blocks of "Business Model Canvas" (BMC) [21].

Key partners	Key activities	Value propositions	Customer relationships	Customer Segments
Who are out key partners? Who are out key suppliers? Which key resources are we acquiring from our partners? Which key activities do partners perform?	What key activities do our value propositions require? Our distribution channels? Customer relationships? Revenue streams?	What value do we deliver to the customer? Which one of our customers' problems are we heling to solve? What bundles of products and services are we offering to each segment? Which customer needs are we satisfying? What is the minimum viable product?	How do we get, keep and grow customers? Which customer relationships have we established? How are they integrated with the rest of our business model? How costly are they?	For whom are we creating value? Who are our most important customers? What are the customer archetypes?
	Key resources		**Channels**	
	What key resources do our value Propositions require? Our distribution channels? Customer relationships? Revenue streams?		Through which channels do our Customer segments want to be reached? How do other companies reach them now? Which ones work best? Which ones are most cost efficient? How are we integrating them with customer routines?	

Cost structure	Revenue steams
What are the most important costs inherent to our business model? Which key resources are most expensive? Which activities are most expensive?	For what value are our customers really willing to pay? For what do they currently pay? What is the revenue model? What are the pricing tactics?

The main elements of the business model already indicated in Table 1 can be represented in a scheme that works also as a visual tool, as shown below in Fig. 1.

The enterprising designer must consider each element of this Model, analyze and discuss it. For example, in the study of customer segments, design entrepreneurs need to define which customer groups to meet and which to ignore. Also, the value proposition must either solve the problem of a customer segment or satisfy a need. Different sources of revenue, such as the sale of goods, usage fee, monthly subscription, rent, advertising, brokerage, likewise, should be considered. The same as the main

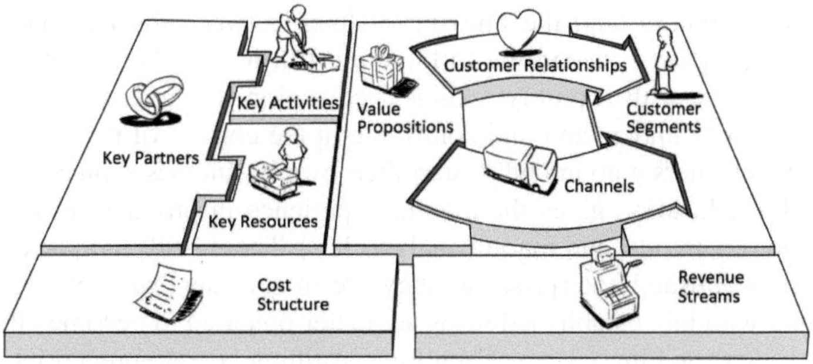

Fig. 1. "Business Model Canvas" in sketch (source [21])

resources for a business to operate, namely physical (factories, buildings, vehicles, stores, stocks…), intellectual (brands, patents, database, expertise), human (especially important for industries that depend on innovation and specialized knowledge), and financial (capital needed to start a new project).

2.4 Entrepreneurship in Future Designers: Learning Dynamics in Design Degree

Skills in Design and creativity are increasingly preferential targets in the business world. From big companies to small corporations, design and Designers are becoming more involved in strategic corporation decisions and they bring a lot of fresh air into the company structures either through creativity, design of projects and new products, or in discussions that help the company thrive. To consider Design as an economic driver is nowadays very clear to companies in different business sector such those presented in Table 2.

Table 2. Design contribution at tactical and operational level

	Goals	Helping	Result	Example
Use design at the **tactical** level	• Achieve goals of certain business activities	• Help plan and design for business activities	• New product lines, new business plans	• Dell • Calvin Klein • IKEA • Nike
Use design at the **operational** level	• Achieve goals of individual products/ services	• Help plan and design for new business offerings	• New products, services	• Levi Strauss • Miele • Corticeira Amorim • Apple

Only the best designers can stand above the crowd in a world where there is a rapid rise in consumer expectations driven by the likes in social networks, a world in which almost everyone can have immediate access to news and information, becoming creators and contributors, and the blurring of lines between the fashionable, usable, trending, stylish products. According to literature [6], it is no longer enough to come out of the university with a purely technical education. Designers need to be entrepreneurial in order to understand and contribute in the context of market and business pressures. For Designers who intend to start their own businesses soon after graduation, entrepreneurship education gives them solid experience in product design and development, technology trends, and market analysis [17]. These skills are just as significant for success in established enterprises as they are in start-ups; students with entrepreneurial training who join established firms are better prepared to become effective team members, designers and innovators. Therefore, instilling an entrepreneurial mind set in design students is essential if we want them to innovate and produce those innovations to better people's lives. But to achieve this new mindset, it is required to restructure the Curriculum in Design formation and install new learning dynamics in Design Degree.

As we previously conclude in previous research [6] "In design, the future professionals should be capable of merging the potential of innovation, competitiveness and commerce to leverage the value of both the design objects and related services either inside or outside the closest scope of Design. This is a gap to overcome in the design curricula of higher education courses by carefully studying the essential requirements".

In the present work, we will be concerned with the dimensions to ponder in the integration of new dynamics in Design curriculum considering the integration of NFD in the context that characterizes today's business environment in a diverse, global and interdependent world. In this article, we are intending to look and analyze the value of improving entrepreneurship and business knowledge management in both, Design degrees and post graduations.

3 Research Methodology and Data Collection

The data collected to this study came from the initial diagnosis phase of the Project "*The Value of Transdisciplinar Projects: A Sustainable way in Academic Research*" (VTP.SAR) and it is the second part of the data obtained in a survey applied to DS in the year of 2018. The first part of the data was studied, and the results obtained were already published [6]. This exploratory study follows a qualitative research approach.

A questionnaire was applied online to a purposive sample of 60 participants (Students and Graduated in Design areas) and the undertaken questions address issues such as sample sociodemographic characteristics, the DS Academic graduation, the projects development and implementation, startups, entrepreneurship, business and design. We seek out with this questionnaire to find the skills acquired during training and graduation that enables DS to accomplish tasks in management project design, in materials and technologies of design, and in marketing and project stewardship and communication. In other hand, the questionnaire also addressed the awareness of the DS (previously enrolled in any Design project work) in relation to the subjects that are not considered in the College curriculum and, accordingly DS, could be relevant for the development of complementary skills and competences, crucial for a better understanding of the Business world.

The questionnaire was composed of 17 questions, 13 of them being closed. The former questions were given in a Likert type scale with five levels of response, and dichotomous questions. In the first case, DS should classify the presented suggestions with (1) being not at all relevant to (4) being very relevant, with intermediary classification (2 and 3). The option of *I do not know/I cannot answer* was also considered.

4 Findings

4.1 Profile of Respondents

The sample (60 respondents) was made up predominantly of women (70%) and no answers were considered invalid. Considering the age, 44% of respondents were between 18 and 22 years old, 30% fit in the range of 23 and 32 years, and 26% were older than 33 years. Considering the occupational status, 57% were students, 27% were professional workers, 16% were working students. About the academic qualifications, DS constitutes most of the sample (46%), 34% has already finished the graduation in Design, 30% were MSc students, and 20% had a PhD in the area of Design.

4.2 DS Awareness of Design Graduation Fitness to the Business World

The data obtained with the survey were quantitative, but the assessment was made under a quantitative approach since we did not consider statistical significance. So, all the registered average values were interpreted under a qualitative paradigm.

Respecting the acquisition of appropriate design skills during the Academic formation, a large majority of DS has a very positive perception (70%). Most respondents (93,3%) consider that they have "very good" and "good" skills in the area of project development, 86% of which consider this to be the most important academic area, devaluing other areas such as materials, technologies design or even the area of communication. But in what concerns the skills that DS considered as crucial to accomplish a project of design development, bearing in mind the business and management areas associated, the findings pointed out that 33% of respondents refers to have fair knowledge and skills, almost half (16%) considers having excellent skills, while 10% considers do not having those skills. *Circa 7%* did not answer the questions. If the questions pointed to the competencies that have been given during the DS studies and considered as needed for the creation of a business plan or give them entrepreneurial competencies, the results reveal a notorious perception of a deficiency in proper qualifications. In view of that, 30% of DS remark to have a weak knowledge in this issue, 30% believe having a fair knowledge, and only 34% refer having a right acquisition of the beforehand referred skills.

Regarding the need to adapt the project to consumer demands, 73.3% consider it very important and 26.7% important, which reveals that all the respondents reflect the need to know about tools and strategies that allow the entry and permanence of their products on the market. The same percentages are obtained in the answers to the questions about the need to know the strategies and tools leading to business success.

Finally, it was interesting to know if the respondents had already accomplished some work experience in the field of design project development and only 30% stated that they had integrated a work team. Of this group, 78% participated in the project as a designer, from research, concept, and ideation, accompanying its development. However, only 11% followed the development of the project in the implementation phase of the business and 11% did not respond. Regarding these projects, 45% were developed in the area of clothing/textiles, 22% in the product area, 17% in the area of design thinking, 5% in the advertising area and 11% did not respond. These results allowed the recognition of gaps, weaknesses and strengths in the existing DS Design curriculum. This study also enabled us to identify specific entrepreneurial training needs in order to align new DS skills for development/implementation of business projects.

5 Discussion and Final Remarks

The importance of studying the new paradigm of including business knowledge and entrepreneurial learnership in Design curriculum at Design graduation [23], is due not only to transformations in the industry, but also to the emergence of a new generation of designers, who wish to take advantage of new business experiences and professional opportunities and turn themselves entrepreneurs. In our literature review we notice that several authors [4, 6, 7] and reports [1, 11] on entrepreneurship stress not only the importance and need to start doing business as early as possible in areas from Engineering to Technology, from Business to Architecture and Design, but also to quickly learn what entrepreneurship really means.

Considering our case study, it is possible to observe several failures in teaching, in terms of contents and in terms of entrepreneurial thinking and attitudes in Design graduations. This reality contradicts the evolution that has occurred in other countries, where there are several subjects related to entrepreneurship that integrate the training curricula [12, 14, 21, 24].

Our research assumes the need of a multidisciplinary logic in Designers formation and seeks to bridge the lack of knowledge and scientific research on NFD, as they try to find answers to the difficulties that face when dealing with projects design or management situations. Difficulties imposed, for instance, by the textile and clothing industry, and the need to create innovative and value-added products capable of differentiating themselves from luxury products on the one hand and low-quality products produced in third world countries on the other are difficult to overtake [7, 23, 25].

In our work, we try to identify some of the most relevant issues in the context of the implementation of entrepreneurship and business notions in design and contribute to the discussion of fundamental aspects of how this may be an asset to Design formation. The surveys, directed to finalists of the courses of Design, about enter in the professional world, and to former students of the area who are already in the labor market, allow us to understand the expectations of the future design professionals, their perspectives and the extent to which they understand entrepreneurship and project design as a means of breakthrough in an increasingly competitive market.

We need to further explore the relationship between Design formation (either in the 1st Cycle or MSc), entrepreneurship & projects design. The articulation between these three fields could be a solution for the future of Designers in a constantly changing industry. Thus, in order to better understand this relationship, more research must be made. With the collection of information from DS, NFD and professionals of the area, this work is also a repository with opinions about the perception of the needs and skills that should be learnt in Design courses that will allow the NFD to be entrepreneurs. In this study we find out an absence of support from entities that can help in the creation of entrepreneurial projects and design-related ventures, in addition to a clear lack of entrepreneurship reflections and business thoughts in design courses, reflected by several problems that NFD face while trying to take the business path and become entrepreneurs.

In this sense, and to conclude, it seems to be relevant the development of an entrepreneurial platform, which should be understood as a model, and whose objective would be (i) to fill the lack of thinking and entrepreneurial concepts in the courses of Design, (ii) give helpful tools, such as the Canvas Model, for analysis and creation of new business models, and (iii) to contribute to the creation of value in the design industry. All situations evidenced in the responses of DS, NFD and young designers' professionals. It is understood that there is a real need to create a solution that can help these young entrepreneurs. Considering the results obtained in the present work, the model to be developed for the creation of a platform, intends to be a first step to improve the Design formation of DS, to take the Design industry to gain added value and to introduce business competencies, entrepreneurship skills and learning dynamics in Design Degrees.

References

1. European Commission, Communication from the Commission to the Council, the European Parliament, the European Economic and Social Committee and the Committee of the Regions: Fostering entrepreneurial mindsets through education and learning (2006)
2. Liñán, F., Fayolle, A.: A systematic literature review on entrepreneurial intentions: citation, thematic analyses, and research agenda. Int. Entrepreneurship Manage. J. **11**, 907–933 (2015)
3. Thompson, E.R.: Individual entrepreneurial intent: construct clarification and development of an internally reliable metric. Entrepreneurship Theory Pract. **33**(3), 669–694 (2009)
4. Lans, T., Gulikers, J., Batterink, B.: Moving beyond traditional measures of entrepreneurial intentions in a study among life-sciences' students in the Netherlands. Res. Post-Compulsory Educ. **15**(3), 259–274 (2010)
5. Paço, A., Ferreira, J., Raposo, M.: Development of entrepreneurship education programmes for HEI students: the lean start-up approach. J. Entrepreneurship Educ. **19**(2), 39–52 (2006)
6. Almeida I.D., Delgado M.J., da Silva J.M.V.-B., Montagna G., Monteiro J.P.: Design and business: growing up as a separate couple. In: Chung W., Shin C. (eds.) Advances in Interdisciplinary Practice in Industrial Design. AHFE 2018. Advances in Intelligent Systems and Computing, vol. 790. Springer, Cham (2018)
7. Hitchings, M.G.: Career Opportunities: Connecting Design Students with Industry. Procedia Soc. Behav. Sci. **228**, 622–627 (2016)

8. Alberti, F., Salvatore Sciascia, S., Poli, S.: Entrepreneurship education: notes on an ongoing debate. In: Proceedings of the 14th Annual IntEnt Conference, University of Napoli Federico II, Italy, vol. 4, no. 7 (2004)
9. Alvarez, S.A., Barney, J.B.: Entrepreneurial opportunities and poverty alleviation. Entrepreneurship Theory Pract. **38**(1), 159–184 (2004)
10. Bae, T.J., Qian, S., Miao, C., Fiet, J.O.: The relationship between entrepreneurship education and entrepreneurial intentions: a meta analytic review. Entrepreneurship Theory Pract. **38**(2), 217–254 (2014)
11. European Commission, Entrepreneurship in Higher Education, especially within Non-Business Studies. Brussels: Final Report of the Expert Group (2008)
12. Collins, O.F., Moore, D.G.: The Enterprising Man. Michigan State University Press, East Lansing (1964)
13. Gibb, A.A.: Enterprise culture: its meaning and implications for education and training. J. Eur. Ind. Training **11**(2), 3–38 (1987)
14. Cooney, T.: Entrepreneurship skills for growth-orientated businesses. In: Workshop on 'Skills Development for SMEs and Entrepreneurship', Copenhagen (2012)
15. Drucker, P.: Innovation and entrepreneurship. Harper & Row Publishers, New York (1985)
16. Mitchelmore, S., Rowley, J.: Entrepreneurial competencies: a literature review and development agenda. Int. J. Entrepreneurial Behav. Res. **16**(2), 92–111 (2010)
17. Bird, B.: Towards a theory of entrepreneurial competency. Adv. Entrepreneurship Firm Emergence Growth **2**, 51–72 (1995)
18. Man, T.W.Y., Lau, T., Chan, D.F.: The competitiveness of small and medium enterprises: a conceptualization with focus on entrepreneurial competencies. J. Bus. Ventur. **17**(2), 123–142 (2002)
19. Robles, L., Zárraga-Rodrígueza, M.: Key competencies for entrepreneurship. In: 2nd Global Conference on Business, Economics, Management and Tourism, Prague, Czech Republic (2014)
20. Chell, E.: Review of skill and the entrepreneurial process. Int. J. Entrepreneurial Behav. Res. **19**(1), 6–31 (2013)
21. Osterwalder, A., Pigneur, Y., Clark, T.: Business model generation: a handbook for visionaries, game changers, and challengers. Wiley, Hoboken (2010)
22. Zeithaml, C.P., Rice, G.H.: Entrepreneurship/small business education in American universities. J. Small Bus. Manage. **25**(1), 44–50 (2005)
23. Branagan, A.: The Essential Guide to Business for Artists and Designers. Bloomsbury, London (2017)
24. Odora, R.J.: Integrating product design and entrepreneurship education: a stimulant for enterprising design and engineering students in South Africa. Procedia Technol. **20**, 276–283 (2015)
25. Günzel, F., Holm, A.B.: One size does not fit at all – understanding the frontend and back-end of business model innovation. Int. J. Innov. Manage. **17**(1), 1340005-1–1340005-34 (2013)

Learning as a Systemic Activity

Olga Pinchuk, Oleksandr Burov(✉), and Svitlana Lytvynova

Institute of Information Technologies and Learning Tools of National Academy of Educational Sciences of Ukraine, 9 M. Berlyns'koho St., Kiev 04060, Ukraine
opinchuk100@gmail.com, ayb@iitlt.gov.ua,
s.h.lytvynova@gmail.com

Abstract. This paper describes theory (systemic structural model) of learning activity, as well as demonstrates (on experimental data) changes in intelligence structure on "micro-age" intervals. They are compared specific changes of intelligence structure of pupils of grades 10–11 and students of the 1st year of Kiev University of Economics, masters of psychophysiology, physicians (psychophysiologists), and patent experts. Principles of ergonomic design for learning workplace are proposed.

Keywords: Human factors · Functional system of activity · Intellect structure · Learning

1 Introduction

To date, it is recognized that we live in the knowledge society and a human individual cognitive and creative abilities become crucial for the humanity development [1]. Children of the twenty-first century were born, grow, study, master the profession live and work in the information age, where the surrounding world is increasingly losing the features of the material world and turning into the world of information and knowledge [2]. Correspondingly, the means of training are changing in time [3], the synthetic part of the educational process [4] replaces increasingly the interaction between "human-human" and "human-material carrier of information".

Throughout history, advances in information and communication technologies, energy and transport technologies, biotechnology and natural sciences, agriculture and industry, military technology, as well as in other areas of science and technology play an important role in the process of radical changes in how people deal with their affairs. Today's society lives de facto and *de jure* in the information age, as evidenced by a number of international instruments adopted at international forums and materials from the World Economic Forum in Davos (2017–2019). As a result, digital space is increasingly embracing all areas of our lives, first of all, education and training that are the basis of a human capital.

Modern education needs to be more and more individual-oriented, securing individual psychophysiological abilities and development. Effectiveness of the collaboration of humans and technical, didactic and organizational subsystems could be achieved, if learning activity is considered and designed as activity of a system which structure includes a human (learner), learning environment and learning tools/means. Each from

W. Karwowski et al. (Eds.): AHFE 2019, AISC 963, pp. 335–342, 2020.
https://doi.org/10.1007/978-3-030-20135-7_33

these three components has its own structure and functions, and general system performance can be described in terms of the systemic-structural activity theory [5].

Purpose. To analyze learning activity from point view of systemic organization including information and cognitive components.

2 Discussion of Results

Three things are needed to succeed in responding to quick changes in life and technology: we must notice and recognize a certain change; we must understand the consequences of this change; we need to make timely and effective decisions about it; it must be understood that not only new means and opportunities arise, but a digital transformation of the human life and human activity that was actively discussed at the Davos Forum in 2019. First of all, it concerns new technologies that accompany us in all areas of our lives and activities - information and communication technologies (ICTs) - and also directly affect not only our present, but also the future.

The transformation of the role of information networks and their place in human life has led to shift the emphasis of network designers towards the anthropocentric nature of their construction and existence, the emergence of the need to apply the concept of not only "integration of networks", but also the concept of "integrated human-centric network." A new type of integrated network becomes not a passive element of innovative processes, but is active as it changes significantly the character of contemporary creation, which becomes a "currency" with its laws of formation, development, circulation and the need for protection. All these types of human activities are related to the creation of new ideas, new intellectual products, which are not in the process of creation strictly localized in the place of their creation, that is relatively open and not protected from external expansion.

The experience of ergonomic science makes it possible to optimize the interaction of a person with the environment and means of activity, including in the information environment. Ergonomics as a scientific and practical discipline is aimed at ensuring high efficiency of human activity, its safety and comfort. One of the ways to achieve such a triple task is to create an effective psycho-physiological support for the ability to work in the process of both work and learning. Macroergonomic approach involves the systematic solution of issues of analysis of a certain type of activity, designing its optimal conditions, selecting and adapting a person to this activity, solving technical and organizational issues of providing effective and safe education and labor.

Activity is a main substance of our everyday life. But its scopes become relative, not clearly defined because of:

- changes in our goals and needs (earning, cognition, leisure),
- tools and facilities (electronic equipment in workplace, in office, at home),
- workplace design and construction (more ergonomic and comfortable),
- location (outsourcing job),
- increasing of mental component work (in contrast with physical one),
- day time span of the same or similar type of activity.

Especially it is actual for education, because today's learning can be carried out anywhere and at any time thanks to different gadgets including mobile ones. From the other side, modern jobs need skills not so specialized, as general. In other words, Requirements for a highly qualified specialist are not of narrow profile, but of a systemic nature. Accordingly, education process should form a specialist from systemic positions: vocational, behavioral, communicative, personal and social. Environment of a human activity should be evaluated as natural (physical, biological), informational, organizational, technical and social. In other words, education is in focus of ergonomics, which combine a human, tools of activity and environment by their interaction to satisfy efficiency of the general system, safety and comfort of activity (Fig. 1).

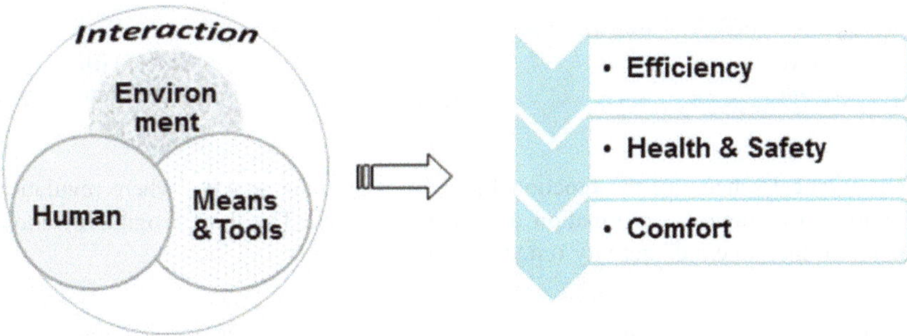

Fig. 1. Ergonomics objects and goal.

Learning is a system of activity that forms a future specialist in his/her interaction with external environment and internal state.

Models and appropriate ICT tools were developed for the system "Learner-technology/tools-environment" (SLTE) accounting a learner psychophysiological individual features and stages of learning [6]. Functional system of cognitive activity (FSCA) in digital environment has been developed by authors as a refinement of the previous general functional system of activity after Anokhin [7]. In general case, the conceptual model is considered as an information stratum of professional activity, and physiological chain "afferent inputs – activity acceptor – physiological control – effectors - act" is an energetic stratum in this context (Fig. 2). According to authors view, the goal of professional learning and training is to form the conceptual model of activity of the particular type, carrying out particular tasks. The FSCA was proposed as a structural functional system representing psychophysiological model of cognitive activity, and as a hyper-complex dynamic (HCD) system. Considering student's ability to learn (SAS) as a three-level closed hyper-complex dynamic system, in which factors of influence on the efficiency of SLTE are the elements of the HCD that interact with each other, "the general suitability for the profession can be represented as the first level HCD, micro-age - as the second level HCD, current - as HCD of the third level".

In addition, the authors propose to single out an independent circuit "cognitive contour" for FSCA in the SLTE, because learning activity can occur without external object, f.e., in VR/AR environment [8], in adaptive learning systems [9]. In the digital

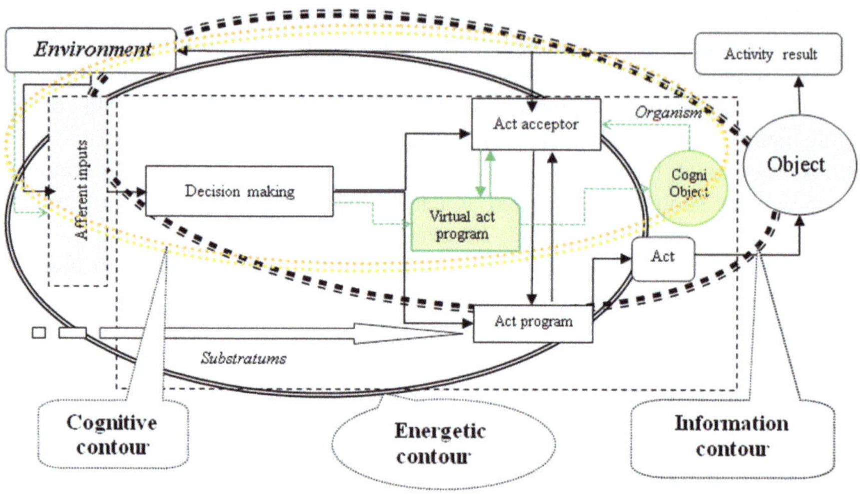

Fig. 2. Theoretical scheme of the functional system of learning activity, where regulation was divided into two contours – information and energetic ones. The third (cognitive) contour is associated with the "internal" activity (cited by [6])

world and synthetic learning the external object may not exist. "Instead of it, the virtual object can appear ("cogni object") that is produced by the virtual act program and can be not a result of training and experience, existing in time of the particular activity (f.e., during the game). This object and interaction with it can be created by sensors (information for them is produced by artificial system, virtual one, simulating real world) and the act acceptor compares virtual result with virtual sensors signals, creating the cognitive model of the synthetic object" [6]. In such a case, an illusion of the object as well as knowledge regards it can appear. But it should be specified that didactics of teaching process can influence to both "Act acceptor" and "Object" in this model.

This model of the learning activity helps to imagine how the system of educational activities is formed.

At the same time, we would like to articulate two sides and three levels of psychophysiological maintenance of learning activity (Fig. 3).

Two sides are: (1) internal, described above, that is associated with physiological maintenance and functional state of the learner in a particular time; (2) external, behavior, that is associated with his/her performance in learning.

Three levels of psychophysiological maintenance: (a) principal ability of learner to learn in general or to study a specific profession; (b) psychophysiological and cognitive changes over a learning time; (c) current state of the learner and his/her ability/readiness to accept effectively the proposed kind of learning tasks (take lectures, laboratory exercises, tests performance etc.). This could be especially important for individual-oriented education.

This model explains relationship between external and internal organization of e learner features and parameters that could be measured to assess his/her learning performance system and its success or degradation. Existed theoretical basis did not

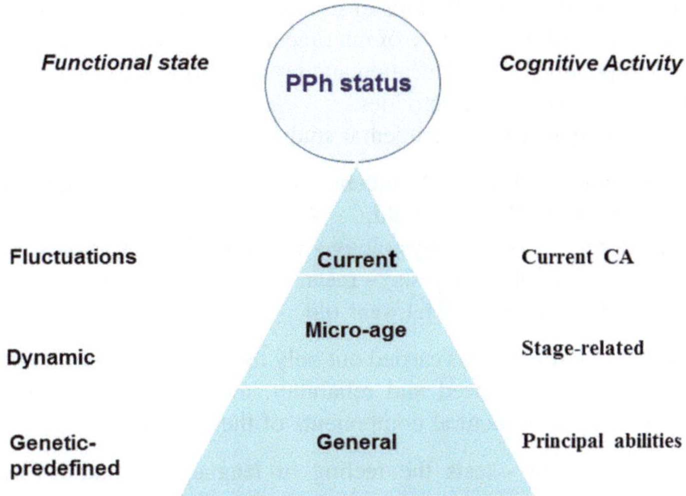

Fig. 3. Sides and levels of psychophysiological maintenance of learning activity

allow to answer questions *What? Where? Why? When? In what way?* should be measure to provide a high accuracy of the human performance prediction. But the model could help us.

A human (learner) activity is accompanied by creation and maintenance of functional system that:

- activates dominant *brain structures*,
- activate *corresponding activity* of one or another physiological system,
- is *quite stable for particular type of the human activity.*

Actual FSA depends on a particular type of learning activity and can activate different mechanisms of performance. Depending on this, a student can be considered as an operator-watcher, operator-researcher or operator-manipulator, and his/her activity can be measured and assessed from viewpoint of ergodynamics.

Human lives and acts in digital space (DS). Young children born, grow, learn and will work with gadgets being linked in networks as a natural environment. Their lives are influenced by DS with old and new hazards, where their performance depends on more cognitive factors (interface, content, models of behavior) and is a result of safety, efficiency, health (HF/E domain). They are discussed challenges evoked by life&activity (L&A) in DS in relation to influenced factors, ways of their avoiding and appropriate tools. Analysis is based on experience of HF/E findings for adults (emergent and military operators) and features of cognitive abilities of high school students. According to research data regards giftedness and its relationship with intellect structure, it is recognized that intellect does not explain the whole giftedness and general abilities of people, but is a background of any ability and its structure can be considered as the systemic characteristic of a human mental performance, as well as professional suitability.

In order to clarify these peculiarities of modern education, the research of common and/or special features of the structure of intelligence among representatives of different age groups (correspondingly, by education and profession) was carried out of groups of people motivated by intellectual activities.

Volunteers participated in experimental studies:

36 pupils of grades 10–11 and students of the 1st year of Kiev University of Economics - mean age 16.7 year old,
27 military masters of psychophysiology - mean age 23.2 year olds,
28 physicians (psychophysiologists) - mean age 32.1 year old,
42 patent experts - mean age 41.1 year old.

The analysis of the results was carried out only for those subjects who performed all tests with a given level of speed and reliability, the intelligence structure after R. Amthauer test was used. Structural components of the intellect are assesse:

1. LS - logical selection - tests the feeling of language, the ability to formulate judgments.
2. GE - definition of common features - tests abstract ability.
3. AN - definition of similarity - tests combinatorial abilities, dynamic thinking.
4. RA - computational and mathematical - tests the ability to solve computational problems of a practical nature.
5. ZR - detection of regularities - tests logical and mathematical thinking.
6. Fs - the choice of figures - tests spatial thinking in terms of geometric combinations on a plane in the formation of an integer shape of its fragments.
7. Wv - task with cubes - tests spatial thinking.
8. Me - memory, attention - tests volume and concentration of attention, as well as operational memory.

Results of comparative analysis have demonstrated that the verbal intellect (VI) is higher than non-verbal one (NI) for all groups of subjects (Fig. 4).

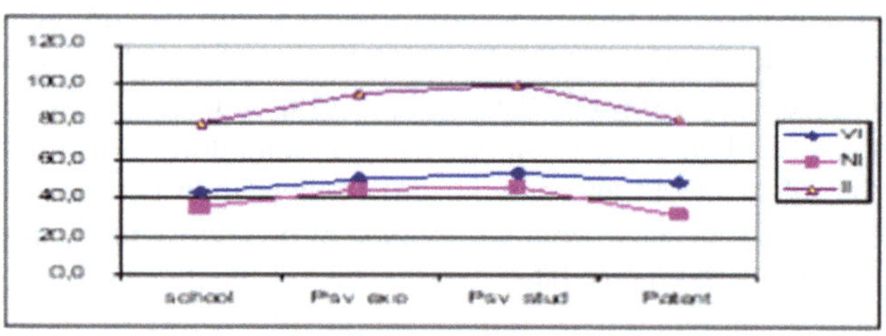

Fig. 4. Verbal and non-verbal intellect of subjects

But the intellect structure differed in those groups (Fig. 5).

As it was expected to some extent, verbal components of the intellect was lowest in schoolchildren, but special thinking and memory were higher than in the patent experts' group. In general, one can make a decision that any type of vocational learning develops intellect as a general thinking system.

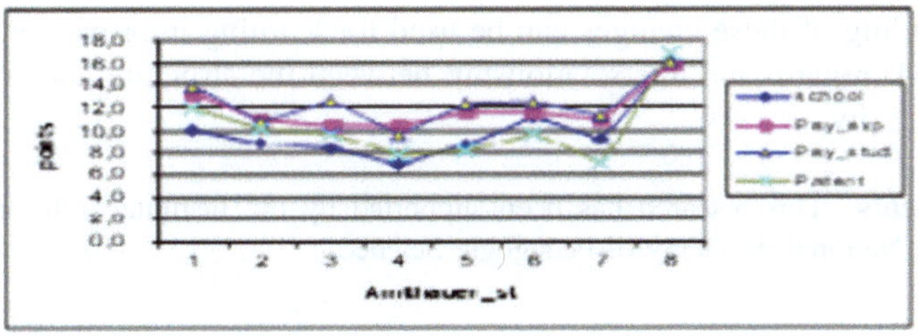

Fig. 5. Intellect structure of the same groups' subjects

Possible question is: what are requirements for the learning in ergonomics of the digital age [10]? What principles of the information and communication technology' system for learning are?

We propose principles of ergonomic design for learning workplace as follows:

1. Subject-oriented design of workplace.
2. Context of used approach.
3. Adequate learning/working tools and means.
4. User-friendly learning/working environment.
5. Developing activity.
6. Effectiveness/reliability as a goal of learning/working process.
7. Safety of user's mental and physical health.
8. Comfort of user's learning/working process.
9. User resilience under possible negative impact from the network and ICT in general.

In addition, we have to articulate the necessity to take into account that digital life and activity gives new opportunities for people and new problems for HF/Ergonomists, lists of draw-backs and risks can and will be extended according to new experience to use eWorld, because in network-oriented space, where "node (human/other agent) – interface – link – network" is a space of information existence [11]. As a result, nea features of ergonomics/human factors appear: in "*hard*" (material) workspace human-produced, "pushed-out", result can be always identified (localized) in space and/or in time, but result of human activity in "*soft*" SLTE or human-aided agent (information) can be not always localized both in time and space.

Such an environment could produce new hazards from the domain of cyber-security [12], that faces new features of the learning as a system activity.

3 Conclusion

This paper describes theory (systemic structural model) of learning activity, as well as demonstrates (on experimental data) changes in intelligence structure on "micro-age" intervals. They are compared specific changes of intelligence structure of pupils of grades 10–11 and students of the 1st year of Kiev University of Economics, masters of psychophysiology, physicians (psychophysiologists), and patent experts.

Understanding of these changes can be used for learning improvement using basic points of SLTE regards the interrelationship between the structure and self-regulation of learning activity.

Acknowledgments. This research has been supported by the Institute of Information Technologies of the National Academy of Pedagogic Science.

References

1. Bykov, V.Ju.: Innovative development of society and modern technologies of open education. Problemy ta perspektyvy formuvannja nacionaljnoji ghumanitarno-tekhnichnoji elity: P78 zb. nauk. pracj/ za red. Tovazhnjansjkogho, L.L., Romanovsjkogho, O.Gh. – Vyp. 23–24 (27–28), Kharkiv: NTU "KhPI", 24–49 (2009)
2. Naden, G.: The great learning curve: how to improve your study habits. The Guardian, 22 March 2018. https://www.theguardian.com/education/2018/mar/22/the-great-learning-curve-how-to-improve-your-study-habits
3. Seel, N.M. (ed.): Encyclopedia of the Sciences of Learning. Springer, Boston (2012)
4. Pinchuk, O., Lyvynova, S., Burov, O.: Synthetic educational environment – a footpace to new education. In: Pinchuk, O., Lyvynova, S., Burov, O. (eds.) Informacijni tekhnologhiji i zasoby navchannja: elektronne naukove fakhove vydannja, vol. 60, № 4, pp. 28–45 (2017). (in Ukrainian)
5. Bedny, G.Z., Karwowski, W.: A systemic-structural activity approach to the design of human-computer interaction tasks. Int. J. Hum. Comput. Interact. **16**(2), 235–260 (2003)
6. Spirin, O., Burov, O.: Models and applied tools for prediction of student ability to effective learning. In: 14th International Conference on ICT in Education, Research and Industrial Applications. Integration, Harmonization and Knowledge Transfer, vol. 2104, pp. 404–411 (2018)
7. Anokhin, P.K.: Principal questions of the general theory of functional system. In: Principles of the System Organization of Function. Moscow: Science, pp. 5–61 (1973)
8. Alexander, T., Ripkens, A., Westhoven, M., Kleiber, M., Pfendler, C.: Virtual tele-cooperation: applying AR and VR for cooperative tele-maintenance and advanced distance learning. In: Andre, T. (eds.) Advances in Human Factors in Training, Education, and Learning Sciences. AHFE 2017. Advances in Intelligent Systems and Computing, vol. 596. Springer, Cham (2018)
9. Scerbo, M.W., Freeman, F.G., Mikulka, P.J.: A brain-based system for adaptive automation. Theor. Issues Ergon. Sci. **4**, 200–219 (2003). https://doi.org/10.1080/1463922021000020891
10. Wilson, J.R.: Carayon pascale: systems ergonomics: looking into the future – editorial for special issue on systems ergonomics/human factors. Appl. Ergon. **45**(1), 3–4 (2014)
11. Burov, O.: Virtual life and activity: new challenges for human factors/ergonomics. In: Symposium Beyond Time and Space STO-MP-HFM-231, STO NATO, pp. 8-1–8-8 (2014)
12. Burov, O.Ju.: Educational networking: human view to cyber defense. Inst. Inf. Technol. Learn. Tools **52**, 144–156 (2016)

Converge of Goals on Learning Operations Research, Case of a Mexican Institute

Guadalupe Hernández-Escobedo[1(✉)], Arturo Realyvázquez-Vargas[1], Karina Cecilia Arredondo-Soto[2], Daniel Acosta-López[1], and María Elena Guevara-Hernández[1]

[1] Departamento de Ingeniería Industrial, Tecnológico Nacional de México/Instituto Tecnológico de Tijuana, Calz. Tecnológico s/n, Fracc. Tomás Aquino, C.P. 22414 Tijuana, Mexico
ghernan@tectijuana.mx, {arturo.realyvazquez, christian.acosta18}@tectijuana.edu.mx, m-e.hernandez@yahoo.com

[2] Universidad Autónoma de Baja California, Calzada Universidad #14418, Parque Industrial Internacional Tijuana, 22390 Tijuana, B.C., Mexico
karina.arredondo@uabc.edu.mx

Abstract. The purpose of this paper is to explore the shared goals of those involved in the learning process of Operations Research at undergraduate level in a Mexican institute. It is because some contradictions and disturbances were discovered within that process affecting mainly the relationships between involved individuals. Also, these can produce change and innovation in that process. For this, there is particular relevance on uncovering which elements actively participate and affect that process. Activity Theory as a novel and systematic approach and a qualitative and interpreting parading were chosen to gain a rich picture of these elements. Three activity systems were developed catching the material and abstract tools and other activity elements. Diverse links between these activity systems and their elements were revealed in the light of contextual features making sense of the relationships between the individuals. It was discovered that their clash produce innovation.

Keywords: Operations Research · Activity Theory · Human activity system

1 Introduction

The learning process involves a set of elements that are connected to each other for achieving the goal of changing the behavior of individuals as a consequence of their experiences [1]. This is seen under the Engineering viewpoint so that it involved internal and external elements affecting that process. The internal elements involve the individuals being the principal reason of it. In this line, it is important to study in detail cognitive processes, which can affect the learning process. Here, detailed studies of those cognitive processes can aid with the discovery of better strategies to produce changes on behavior. In current literature, there are multiple examples of models presenting diverse elements that can be considered for achieving the behavioral

W. Karwowski et al. (Eds.): AHFE 2019, AISC 963, pp. 343–352, 2020.
https://doi.org/10.1007/978-3-030-20135-7_34

changes. The strong relationships between these elements can help on producing the expected individual alterations [2]. From this, it can be inferred that multiple combinations can be done between the elements increasing complexity in the achievement of the aims. These aims are implicitly indicated in the models. Specifically, it is to evaluate the effectiveness of the diverse combinations showed in the models.

On the other hand, the external elements are related to those that can affect the process and are contained in the environment in which the learning process is carried out. Some of them are those that mediate the process into its context. In this line, diverse elements are considered such as technology [3], presentations on specialized software [4], solving problems [5], among others. These investigations are aligned to measure the impact at individual level. They also permit to measure at the same time the effectiveness of combinations of internal and external elements within the learning process.

Moreover, various forms emerged on considering both elements: internal and external. Most of these studies are carried out in different contexts in order to measure their effectiveness. A recurring context used to research the learning process is the university. Diverse elements and combinations are studied [6]. For instance, some of the elements are processes to learn [7], strategies to increase the effectiveness [8], recognition of student paradigms [9], social media [10], among others. However, it is inferred that the majority of these investigations are done in controlled environments, so there is an open gap in the knowledge related to research in natural contexts such as university is. In addition, the consideration of internal and external elements affecting the learning process can be a great opportunity to uncover other elements affecting that process.

This paper is focused on covering mentioned issues to fill this gap. It starts by presenting the research context, which is the learning process of the Operations Research subject carried out in the Industrial Engineering program in a Mexican institute. Moreover, the paper covers the following. First, a methodology is presented. After, the research results are depicted. Finally, the conclusions and recommendations are set out.

2 Methodology

In order to gain a rich picture of the learning process, a qualitative and interpretive paradigm is used. A methodology is developed so that each involved individual in the process can make available information about the process. Activity Theory is chosen such as a conceptual and an analytical tool to explore shared goals of individuals implicated. To do so, an institute located in the northwest of Mexico is selected as case study. Additionally, three methods are employed to make available information of the individuals involved in that process is presented below.

2.1 Activity Theory

This theory is elected because it sees the learning process such as a motivated activity in which diverse linked individuals exhibiting their role in that process. With this tradition, the cognitive and social constructionist approaches are caught considering that individuals reside in a subjectively world, which can signify social understanding [11]. It is covering the internal and external elements of the learning process. Also, this approach permits to know more about the consciousness of individuals who are in continuous interaction with other individuals performing the activity in study [12]. Moreover, the uses of abstract and material tools are included into the same approach. Similarity, the approach helped to appreciate how reality of individuals are united with the activity as well as to understand how the activity is object oriented but not detached from same individuals. This aids with the interpretation of the activity that can be described in social and cultural properties. It is because culture shaped in a broad sense the individuals accomplishing activities. These activities are seen as the unity of analysis allowing expanding them at individual and collective levels. In the same form, those activities are constituted by activities, actions and operations focused on the achievement of objects, goals and conditions respectively [13]. Particularly, the activity in study is the learning process of Operations Research at undergraduate level as mentioned above.

This process is seen as a collective, artifact-mediated and object oriented activity system [14]. Here, it is recognized as a multi-voicedness system as well as a product of transformations accumulated over time. These are consequences of changes and developments of the system are caused by contradictions and tensions, which consequently originate extended series of qualitative adjustments. It is important to recognize that contradictions and tensions are added structural tensions in activity systems and between other activity systems causing innovation on them [15]. They are four types. The first ones can be found within each element of the activity system. The second ones can be discovered between the elements composing the activity system. The third ones can be uncovered between the central activity system and a more advanced activity system. The last ones can be located between the central activity in study and its neighbor activities. Consequently, this frame permits to study the activity in study in a natural context as the learning process of Operations Research at undergraduate level is. This also enables to contemplate the purposes of individuals, associations between individuals, cultural and historical progresses of the learning process, among other aspects. Moreover, this gives the chance to understand how the diverse elements of the activity system are integrated as a whole. Here, the elements are subject(s), object(s), tools and artifacts, communities, rules and norms, and division of labor of the activity system.

2.2 The Case Study

In Mexico, there are diverse types of universities and each of them tries to achieve diverse types of goals solicited by their central bodies. They are principally categorized as public and private. An administration council administrates the private universities and these universities are principally focused on generate benefits to the members of the councils. On the other hand, diverse governmental bodies manage the public universities and define the objectives that have to be achieved public by the universities. The principal objective is to develop competencies for work to the students at undergraduate level. This is the case of the institute that participates in the study. The institute is a campus of a decentralized organization of the federal government that clusters 254 institutes, four centers developing equipment and two research centers. These are located in the main cities of Mexico. Moreover, this organization regulates the contents of the diverse subjects that mainly compose the academic programs in the engineer training. It is important to remark that the participating institute is located in a city placed in the northwest of Mexico perceiving diverse a different culture in comparison to others situated in other cities in Mexico.

Particularly, the subject in study was Operations Research that is included in the Industrial Engineer academic program. Its objectives are clearly exposed in the syllabus following the program, which is based on the development of competencies by using the acquired knowledge in the practice. Additionally, these objectives define three individuals immersed in the process: the students, the lecturer and the institute. They define what are the objectives that have to be achieved by each of them and what abstract and material tools can be employed in that process. The syllabus contained 11 points declaring the name of the subject, its presentation and objectives, previous programs, competencies to achieve, previous competencies, themes to be developed, activities in the learning process, suggested practices, suggested application project, suggested forms to evaluate the competencies, and suggested references.

On the other hand, the institute, as a crucial representative of the governmental organization, has to provide the place and environment in which the objectives have to be achieved linking students and lecturers. In this line, the lecturer takes the role of mediator between the institute and students. Similarly, the lecturer gives particular forms in which the objectives of the subject have to be achieved. This is stated in the diverse norms regulating the learning process in the institute and the aim is to satisfy the institute and students. In order to verify compliance of the diverse tasks on the process achieving the objectives, the lecturer has to plan and implement the particular objectives stated in the syllabus. To do so, the lecturer presents at the initial stage of the semester the sketch and it is verified three times along the semester in order to confirm that achievement. Lastly, the program defines the required background of the student who can study the mentioned subject. This background is connected with the knowledge obtained from previous subjects. As a matter of detail, the student has to approve previous courses as Algebra, Differential and Integral Calculus, Linear Algebra, among other subjects linked with the course included in the study. Most of the students taking this course attend the fourth, fifth, sixth, and seventh semester of their studies. Besides, they are between 19 and 22 years old.

2.3 Data Collection

Three methods are employed to gather data during the fieldwork. Diverse documentation related to the learning process is revised and the institute provides relevant information to the research. Data include the subject contents, organizational handbooks, information generated during the semester per each lecturer teaching the subject matter of study, norms and regulations, among other documents used in the routine operation of the institute. 35 documents are obtained from the institute. Similarly, one lecturer provides all materials used in the learning process. During 2018, three lecturers teach the subject, but only one teaches both semesters. This lecturer participates in the study and attends 120 and 160 students per semester respectively. The obtained documents contain the sketches; information generated during the semester; copy of the tests; a report of the final application project, and solicited questionnaires connected with the learning process. The second method is an interview done to that lecturer. Here, the Critical Incident Technique [16] and its variant [17] are used to obtain additional information about his current practices. The interview lasts about one hour and a half and the lecturer provides 10 documents connected with his practices. The third method is a questionnaire containing seven questions about the perception of the practices in the learning process. The students assisting to the mentioned course answered this questionnaire. The questions are: (a) What do you like and dislike of the subject? (b) What contributions and knowledge have been obtained during the subject? (c) What would you propose to improve the teaching of the subject? (d) How would you apply the knowledge obtained in your daily life and professional performance? (e) Until now, what have been your greatest knowledge and experience acquired in your Industrial Engineering studies? (f) What do you think you need to learn to develop as Industrial Engineer? (g) What do you hope to contribute to society once you finish your Industrial Engineering studies? During the first semester of 2018, 120 questionnaires are delivered by the students, and during the second semester of 2018, 160. Each answer contains at least 500 words.

2.4 Analysis

Activity Theory as an analytical tool allows the understanding of the relationships between individuals, object of the activity and mediation of components at individual and collective levels. This is in the learning process discovering that each individual can be considered separately conforming each part of the whole. In other words, each individual is linked with the learning process, but they exhibit shared goals achieving the object of the activity. Its deconstruction offers some advantages on gaining important insights in how motivations are moved to achieve goals in the learning process integrating social and cultural residues in context. It is possible because the sequences of activity are framed using Activity Theory. The general form of the activity system is presented in the Fig. 1. However, activity as unity of analysis gives advantages and disadvantages to study what is happing in the learning process. These are primordially in the form in how the activity has to be analyzed. For this, three generations of Activity Theory are developed. The first one is centralized in the idea of mediation by abstract and material artifacts or tools. The second one is based on the concept of collective mediation by rules and norms; community and division of labor. The Fig. 1 shows both the first and second generations in the form of triangle. The first generation includes the motivation, subject(s), tools, object and outcome. The second generation incorporates the rest of elements showed in the figure. The last generation presents the opportunity to study questions related to dialogues and diversity between diverse perspectives or traditions [15].

An interpretive approach is chosen as feasible approach to gain in-depth understanding of the diversity of perspectives on the learning process as a naturalistic context. To do so, multiple methods are used to capture that is little acknowledged [19], as mentioned above. Case study is seen as a viable approach because the research includes and is focused on a naturalistic context. This also provides advantages over

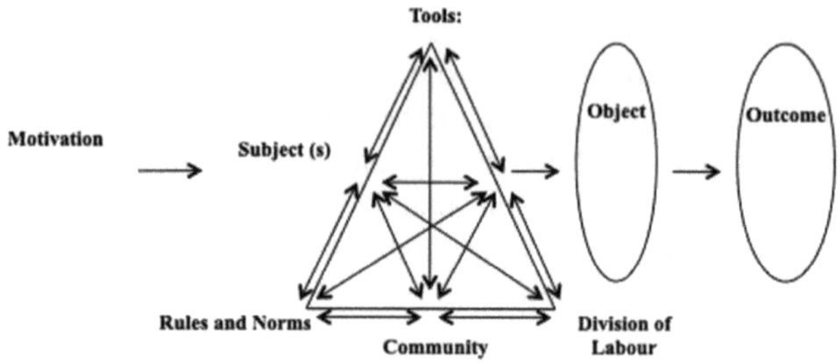

Fig. 1. The structure of a human activity system (adapted from Engestrom [18])

other approaches to investigate the nature of the learning process in context. For these reason three methods are utilized to gather data as mentioned before. All data are transcribed verbatim and an open, axial and selective coding approach is applied to analyze data. Data are examined until categories are saturated referring to the elements of the activity system of each individual participating in the learning process. Three activity systems are created and after they are linked in one exhibiting the shared goals. It is particular emphasis on finding the contradictions and tensions mentioned in the Sect. 2.1. In addition, the discovered elements are employed to uncover links in the light of contextual characteristics making sense of the relationships between its elements and activity systems. Motivations and uses of tools are two elements opened the discussion referring to affecting directly or indirectly the learning process. The encounter of contradictions and tensions puts on manifest the clash of the elements but similarly reveals gaps for innovating the learning process.

3 Results

According to the analysis of the gathered information, three human activity systems are developed. Each of them is related to the involved participants in the learning process. In order to make a rich picture in how the individuals converge in the process, the systems are united representing the shared objectives or goals. The Fig. 2 shows that structure exhibiting the activity elements of the developed systems.

3.1 Administrators

According to the regulations of the central organization, administrators have authority and responsibility to achieve goals signaled by them. Specifically, one of those goals is related to the learning process in terms of facilitating that process. It includes the provision of infrastructure and equipment. It also incorporates certain regulations in order to control the behavior of students. These administrators are also responsible for seeking new ways to perform the activities focused on providing diverse elements to facilitate that process.

However, administrators have evolved the ways in the characteristics of these elements. For instance, they usually change the rules and norms from their perspective

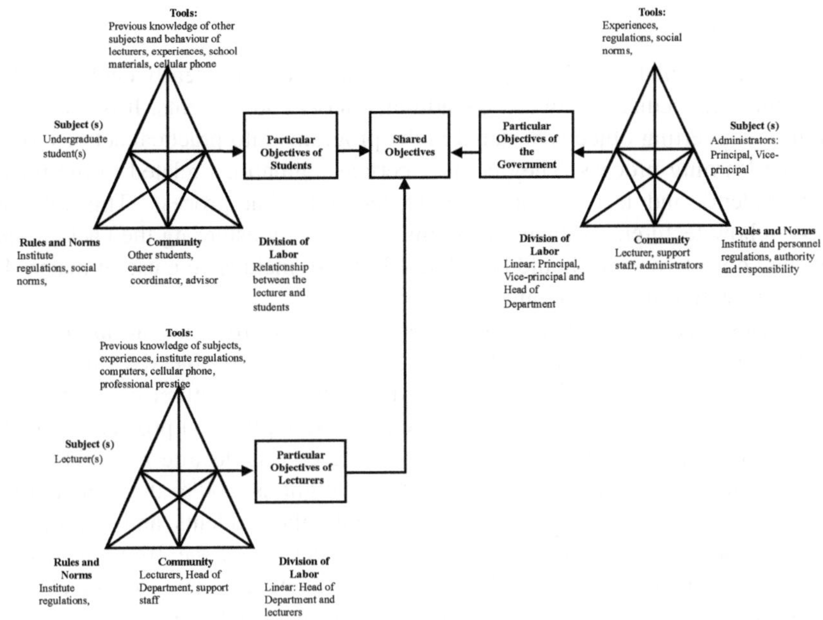

Fig. 2. The structure of shared objectives between students, administrators and lecturers

in order to increase the number of students per classroom. This action generates tensions and contradictions in relation with the rules exhibited in the institutional regulations that clearly signaled the maximum number of students per classroom. Sometimes, the surplus of students is near of 30%. It is important to remark that the number of students per classroom is limited in a basis of the construction regulations and Civil Protection legislations.

Moreover, administrators use diverse material and abstract tools to achieve the goals. It includes the employment of the regulations that are not disseminated among personnel and students. In occasions, these actions provoke diverse tensions with both students and personnel. Referring to the students, they have expressed their frustration in how new regulations stopped certain actions included into the learning process. For example, the inhibition in the use of material tools such as computers and cellular phones within the learning process trough decreasing of the bandwidth used on the Internet. The argument presented for this action was based on regulations concerned with the use of funds provided by the central organization, which clearly limits the costs linked with Internet services. On the other hand, the lecturers exposed some tensions in the use of computers such as material tools within the learning process. It is because the computers do not have technical capacity to support the current software used in the subject, or the computers do not have that software. This inhibits the use of material tools within that process.

3.2 Lecturers

Referring to the regulations, the objectives of the lecturers are clearly declared. One of them is being the mediator between the administrators and students. It is seen such as the agent of converting objectives stated in the programs into practical actions to make feasible the learning process. They employ certain regulations related to the relations between students and lecturers in order to facilitate that process. This relation is observed in how lecturers evaluate the learning process considering the background of students and objectives that should be achieved. It also includes the use of material and abstract tools stated in the program.

Nonetheless, lecturers stated the interpretation of objectives according to personal and professional goals. It includes diverse forms in how the objectives are weighted taking the form of evaluation indexes. This means that each lecturer considers diverse criteria such as attendances, exams, solving exercises, application projects, understand current literature, among others. Once they chose the criteria, lecturers weight them and assign percentages to them. The basis of this weight is 100% through a formula exhibiting these criteria. Some tensions arise with the students and administrators because they can consider it is not stated in the objectives of the program. However, lecturers used academic freedom such as the main argument to reinforce the evaluation criteria. It is important to remark that this argument is stated in the institutional regulations permitting to its application in the context.

On the other hand, the use of material and abstract tools within the learning process is a source of tensions and contradictions between the diverse individuals involved in the process. It is because some of the tools do not meet the minimal technical specifications to employ them on that process. For instance, the employment of computers and specialized software is signaled in the programs such as necessary tools to solve examples during the classes. But, these tools sometimes do not run the software. As being noticed of these situations, some lecturers opt to use free software in cellular phones and computers of students; however, the decreasing of the bandwidth of the Internet limits these uses.

3.3 Students

Respecting to the institutional regulations, the students are the most important element within the learning process. The entire system is aligned to consider the necessities of them in reference to the context including their elements. It is stated that organizational objectives can provide knowledge, experiences and competencies to students in order to prepare for work in the near future. To do so, students become acquainted of these regulations to facilitate the learning process. In addition, they notice the types of abstract and material tools that can use into the learning process.

Nevertheless, students employ the institute regulations in your benefit. It includes the pondering of those sections related to benefits in operational terms of the learning process. For instance, it is usual that students exploit the offered opportunity on choosing the lecturer who teaches the subject on study. This provokes imbalances in the number of students per classroom so that the authorities permitted increasing the number of them per classroom. These imbalances are not only seen in the number of students per classroom but also in the disequilibrium of the workloads per lecturer. To do so, it is normal that regulations are changed in order to allow them, as was mentioned in Sect. 3.1. Students can produce unusual situations enclosing additional risks in terms of Civil Protection issues.

Another theme is concerned with the use of material and abstract tools within the learning process. Sometimes, the students have the knowledge of the fact that computers do not have the technical capacity to run the software required to solve the exercises in class so that they use their own computers. However, administrators put some restrictions on connecting those computers to the institutional net. This inhibits the use of mentioned software. In addition, the students cannot connect their cellular phones to Internet because the bandwidth is insufficient to attend the number of users at the same time. Consequently, this directly affects to the learning process.

4 Conclusions and Recommendations

The results revealed an interesting rewrite and understanding of rules and norms and unusual use of some abstract and material tools in the learning process. Related to the norms and rules, each individual used in terms of facilitating the achievement of their objectives; however, they cause tensions and contradictions that produce diverse actions exhibiting innovation on them. This is seen in ways in how these norms and rules are rewritten affecting the learning process. For instance, in the case of administrators, the rewrite of those similarly remarks rule weighting on putting the learning process above of Civil Protection legislations.

Moreover, the uses of abstract and material tools are clearly denoted such as a crucial factor affecting the learning process. It is seen in the views in how they are considered within that process. If this is observed from an economical perspective, their use is limited to the minimal expression so that diverse actions are done in order to minimize that affectation. Here, the students and lecturers make changes in the activities focused on achievement the learning process discovering innovation in these activities.

However, it is important to remark that innovation can be seen as a way to achieve objectives enunciated by a body that is clearly away from the learning process. For instance, the reduction of funds used to pay the Internet services is a restriction that should be studied in detail in order to uncover benefits on increasing the funds for rise the bandwidth used on Internet. Consequently, this action can change some parameters in how the learning process is carried out in the institute.

Acknowledgments. We would like to acknowledge to the Tecnológico Nacional de México/Instituto Tecnológico de Tijuana and the Universidad Autónoma de Baja California for partially funding this study. We would also like to thank to the participants for contributing with the study. The interpretations and views in this paper, nevertheless, are solely those of the authors.

References

1. Wilder, B.: Learning is a process. https://www.reliableplant.com/Read/13812/learning-is-a-process. Accessed 15 2019 (n.a.)
2. Kolb, D.A.: Experiential Learning. Prentice Hall, Englewood Cliffs (1984)
3. Park, S.Y.: An analysis of the technology acceptance model in understanding university students' behavioral intention to use e-learning. Educ. Technol. Soc. **12**(3), 150–162 (2009)
4. Thomas, P., McKay, J.R.: Cognitive styles and instructional design in university learning. Learn. Individ. Differ. **20**, 197–202 (2010)
5. Yew, E.H.J., Goh, K.: Problem-solving learning: an overview of its process and impact on learning. Health Prof. Educ. **2**, 75–79 (2016)
6. Artigue, M.: The teaching and learning of mathematics at the university level. Crucial questions for contemporary research in education. Not. AMS **46**(11), 1377–1385 (1999)
7. Bennett, S., Agostinho, S., Lockyer, L.: The process of designing for learning: understanding university teachers' design work. Educ. Technol. Res. Dev. **65**(1), 125–145 (2017)
8. Biggs, J.: Teaching for Quality Learning at University, pp. 165–203. SRHE and Open University Press, Buckingham (1999)
9. Yang, F.Y., Tsai, C.C.: Investigating university student preferences and beliefs about learning in the web-based context. Comput. Educ. **50**, 1284–1303 (2008)
10. Madge, C., Julia Meek, J., Wellens, J., Hooley, T.: Facebook, social integration and informal learning at university: 'it is more for socialising and talking to friends about work than for actually doing work'. Learn. Media Technol. **34**(2), 141–155 (2009)
11. Jacob, E.K., Shaw, D.: Sociocognitive perspectives on representation. Ann. Rev. Inf. Sci. Technol. (ARIST) **33**, 131–185 (1998)
12. Kaptelinin, V., Nardi, B.: Acting with Technology: Activity Theory and Interaction Design. MIT Press, Cambridge (2006)
13. Leont'ev, A.: Activity, Consciousness, and Personality. Prentice Hall, Englewood Cliffs (1978)
14. Engeström, Y.: Innovative learning in work teams: analysing cycles of knowledge creation in practice. In: Engeström, Y.R.M., Punamäki, R.-L. (eds.) Perspectives on Activity Theory, pp. 377–406. Cambridge University Press, Cambridge (1999)
15. Engeström, Y.: Expansive learning at work: toward an activity theoretical reconceptualization. J. Educ. Work **14**, 133–156 (2001)
16. Flanagan, J.: The critical incident technique. Psychol. Bull. **51**(4), 327–358 (1954)
17. Chell, E.: Critical Incident Technique. Essential Guide to Qualitative Methods in Organizational Research, pp. 45–60 (2004)
18. Engeström, Y.: Learning by Expanding: An Activity Theoretical Approach to Developmental Research. Orienta-Konsultit, Helsinki (1987)
19. Denzin, N.K., Lincoln, Y.: The Discipline and Practice of Qualitative Research. Handbook of Qualitative Research. 2nd edn., pp. 1–28 (2000)

Usability of Learning Technology

Analysis of Types, Positioning and Appearance of Visualizations in Online Teaching Environments to Improve Learning Experiences

Jessica Brandenburger(✉), Manfred Constapel,
Horst Hellbrück, and Monique Janneck

The Center of Excellence CoSA, Department of Electrical Engineering,
Luebeck University of Applied Sciences, Lübeck, Germany
{Jessica.Brandenburger,Manfred.Constapel,
Horst.Hellbrueck,Monique.Janneck}@th-luebeck.de

Abstract. In this paper we investigate different visualizations of learners' data related to collaborative online learning in terms of suitability and attractiveness to students. Furthermore, we analyze whether positioning and color appearance of these data visualizations might have an effect on learners' behavior. To that end, we conducted an online study (n = 120) as well as an eye tracking study (n = 20) to compare different types of visualizations. Results show that students prefer classical data visualizations like bar charts. Visualizations placed in the sidebar of a two column web interface get less attention than visualizations in the header of the main content area. Color schemes do not seem to influence the perception of visualizations. We discuss possible explanations and implications for designing data visualizations in learning environments.

Keywords: Visualizations · Eye tracking · Online teaching environments · Human computer interaction · Learning analytics · Visualization methods

1 Introduction

Collaborative learning has been identified as a valuable approach to encourage interaction in online learning environments and provide more valuable learning experiences [1, 2]. However, group work is also often associated with typical difficulties, such as social loafing and generally unbalanced participation [3]. Visualization of intermediate results of group performance and individual behavior may support students in collaborative online teaching environments. The aim of our research is to improve group awareness in computer-supported collaborative learning scenarios by intelligent automated interventions. To that end, we automatically track and analyze activities of group members in learning management systems (LMS) according to a set of rules derived from research on successful group behavior. Results are utilized to implement visualizations for mirroring (problematic) learning behavior and guiding group members towards a more constructive behavior [2, 3]. In this approach, *mirroring* means that group (members') behavior – e.g. distribution of work load, participation of individual

W. Karwowski et al. (Eds.): AHFE 2019, AISC 963, pp. 355–366, 2020.
https://doi.org/10.1007/978-3-030-20135-7_35

group members – is simply made visible without further interpretations and suggestions, while *guiding* includes instructions for group members (e.g. alerting them to an upcoming deadline) [4]. To select suitable visualizations for that purpose, we conducted an online study (n = 120) to examine and compare various types of visualizations in terms of user experience, aesthetics and overall impression.

An important prerequisite for the effectiveness of data visualizations is that they catch users' attention and are clearly discernible among the multitude of information within a learning management system. Most LMS feature a two or three column web interface, therefore only limited possibilities for positioning visualizations of learning analytics data are available. To analyze the possible effects of positioning visualizations on different parts of the screen and capture students' first impressions of these visualizations, we conducted an eye tracking study with n = 20 participants. In addition to positioning, the effects of different color schemes were explored.

In summary, this paper addresses what types of visualization are attractive and easy to understand and interpret and investigates where and how these visualizations should ideally appear in a LMS.

2 Background and Concept

There are numerous visualization methods. Likewise, there are several tools, frameworks and libraries [5] to simplify the development of interactive visualization of various types of data. Most data comparisons are based on pie, bar, column, line, and dot charts [6]. An overview of 100 comprehensively documented and evaluated visualization methods is obtainable by the so-called *periodic table of visualizations* [7]. Methods for data, information, concept, strategy, metaphor and compound visualizations are illustrated in that work. Other solutions for the representation of relations, comparisons, and distributions are classified by Abela's *chart chooser* [8]. The *periodic table*, the *chart chooser* and visualization components of the popular JavaScript Library D3.js [5] and Crossfilter.js haven been considered in our development of suitable visualizations for online teaching environments.

Visual variables – including position, size, brightness, texture, color, orientation, shape, movement, saturation etc. – are used to encode and represent data. The complexity of a visualization is highly influenced by the visual variables involved [9, 10]. Especially color choice plays an important role for visualizations [11].

In our work we use data visualizations to provide feedback to students regarding their learning behavior in a collaborative online teaching environment. As a basis, we looked at typical problems regarding collaborative work in small groups (between 3 to 5 members) identified and classified by Strauss et al., 2018 [3]. Namely, we aimed to develop suitable support mechanisms for the following situations:

1. *Imbalanced participation*; i.e. workload of different group members varies heavily.
2. *Missing/late feedback to forum posts*, i.e. group members reply to their fellow group members' requests belatedly or not at all.
3. *Missing awareness of task progress*, i.e. it is hard for group members to judge the overall progress of their group, for example because feedback or individual results are missing.

To support overall group awareness, we designed data visualizations (*mirroring*) and interventions and recommendations (*guiding*) for these three problem categories to be displayed in the LMS (in this case, Moodle). Figure 1 illustrates the general structure of the Moodle plug-in, which was designed to be displayed either in the Moodle sidebar or in the page header.

For each problem category we designed a variety of visualizations using different forms (see Fig. 2), such as classical data and information visualizations as well as metaphor and concept visualizations, which were subsequently evaluated (mirroring component) by students in an online study (study 1, Sect. 3) and an eye tracking study (study 2, Sect. 4).

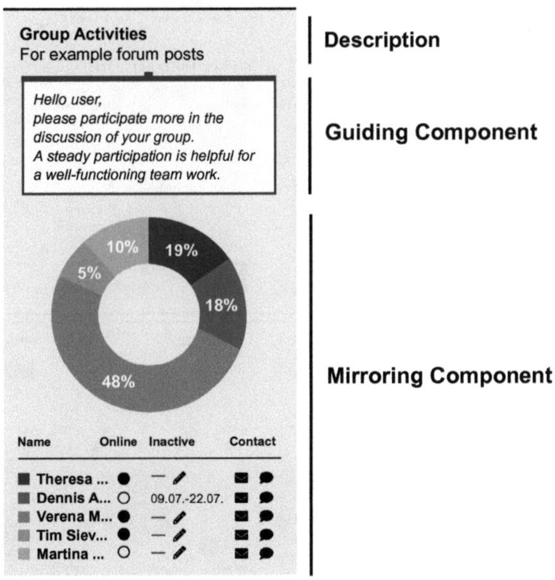

Fig. 1. Modular structure of an exemplary visualization to be displayed in the Moodle sidebar.

3 Study 1 – Comparison of Visualization Forms

3.1 Method and Sample

A total of 23 different visualizations for the three problem categories described in Sect. 2 [cf. 3] were evaluated by means of an online study (Table 1). The participants were recruited via university newsgroups and social media. Sole criterion for inclusion was that participants were currently studying at a university or had already obtained a university degree.

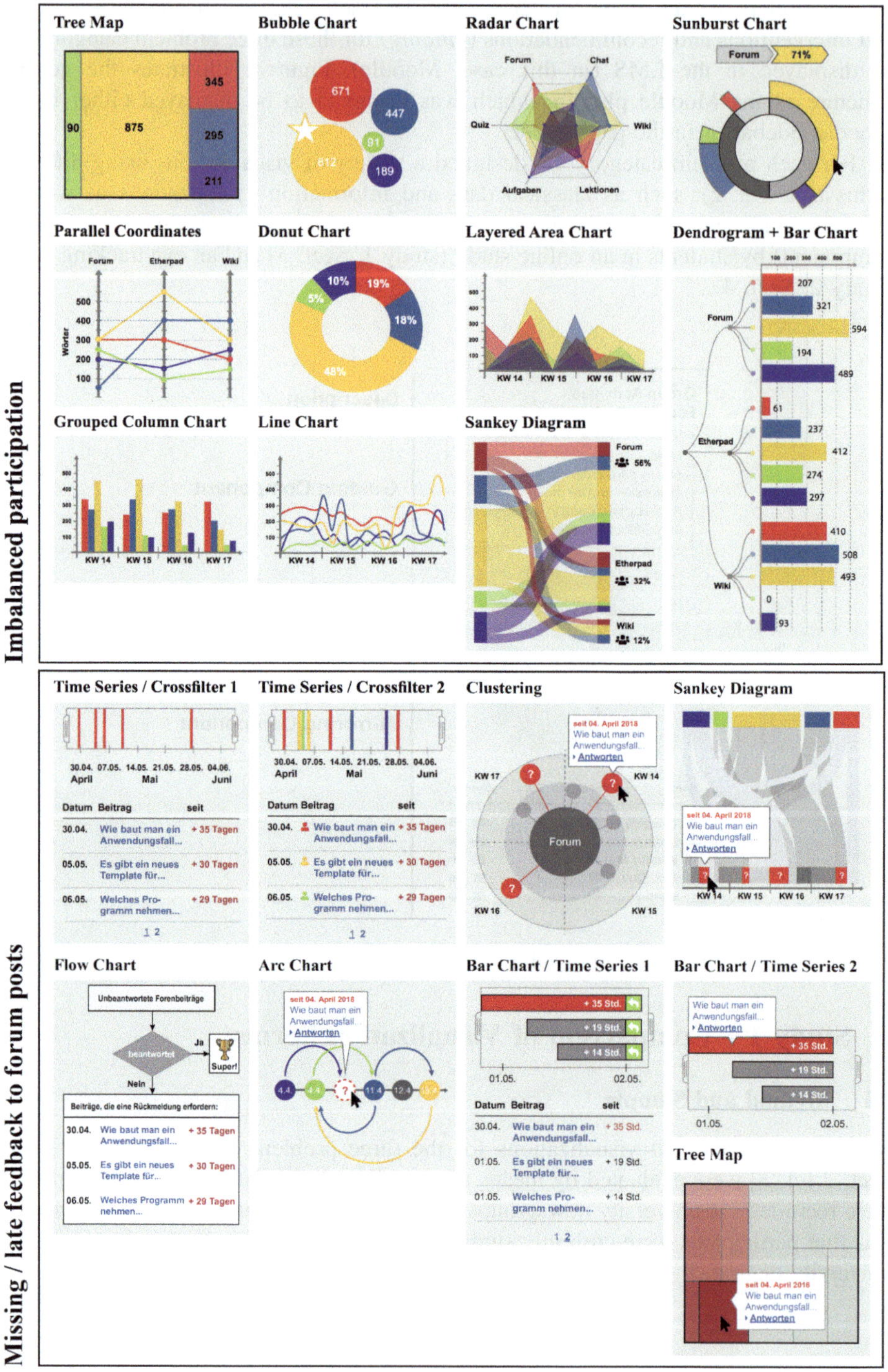

Fig. 2. Visualization forms/methods for mirroring focussed typical problems

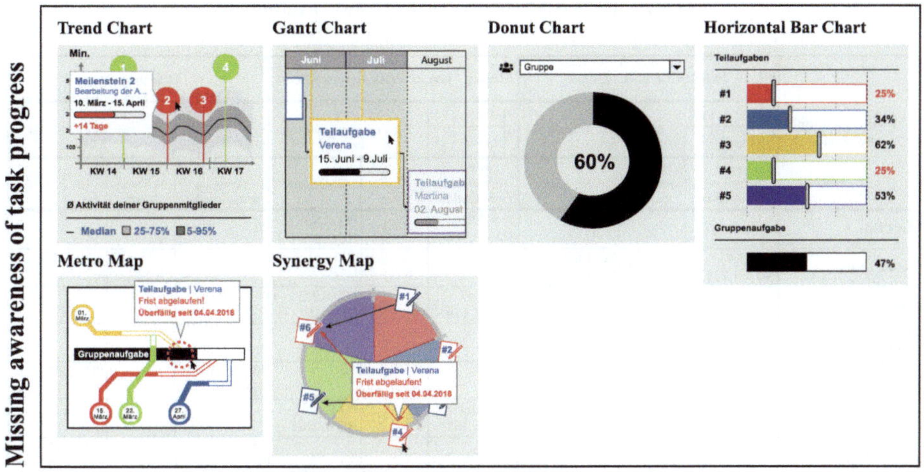

Fig. 2. (*continued*)

A total of 120 students/alumni (female = 38, male = 51, without indication = 31; mean age 27.65, range 20 to 55 years) took part in the online study. 55.8% of participants indicated a positive attitude towards team work in general, 30% a negative attitude (14.2% neither positive nor negative).

Participants were introduced to the subject of online team work by a short scenario. Afterwards they were presented with the different charts (see Fig. 2) in random order and were asked to rate it on a five-point likert scale (ranging from 1 – I do not like it to 5 – I like it very much). Afterwards, a randomly chosen visualization from each problem category was presented for a more thorough evaluation in terms of aesthetic appearance and user experience, using the standardized VisAWI-S [12] and UEQ-S [13] questionnaires.

Furthermore, participants evaluated 23 different harmonic color schemes, such as monochromatic, analog, complementary, partially complementary, triadic and tetradic color compositions.

3.2 Results

Comparison of Visualizations. For all three problem categories, students preferred classical data or information visualizations like *donut chart, bar (grouped column) chart, bubble chart and timelines,* both regarding usability and aesthetics (see Table 1). Over all categories, the *bar chart* reached the highest general aesthetic factor (5.8), while the *synergy map* as a concept visualization received the lowest rating (2.9).

Moreover, participants were able to indicate whether the visualizations were difficult to understand. In this regard, data visualizations performed better than other forms of visualizations like the *sunburst chart*, *sankey diagram* or *synergy map,* which were rated as difficult to understand by more than 20% of participants.

Table 1. VisAWI-S and UEQ (short) ratings

Visualizationform/ -method	Imbalanced participation			Missing / late feedback to forum posts			Task progress		
	n valid 120, missing 0	10 to 12 (after equally distributed draw)		n valid 95, missing 25	9 to 12 (after equally distributed draw)		n valid 90, missing 30	14 to 17 (after equally distributed draw)	
	overall impression (attractiveness)	VisAWI-S (Aesthetics factor)	UEQ (short)	overall impression (attractiveness)	VisAWI-S (Aesthetics factor)	UEQ (short)	overall impression (attractiveness)	VisAWI-S (Aesthetics factor)	UEQ (short)
Arc Chart				M = 2,07 SD = 0,997	M = 3,7	PQ -1,438 (--) HQ 1,104 (+)			
Bubble Chart	M = 3,43 SD = 1,228	M = 4,7	PQ 1,600 (++) HQ 0,075 (--)						
Clustering				M = 2,47 SD = 1,138	M = 4,4	PQ -0,225 (--) HQ 1,150 (+)			
Dendrogramm + Bar Chart	M = 2,78 SD = 1,161	M = 4,4	PQ -0,389 (--) HQ 0,639 (--)						
Donut Chart	M = 3,83 SD = 0,973	M = 4,9	PQ 1,875 (+++) HQ -0,083 (--)				M = 3,56 SD = 1,072	M = 4,6	PQ 1,867 (+++) HQ -0,767 (--)
Flow Chart				M = 2,69 SD = 1,168	M = 3,7	PQ 0,464 (--) HQ 0,036 (--)			
Gantt Chart							M = 2,8 SD = 1,144	M = 3,7	PQ 0,044 (--) HQ -0,206 (--)
Grouped Column Chart	M = 3,49 SD = 1,092	M = 4,6	PQ 1,722 (++) HQ -0,972 (--)						
Horizontal Bar Chart							M = 4,3 SD = 0,8	M = 5,8	PQ 2,518 (+++) HQ 0,589 (--)
Horizontal Bar Chart + Time Series 1				M = 3,16 SD = 1,075	M = 4,4	PQ 0,727 (-) HQ 0,250 (--)			
Horizontal Bar Chart + Time Series 2				M = 3,01 SD = 1,096	M = 3,4	PQ 0,318 (--) HQ -0,841 (--)			
Layered Area Chart	M = 2,44 SD = 1,063	M = 4,2	PQ -0,125 (--) HQ 0,750 (-)						
Line Chart	M = 2,83 SD = 1,234	M = 3,9	PQ 0,025 (--) HQ -0,700 (--)						
Metro Map							M = 2,59 SD = 1,189	M = 4,1	PQ -1,143 (--) HQ 0,911 (-)
Parallel Coordinates	M = 2,48 SD = 1,171	M = 4,4	PQ 0,075 (--) HQ 0,775 (-)						
Radar Chart	M = 2,37 SD = 1,173	M = 4,2	PQ -0,625 (--) HQ 0,906 (-)						
Sankey Diagram	M = 1,9 SD = 1,126	M = 3,7	PQ -1,167 (--) HQ 1,542 (+)	M = 1,84 SD = 0,926	M = 3,8	PQ -1,500 (--) HQ 0,667 (-)			
Sunburst	M = 2,08 SD = 0,94	M = 3,5	PQ -1,409 (--) HQ 0,250 (--)						
Synergy Map							M = 1,94 SD = 0,916	M = 2,9	PQ -1,446 (--) HQ 0,857 (-)
Time Series + Crossfilter 1				M = 2,67 SD = 1,106	M = 4,2	PQ 0,159 (--) HQ 0,205 (--)			
Time Series + Crossfilter 2				M = 2,6 SD = 1,244	M = 4,4	PQ -0,250 (--) HQ 0,611 (--)			
Tree Map	M = 2,54 SD = 1,166	M = 3,8	PQ 0,614 (--) HQ -0,432 (--)	M = 1,81 SD = 0,812	M = 3,6	PQ -0,750 (--) HQ 0,295 (--)			
Trend Map							M = 2,63 SD = 1,022	M = 3,7	PQ 0,100 (--) HQ 0,283 (--)

Comparison to UEQ Benchmark (short)
Excellent +++ Good ++ Above average + Below average - Bad --

Evaluation of Color Schemes. Regarding color schemes, participants clearly preferred tetradic color constellations (see Fig. 3). The most popular combination was chosen by 41.6% of participants, showing a clear preference. There were no correlations with gender or field of study, respectively.

Fig. 3. Preferred color constellations (left: most popular).

4 Study 2 – Effects of Positioning and Color

4.1 Method and Sample

Eye tracking is a well-tested method for measuring attention processes and interface designs in Human-Computer Interaction research [14]. An eye tracker records the reflection of infrared light from the retina and the cornea [15]. Prior eye tracking studies with Moodle show that content placed in the left top corner gets the most attention [16]. Thus, we expected that visualizations placed in the sidebar would receive less attention than visualizations placed in the header.

We conducted an eye tracking study to measure attention processes while looking at LMS pages with varying types of data visualizations. The eye-tracking study took place in the usability lab of our university. For analysis, we used NYAN 2 Architect.

20 students (50% female, 50% male; average age 24.75 years, range 20 to 35 years) participated. It was noted whether they wore glasses or contact lenses. Besides they were asked to confirm that their capacity of reaction was not impaired, e.g. by taking medication.

Only static screens with no interactive elements were used as stimuli. The visualizations were embedded in actual Moodle pages for a more realistic experience, showing e.g. start page, forum, wiki, or messages page. In total, we created 24 stimuli with 7 different visualizations all composed of different views and five of them also in different colors. We chose visualizations that had received the highest ratings in the online study presented in Sect. 3. The switch interval was set to five seconds with two seconds of black blank screen in between. While no absolute standards for interpreting eye tracking results exist [15], prior studies have shown that users' first visual impression is crucial for their evaluation of an interface; and first impressions are formed in about 50 ms [17]. Regarding *positioning*, we compared placing the visualizations within the sidebar vs. in the header of the Moodle page. We defined two different areas of interests (AOIs) for analysis (Fig. 4). Measures included fixation duration, fixation count, and gaze duration on the areas of interest (AOI).

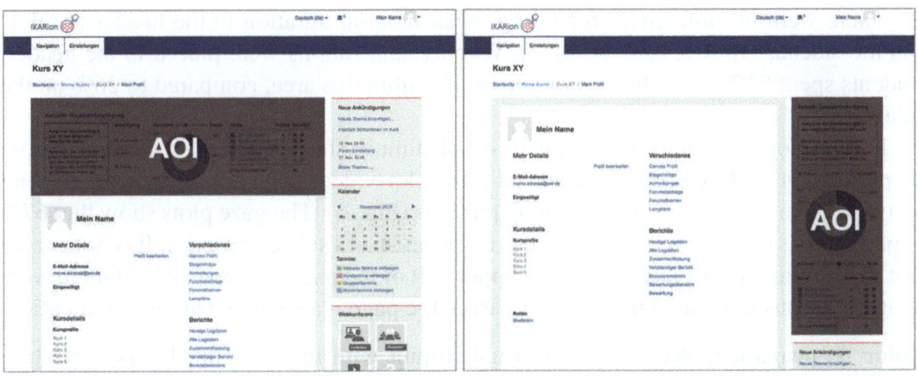

Fig. 4. Sample stimuli *(donut chart)* with AOIs used for the eye tracking study

Furthermore, we investigated whether different *colorings* had an impact on attention processes. Since study 1 revealed a strong preference for a rather bright tetradic color scheme, we expected brightly colored visualizations (i.e. tetradic color combinations) to get more attention than more subdued colors (i.e. analog color combinations).

4.2 Results

Positioning. The fixation count describes how many components of the web interface were detected by the participants [18, 19]. The total gaze duration describes the time people spent on the areas. To compare header and sidebar positioning, a t-test was calculated regarding fixation count, total gaze duration and mean fixation durations. Significant differences at fixation count and total gaze duration ($p < 0{,}001$) were found: All visualizations placed in the sidebar got less attention than the same visualizations positioned in the header (see Table 2), which were perceived by nearly 100% of students. Visualizations placed in the sidebar were ignored by up to 30% of participants (see Fig. 5).

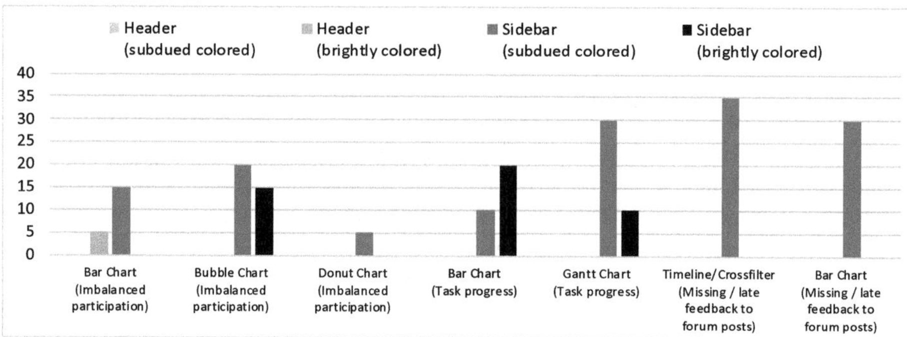

Fig. 5. Unrecognized visualizations by regions and color schemes

There were no differences regarding mean fixation duration in the header (AOI 1) and the sidebar (AOI 2, 0,23 vs. 0,22). When visualizations were placed in the header, students spent 53% of the five second interval within this area, compared to 26% on the sidebar.

Furthermore, we analyzed heatmaps of all stimuli, showing fixation duration across all participants. Moreover, we analyzed gaze plot videos for each stimulus across all participants to get insights into scanning patterns (Fig. 6). The gaze plots show that 67% of participants directed their first glances to the left-hand side of the page. It is noticeable that the participants usually fixated the upper half of the page. Interestingly, if there were images or icons integrated in the lower part of the page it also attracted further attention.

Color Appearance. We compared visual stimuli with more subdued versus brightly colored visualizations (Table 2). Regarding the *bubble* and *bar chart* (category *imbalanced participation*) and the *gantt-chart* (category *task progress*), the mean fixation

Table 2. Mean values and standard deviation (in parentheses) of fixation counts

Fixation Count Mean (SD)	**Imbalanced participation**											
	Bar Chart				**Bubble Chart**				**Donut Chart**			
	Header		Sidebar		Header		Sidebar		Header		Sidebar	
Color scheme	subdued	bright	subdued	bright	subdued	bright	subdued	bright	subdued	bright	subdued	bright
Inside AOI	14,20 (6,01)	13,35 (5,31)	5,55 (3,94)	8,35 (4,03)	12 (5,84)	13,2 (4,01)	5,15 (3,95)	7,10 (5,73)	12,85 (5,01)	12,55 (6,40)	7,8 (3,58)	7,10 (4,36)
Outside AOI	5,10 (3,91)	6,50 (4,50)	13,85 (5,66)	12,30 (4,87)	7,65 (4,67)	6,80 (4,51)	12,65 (4,33)	12,05 (5,18)	6,05 (4,96)	7,30 (5,71)	11,50 (4,58)	11,30 (4,39)

Fixation Count Mean (SD)	**Missing awareness of task progress**							
	Bar Chart				**Gantt Chart**			
	Header		Sidebar		Header		Sidebar	
Color scheme	subdued	bright	subdued	bright	subdued	bright	subdued	bright
Inside AOI	12,15 (5,30)	8,95 (4,65)	5,70 (4,43)	7,45 (5,42)	13,75 (4,67)	12,65 (4,21)	5,55 (5,61)	7,20 (5,21)
Outside AOI	7,85 (5,16)	9,70 (5,36)	13,95 (4,49)	11,05 (6,30)	4,85 (4,04)	5,85 (3,92)	13,90 (5,67)	11,45 (5,26)

Fixation Count Mean (SD)	**Missing / late feedback to forum posts**							
	Timeline / Crossfilter				**Bar Chart**			
	Header		Sidebar		Header		Sidebar	
Color scheme	subdued		subdued		subdued		subdued	
Inside AOI	10,45 (4,95)		3,60 (4,02)		11,10 (4,89)		3,75 (3,60)	
Outside AOI	9 (4,83)		15,50 (4,86)		9 (4,89)		16,90 (3,70)	

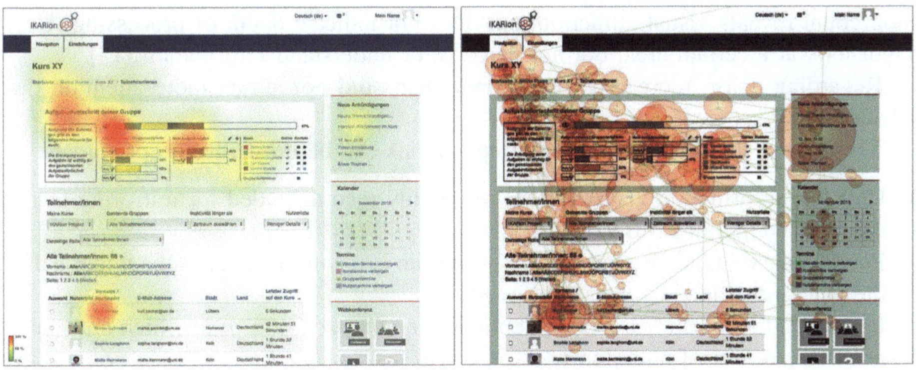

Fig. 6. Left: Heatmap example showing fixation duration across all participants for one stimulus. Right: Screenshot example of overlapping scanpath videos (gaze plots).

counts were higher for brightly colored visualizations, independent of position. The *bar chart* (category *task progress*) and the donut chart *(imbalanced participation)* attracted more attention in subdued colors. No significant differences could be observed.

5 Conclusion

This paper explored different types of visualization for mirroring group activities in online learning environments as well as their positioning and color appearance. Visualizations were developed after extensive research and analysis of visualization methods and evaluated regarding their attractiveness, aesthetics and user experience by means of an online study (n = 120).

Classical data and information visualizations, such as *bar-*, *bubble-* and *donut-chart,* were consistently rated best in all three scenarios of problems that may occur in group work, both regarding aesthetics and usability. These types seem to convey quantitative data in a simple and easy-to-understand form. Furthermore, they are frequently used in both print and online media and therefore probably more familiar to users than other types of visualizations. Other visualization methods use and communicate more complex metaphors, strategies, concepts etc. Metaphor visualizations and also partly concept visualizations received much lower ratings and were more often evaluated as being hard to understand. Thus, for an online teaching environment where information should be quickly understood by students, the use of data and information visualizations is highly recommended.

By means of an eye tracking study (n = 20) we observed that the header area of our two column LMS (Moodle) attracted more attention than the sidebar. Thus, important content should be placed in the main content area.

However, the fixation count depends on the content of the respective area. If the main content area displays a lot of information (e.g. text), the sidebar is not noticed very well. This is due to common scanning patterns [20], as people fixate the first words on the left [16], and not necessarily caused by a lack of information in the sidebar. Furthermore, when interpreting results, one has to take into account that fixation count reveals visual attraction, but does not indicate depth of processing: Users might look at a certain area, but not perceive or understand information [15].

Regarding color, more brightly colored charts did not attract more attention than subdued color schemes.

To sum up, the following design implications can be drawn from our studies:

- Data and information visualizations are highly recommended for mirroring learner data to students in LMS.
- Positioning visualizations at the top of the site is recommended. In the upper half of the page most fixations were observed.
- Images and icons attract users' attention, even on parts of a page which are usually not at the center of attention.
- Bright tetradic color schemes are evaluated most positively by users. However, they do not attract more attention.

We are currently testing the mirroring and guiding components in a field test in real learning settings to analyze their impact on group behavior. In further studies, comprehensibility of the visualizations as well as interactive and dynamic visualizations will be examined. Furthermore, their display on mobile devices should be considered in the design process.

References

1. Chi, M.T.H.: Active-constructive-interactive: a conceptual framework for differentiating learning activities. Cogn. Sci. **1**, 73–105 (2009)
2. Diziol, D., Walker, E., Rummel, N., Koedinger, K.: Using intelligent tutor technology to implement adaptive support for student collaboration. Educ. Psychol. Rev. **22**(1), 89–102 (2010)
3. Strauß, S., Rummel, N., Stoyanova, F., Krämer, N.: Developing a library of typical problems for collaborative learning in online courses. In: Kay, J., Luckin, R. (eds.) Rethinking Learning in the Digital Age: Making the Learning Sciences Count, 13th International Conference of the Learning Sciences (ICLS) 2018, vol. 3. International Society of the Learning Sciences, London (2018)
4. Soller, A., Martinez, A., Jermann, P., Muehlenbrock, M.: From mirroring to guiding: a review of state of the art technology for supporting collaborative learning. Int. J. Artif. Intell. Educ. **15**, 261–290 (2005)
5. Bostock, M., Ogievetsky, V., Heer, J.: D3 data-driven documents. IEEE Trans. Vis. Comput. Graph. **17**, 2301–2309 (2011)
6. Zelazny, G.: Say It With Charts: The Executives's Guide to Visual Communication, 4th edn. McGraw-Hill Education Ltd, New York (2001)
7. Lengler, R., Eppler, M.: Towards a periodic table of visualization methods for management. In: IASTED Proceedings of the Conference on Graphics and Visualization in Engineering (GVE 2007), Clearwater, Florida, USA (2007)
8. Abela, A.: Chart Chooser & Advanced Presentations by Design: Creating Communication that Drives Action. Pfeiffer (2008). http://extremepresentation.typepad.com/blog/2006/09/choosing_a_good.html. Accessed 04 Feb 2019
9. Tominski, C., Aigner, W., Miksch, S., Schumann, H.: Images of time. In: Black, A., Luna, P., Lund, O., Walker, S. (eds.) Information Design: Research and Practice. Routledge (2017)
10. Bertin, J.: Semiology of Graphics: Diagrams, Networks, Maps. The University of Wisconsin Press, Madison (1983)
11. Stone, M.: Choosing Colors for Data Visualization. Business Intelligence Network. https://www.perceptualedge.com/articles/b-eye/choosing_colors.pdf
12. Moshagen, M., Thielsch, M.T.: A short version of the visual aesthetics of websites inventory. Behav. Inf. Technol. **32**(12), 1305–1311 (2013)
13. Schrepp, M., Hinderks, A., Thomaschewski, J.: Design and evaluation of a short version of the user experience questionnaire (UEQ-S). Int. J. Interact. Multimed. Artif. Intell. **4**(6), 103–108 (2017)
14. Jacob, R.J.K., Karn, K.S.: Commentary on section 4. Eye tracking in human-computer interaction and usability research: ready to deliver the promises. In: Hyönä, J. u.a. (Hrsg.) The Mind's Eye. Cognitive and Applied Aspects of Eye Movement Research, Amsterdam, pp. 573–605 (2003)
15. Bojko, A.: Eye-Tracking the User Experience: A Practical Guide to Research, 1st edn. Rosenfeld Media, Brooklyn, New York (2013)
16. Rakoczi, G.: Effektive Kursgestaltung - Erkenntnisse der Eye Tracking Forschung nutzen. MoodleMoot Austria, Linz (2012)
17. Lindgaard, G., Fernandes, G., Dudek, C., Brown, J.: Attention web designers: you have 50 milliseconds to make a good first impression! Behav. Inf. Technol. **25**(2), 115–126 (2006)
18. Goldberg, J.H., Kotval, X.P.: Computer interface evaluation using eye movements: methods and constructs. Int. J. Ind. Ergon. **24**(6), 631–645 (1999)

19. Duchowski, A.: Eye Tracking Methodology. Theory and Practice, 3rd edn. Springer, London (2017)
20. N. N. Group: How People Read on the Web: The Eyetracking Evidence, Nielson Norman Group, Technical report (2013)

Job Interview Game for High Functioning Autist

Johan de Heer[1(✉)], Rafal Hrynkiewicz[2], Thomas de Groot[2], and Emine Poyraz[3]

[1] Thales Research & Technology, Gebouw N, Haaksbergerstraat 67, 7554 NB Hengelo, The Netherlands
Johan.deHeer@nl.thalesgroup.com

[2] TXchange Game Based Learning Solutions (a Thales Company), Gebouw N, Haaksbergerstraat 67, 7554 NB Hengelo, The Netherlands
{Rafal.Hrynkiewicz,Thomas.deGroot}@txchange.nl

[3] Digital Media in Human Development, Leiden University, Rapenburg 70, 2311 EZ Leiden, The Netherlands
e.poyraz@umail.leidenuniv.nl

Abstract. We report on a job interview simulation game developed as part of the BRAINTRAIN project funded by the EU. The on-line accessible single-player 2D narrative game is adaptive to the trainees' (a) answering responses to questions by a virtual interviewer, (b) point of gaze towards the interviewer, and (c) his/her stress levels. Adaptiveness was realized by changing the game world in terms of more or less abstract interviewing questions, increasing or decreasing visual and audio clutter in the game interface. The game is intended to be an intervention instrument aiming to improve social communication and interaction skills. The target group are high functioning Autism Spectrum Disorder (ASD) adults. To test the feasibility of such a tool, we report on findings from a market study in terms of a SWOT analysis, and we show the results from a user study in terms of hypothesized effects on social skills improvements.

Keywords: Game based learning · Social skill development · Autism

1 Introduction

Here we like to report on a job interview simulation game developed as part of the BRAINTRAIN project[1] funded by the EU. BRAINTRAIN will improve and adapt the methods of real-time fMRI neurofeedback (fMRI-NF) for clinical use, including the combination with electroencephalography (EEG) and the development of standardized procedures for the mapping of brain networks that can be targeted with neurofeedback. One of BRAINTRAINs' objectives is to transfer fMRI-NF findings to other feedback-guided technologies that could be applied in everyday settings (e.g., non-intrusive virtual reality coupled or not with EEG).

[1] http://www.braintrainproject.eu/.

W. Karwowski et al. (Eds.): AHFE 2019, AISC 963, pp. 367–376, 2020.
https://doi.org/10.1007/978-3-030-20135-7_36

For that purpose a serious game was developed based on results from the University of Coimbra IBILI institute regarding cognitive and methodological research on social cognition networks, to boost adaptive social decision making in high functioning ASD adults [1]. Recently, several studies showed the feasibility and efficacy of Virtual Reality Job Interview Training (VR-JIT) in randomized controlled trials [2–5]. VR-JIT demonstrated improving job interview skills and self confidence in adults with Autism Spectrum Disorder (ASD). BRAINTRAIN project partners' Thales and University of Coimbra joined their expertise's and prototyped a serious game for training social executive functions, and coupled it with galvanic skin conductance and eye-movement recordings. The on-line accessible single-player 2D narrative game is adaptive to the trainees' (a) answering responses to questions by a virtual interviewer, (b) point of gaze towards the interviewer, and (c) his/her stress levels. Adaptiveness was realized by changing the game world in terms of more or less abstract interviewing questions, increasing or decreasing visual and audio clutter in the game interface. The outline of this paper is as follows. We first describe the Job interview game followed by the results of the market study to test the feasibility of such a product, and a usability study to test its efficacy. Finally, we wrap up with some conclusions.

2 The Job Interview Game

During the scope of the BRAINTRAIN project, an European project to study and optimize neurofeedback approaches for application in several brain disorders, a behavioral change game was created to train individuals with autism spectrum disorders in the process of taking a Job Interview. For more in-depth elaborations why this is relevant we point the interested readers to [1–5]. In this paper we focus more on the game itself, how it is perceived by the target group and the prospective market.

The game setting is an interview for the player's dream job. Several levels of difficulty are available and there is an adaptive system which changes the difficulty level based on the performance of the player. The game also implements a system of incentives (score, badges) and feedback on performance and behavior. See Fig. 1 for a schematic overview.

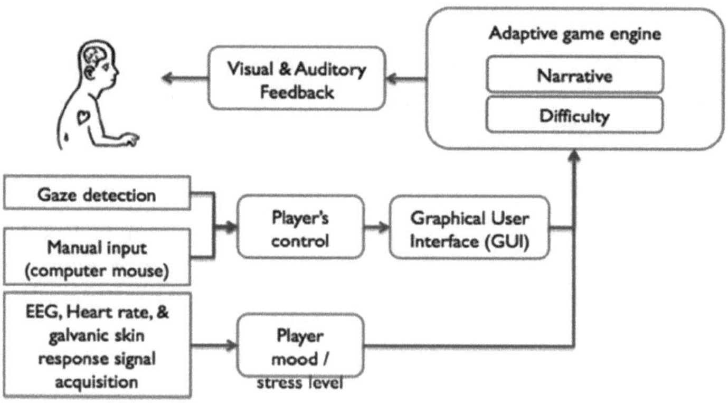

Fig. 1. Single player turn taking 2D narrative game.

Training of the following aspects of the executive function (attention, verbal reasoning, cognitive flexibility, problem solving).

- Dealing with visual and auditory clutter/information density;
- Leading a conversation;
- Comprehension of text/speech;
- Adapting to changing and unpredictable situations.

The learning objectives translated to the following game mechanics and player challenges.

- Visual clutter, sounds and distractions;
- Text clarity/ambivalence;
- Choice of multiple answer options and ability to ask for a "time-out";
- Unexpected events (visual and auditory interruptions);
- Scoring of the answers and behavior of the player to indicate performance.

Players' input.

- Computer mouse to answer questions;
- Point of gaze towards the game interface;
- Player mood based on galvanic skin response measurements.

Feedback during the game.

- Visual cues through the interviewer facial expressions, as feedback on given answers.

Feedback after the game.

- Score, indicating likelihood to get the job you applied for;
- Badges, rewarding the player for various aspects of preferred behavior and stimulating replayability;
- Textual feedback on any number of specific topics related to the scenario.

Screenshots

- Overall description of the game interface (Fig. 2);
- The interviewer presents herself before starting the questions (Fig. 3);
- Example question and answer (Fig. 4);
- Two different feedbacks: positive (left) and negative (right) performed on a clear environment (top) or in an environment full of distractors and visual clutter (bottom) (Fig. 5);
- The participant can request a time-out and give information about what is troubling him (Fig. 6).

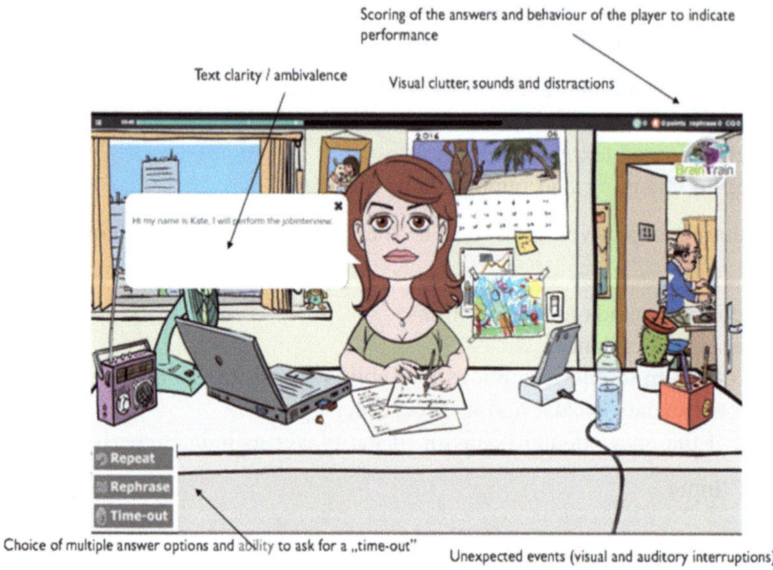

Fig. 2. Overall description of the game interface.

Fig. 3. The interviewer presents herself before starting the questions.

Fig. 4. Example question and answer

Fig. 5. Two different feedbacks: positive (left) and negative (right) performed on a clear environment (top) or in an environment full of distractors and visual clutter (bottom).

Fig. 6. The participant can request a time-out and give information about what is troubling him.

3 Market Study

This additional study was also funded by the BRAINTRAIN project to examine if the Job Interview game could be considered a minimal viable product (MVP). In total 12 interviews were conducted in 2017 and 2018 with representative structures in UK and Portugal. The study was outsourced to an external market research firm, which produced a SWOT analysis capturing their findings.

Strengths:

- An adaptive game with different levels of difficulties;
- A complementary tool to coaching session;
- A continuum between support structures and home training session;
- A larger target group including people with soft skill learning disabilities and not only ASD;
- A remotely use: possible allowing associations to target a larger population;
- The possibility to use the data collected for research purpose.

Opportunities

- Very few competitors on the market;
- Governmental ambition to improve the employment rate of disabled people, including autism;
- People suffering from autism loving IT tools.

Weaknesses

- The necessity to work on a pilot study to have a proof of concept;
- A game not adapted for all levels of autism.

Threats

- Charities and associations are not used to pay to implement tools in their activities;
- Many independent structures in both countries;
- A need to construct an efficient communication plan to spread the use of the game.

4 Usability Study

A user study [6] was conducted to test the efficacy of the game. We hypothesized that in a pretest – posttest experimental paradigm with the game as treatment we expect improvement of social competencies and skills in a representative ASD subject group. We benchmarked the subject group with ASD with a subject group without ASD. Subjects filled in the Screenings Questionnaire for Autism [SRS-A] [7] during pre- and posttest. SRS-A is measuring social responsiveness. Social responsiveness consists of social communication (22 items), social motivation (11 items), social awareness (19 items) and rigidity (12 items). Social communication defines the level of adequate responses on social cues during social interactions. Social awareness defines to what extent social cues are recognized and interpreted. Social motivation measures the level of engagement to be part of social interactions. Rigidity concerns the scale that focuses on the averseness against change. All items are measured using a 4-point Likert scale.

Subjects were acquired via social media. In total 42 Dutch participants diagnosed with ASD (62% men, 38% women) took part in the investigation. Ages ranged from 18 till 62 years (M = 29.84, SD = 10.36). The average educational background was on the level of university of applied sciences, 45% of the subjects was unemployed and 55% employed. Twelve participants without ASD participated (25% men, 75% women) with ages that varied between 18 till 52 (M = 32.64, SD = 10.84); employed 83.3% and 16.7% unemployed. Educational background on same level as ASD group.

The experiment took about 45 min. Filling in the questionnaire during pretest took around 10–15 min followed by game play (38 in-game job interview questions and answering took about 15–20 min), and again the SRS-A in the post test phase took 10–15 min as well. Subjects received an e-mail, including a briefing regarding the experiments and a few links to fill in SRS-A on-line and a link to play the game. Because subjects played the game at home it was played without the eye-movement recording apparatus and galvanic skill response measurement system.

After data inspection 24 ASD and 11 non-ASD subject were taken into account for further statistical analysis. The reduction of the data set was mainly due to the fact that some ASD subjects did not fill in the SRS-A for the second time (Fig. 7).

As expected the difference on the overall SRS-A scores between the two groups differed significantly. ASD subjects scored on average higher as compared with the non-ASD group. A high score indicates lower social responsiveness. This was true for all four subscales: social communication, social motivation, social awareness and rigidity. Although the scoring on SRS-A between pre and posttest is in the expected direction the difference for both groups is not significant. In fact, our hypothesized effect that the game intervention enhances social responsiveness was not supported by

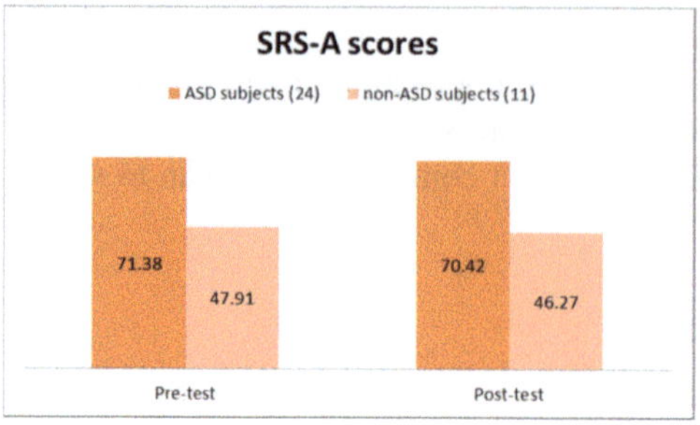

Fig. 7. SRS-A scores for pre and posttest between ASD and non-ASD subjects

the data and statistical analyses (pair t-tests and ANCOVAs) in this study. Further, no significant differences or interaction effects were found when we took employment (yes/no) and gender (men/women) into account as co-variates. At max, our results show a trend in the expected direction, namely that the overall SRS-A scores and scores on all four subscales were lower after game play.

From a methodological perspective this user study could be criticized on several points varying from how subjects were selected (via social media), how data was collected (at subjects' home with no experimental control), no control group available to support the observed trend, the limited number of subjects that participated in the study (24 with ASD and 11 without ASD), et cetera.

However, it is more interestingly to look how the SRS-A scores are interpreted (see Table 1). Since SRS-A is a screening test scores between 40–60 indicate a normal social responsiveness (as seen for our normal subjects); scores between 61–75 indicate mild to moderate shortage in social responsiveness, and scores above 76 indicate serious shortcomings in social responsiveness.

Table 1. Interpretation of SRS-A scores

SRS-A scores	
40-60	normal social responsiveness
61-75	mild to moderate shortage in social responsiveness
> 76	serious shortcomings in social responsiveness

The question therefore is do we see a change in scale position such that the interpretation/qualification is different between pre-test screening and post-test screening? Since our subjects were limited in number we inspected the pre- and post-test scorings on an individual level to examine if a change in interpretation could be found, for example from serious shortcomings to mild or from mild to normal. We found that for 8 subjects (7 ASD and 1non-ASD) scores on the posttest were lower than on the pretest with a change in interpretation as a result (see Table 2). This result was found on various subscale levels. Only subject 7 went from normal to mild in the social communication subscale.

Table 2. SRS-A scores for 8 subjects showing a change in interpretation

		SRS-A overall				SRS-A Social Communication				SRS-A Social Awareness				SRS-A Social Motivation				SRS-A Rigidity			
Subject	ASD	Pre		Post		Pre		Post		Pre		Post		Pre		Post		Pre		Post	
1	ASD	73	Mild	63	Mild	67	Mild	57	Normal	73	Mild	58	Normal	61	Mild	59	Normal	85	High	80	High
2	ASD	70	Mild	66	Mild	69	Mild	65	Mild	78	High	72	Mild	59	Normal	61	Mild	61	Mild	56	Normal
3	ASD	67	Mild	64	Mild	62	Mild	59	Normal	75	Mild	72	Mild	63	Mild	57	Normal	61	Mild	63	Mild
4	ASD	59	Normal	54	Normal	64	Mild	57	Normal	49	Normal	46	Normal	65	Mild	59	Normal	54	Normal	52	Normal
5	ASD	58	Normal	53	Normal	50	Normal	48	Normal	55	Normal	52	Normal	63	Mild	55	Normal	65	Mild	58	Normal
6	ASD	78	Mild	77	Mild	78	High	75	Mild	75	Mild	72	Mild	74	Mild	76	High	74	Mild	74	Mild
7	ASD	65	Mild	63	Mild	60	Normal	65	Mild	66	Mild	61	Mild	66	Mild	63	Mild	61	Mild	56	Normal
8	non	65	Mild	58	Normal	69	Mild	59	Normal	64	Mild	57	Normal	61	Mild	57	Normal	56	Normal	54	Normal

Note, however, that on the overall score no result in interpretation was found except for the non-ASD diagnosed subject number 8. A preliminary conclusion based on this finding may indicate that the game is a fit for purpose for a some people that are ASD diagnosed but not for all, and for some subscales. In our case 7 subjects out of 24, say around the 25–30% range may benefit to improve some aspects of their social responsiveness, that is a positive change in social communication, awareness, motivation and/or rigidity.

5 Conclusion

This study was a result of the BRAINTRAIN project in which we designed, developed and prototyped a serious game for high functioning adults in the autism spectrum. The objective of the game is to improve (aspects of) social responsiveness in the context of a job interview in order to be better prepared on such interviews in real life. The game world is adaptive to player responses, stress levels and point of gaze. The prototype was part of a market study in Portugal and UK indicating the potential for this game aiming to improve social responsiveness. At the one hand, the results of a first user study were not convincing from a statistical point of view, and improvements from a methodological perspective are suggested. However, looking to the data on individual levels and how scoring on the SRS-A scale are interpreted we found that the game leads to change in interpretation for 25–30% of the ASD subjects. The user study showed that this group based on their scores in SRS-A changed in screening position from mild/moderate shortcoming in aspects of social responsiveness to normal. Future work is directed to develop the prototype to the next maturity level and to test for which specific ASD group the game may be a relevant training instrument to enhance social skills.

Acknowledgments. This work part of BRAINTRAIN - a four-year Collaborative Project supported by the European Commission, under the Health Cooperation Work Programme of the 7th Framework Programme, under the Grant Agreement n° 602186. In addition, we like to thank our project partner the IBILI research group from the University of Portugal for providing relevant insights regarding ASD, writing the game scenario and feedback on the game solution. We also like to thank the BRAINTRAIN project management for facilitating and financing the market study. Finally, Emine Poyraz for conducting the user study, in fulfillment of her M.Sc. degree from Leiden University – The Netherlands.

References

1. Simões, M., Bernarsdes, M., Barros, F., Castelo-Branco, M.: Virtual travel training for autism spectrum disorder: proof-of-concept interventional study. In: JMIR Serious Games, vol. 6, No. 1, January–March (2018). https://games.jmir.org/2018/1/e5/2
2. Whyte, E.M., Smyth, J.M., Scherf, K.S.: Designing serious game interventions for individuals with autism. J. Autism Dev. Disord. **45**(12), 3820–3831 (2015)
3. Smith, M.J., Ginger, E., Wright, K., Wright, M., Lounds Taylor, J., Boteler Humm, L., Olsen, D., Bell, M.D., Fleminga, M.F.: Virtual reality job interview training in adults with autism spectrum disorder. J. Autism Dev. Disord. **44**(10), 2450–2463 (2014)
4. Arter, P., Barna, J., Brown, T., Fidiam, R., Fruehan, A., Law, M.: Using Virtual reality to improve job interviewing skills for individuals with ASD. Presented at the international society for technology in education (ISTE) conference (2018). http://conference.iste.org/2018/program/search/detail_session.php?id=110817130
5. Bresnahan, T., Rizzo, A.A., Burke, S.L., Partin, M., Ahlness, R.M., Trimmer, M.: Using virtual interactive training agents with adults with autism and other developmental disabilities. In: Proceedings of the 11th International Conference on Disability, Virtual Reality & Associated Technologies, Los Angeles, California, USA, 20–22 September 2016 ICDVRAT (2016). ISBN 978-0-7049-1547-3
6. Poyraz, E.: BrainTrain: evaluatie van het effect van gamen op de sociale responsiviteit bij Nederlandse individuen met ASS. M.Sc thesis from Leiden University (2019)
7. Roeyers, H., Thys, M., Druart, C., De Schryver, M., Schittekatte, M.: SRS Screeningslijst Voor Autismespectrumstoornissen. Hogrefe Uitgevers BV, Amsterdam (2011)

Antecedents to Training Engagement

Nadiah Alimon[1(✉)], Roziana Shaari[1], Lily Suriani Mohd Arif[1], Shah Rollah Abdul Wahab[1], and Nazry Yahya[2]

[1] School of Human Resource Development and Psychology, Faculty of Social Sciences and Humanities, Universiti Teknologi Malaysia, 81310 Johor Bahru, Johor, Malaysia
nadiahalimon1985@graduate.utm.my, rozianas@management.utm.my, {lily, shahrollah}@utm.my
[2] JP Skills Centre, Johor Port Berhad, 81707 Pasir Gudang, Johor, Malaysia
nazry@johorport.com.my

Abstract. This study aims to discuss the joint relationship of trainer directiveness and trainee learning experiences with training engagement. Survey responses were gathered from 99 employees at Johor Port, Johor, Malaysia using self-administered questionnaires. The result revealed that trainer directiveness has a positive relationship with training engagement. It was also found that trainees' learning experiences are positively correlated with training engagement. This study suggests that organizations take into account the differences in trainer's instructional style and trainee's cognitive process and motivation when implementing training programs since both factors may significantly affect the outcome of the training program as implied by the findings of this study. The study also proposes to the trainers the use of directive style in instructional settings as the specialization of directiveness give a positive influence to training engagement and subsequently, provide benefits to the trainee learning.

Keywords: Trainer characteristics · Trainee characteristics · Training engagement · Training transfer

1 Introduction

Trainer characteristics are less reported in previous studies as trainees will often have favorable attitudes toward the training experiences itself [1]. Therefore, it is not surprising when organizations are still investing heavily in designing and developing training programs only [2]. Nevertheless, of late, there is a significant amount of interest among researchers to study trainer characteristics and how it may affect training engagement. This is in accordance with Baldwin, Ford, and Blume [1], who contended that trainer characteristics have an impact on training engagement. However, the study on the interaction between trainer characteristics and trainee receptiveness is still very much lacking [3]. The role of the trainer's knowledge, skills, abilities and other attributes [4] and their impacts on training engagement demand further attention in the training literature as most of the prior studies regarding this engagement failed to acknowledge the impact of these roles [5]. Moreover, Harris, Chung, Hutchins, and Chiaburu [6] proposed that the definition of trainer characteristics should also include

W. Karwowski et al. (Eds.): AHFE 2019, AISC 963, pp. 377–384, 2020.
https://doi.org/10.1007/978-3-030-20135-7_37

trainer directiveness (i.e. the use of behaviours that aims to structure learning, outline goals and provide feedback). And, this consequently, inspires us to investigate the relationship between trainer directiveness and training engagement. In this study, the relationship between trainee learning experiences and training engagement are also examined. Furthermore, the findings of this study are hoped to help increases our understanding on training transfer. Thus, the purpose of this study will answer the following research questions:

- Does trainer directiveness correlate to training engagement?
- Does trainee learning experiences correlate to training engagement?

2 Theoretical Rationale and Expected Relationships

2.1 Trainer Directiveness

As previously mentioned, the lack of attention given toward trainer characteristics in prior studies was due to trainees' tendency to give a positive response in the post-training measurement [1] and organizations are more focused and invested in the development of training [2]. The influential roles of directiveness (i.e. the use of directive behaviors) which should be included in trainer characteristics are also poorly charted [6]. These behaviours are important in instructional settings, where learners will encounter new concepts, are tasked to learn new skills and rely on more guidance than in typical day-to-day settings [7, 8]. We conceptualize directiveness influences the trainer's instructional techniques. An inquiry on trainer directiveness gives a worthwhile return because these behaviors could be coachable and place more emphasis on learner control and interaction in the learning process [6]. From the perspective of instructional design, trainer with directive style is likely to be involved in facilitating a more structured environment by cueing learners to fulfill task requirements, monitoring learning performance and providing meaningful feedback toward learning goals [7, 8]. Previous studies revealed that when trainers implement the strategies of cognitive rehearsal and behavioral practice during training, participants' learning scores were positive [9–11] and could stimulate learners' engagement [12, 13]. Trainees will become increasingly active in the learning process when they are more engaged [14]. Hence, how to interact with or engage trainees is worth considering by a trainer [5].

2.2 Trainee Learning Experiences

Training outcomes will become well if the interaction between trainer and trainee characteristics is emphasized together [3, 6, 15–17]. Learning experiences entail the experiential learning style (i.e. a form of processing of intuitive, automatic and associated primarily with affect and emotional responses) [18, 19] interact together with motivation, which are segmented to the characteristics in learning. The experiential learning style is originated from the cognitive-experiential self-theory, which differ from each learner in relation to the information processing. The experiential learner requires external and affective cues to effectively activate the experiential system and, subsequently, the information processing. The positive cues may be provided by a trainer who practices directive style that stimulate delivery techniques and they are then motivated to engage in a similar action to re-experience this affect (i.e. motivation to learn) [20]. Trainees who do not give a preference toward the experiential learning are less likely to be affected by the trainer's directive style. Motivation to learn (i.e. an internal state that arouses, directs and sustains learner's behaviors towards achieving certain goals) [21] is an important component for active engagement in the learning process [22, 23]. The use of directive behaviors by the trainer could reduce learners' inhibition and any negative perception [24] & [6], intensify their cognitive interest and motivation to learn [25, 26], further, producing the higher level of engagement [27]. When trainees are more motivated, they will be more engaged [14] and perceive greater utility with the training [28, 29].

3 Methodology

3.1 Participants and Procedures

Data collection in this study was a part of the research initiative aimed to understand training engagement through usage of directive behaviors by the trainer. Respondents were 99 employees at Johor Port, Johor, Malaysia, who have participated in technical training programs provided by the port itself. There were 8 technical training programs involved and was instructed by 5 trainers.

3.2 Measures

Participants were asked to respond to the established various key study variables using 6-point Likert-type agreement scales, which level 1, strongly disagree to 6, strongly agree, together with demographic inquiry approach. Questionnaires were distributed at the end of the training programs and the distribution took an approximately 5 weeks to assess the appropriate average rate of a trainer's directive style and to obtain the desired number of participants.

Trainer Directiveness. This variable was provided with 15 inquiry items and the value of measures for the participants through the use of Cronbach's Alpha analysis is $\alpha = 0.911$, met acceptable reliability statistics (i.e. $\alpha = 0.88$) [26].

Trainee Learning Experiences. 10 inquiry items equipped in this variable and the participants' measure was acquired with the worth of $\alpha = 0.942$, accommodated with a desirable reliability statistic (i.e. $\alpha = 0.87$) [18].

Training Engagement. Participants completed 7 items of inquiry adapted from a scale developed by Schaufeli and Bakker [30] to measure training engagement. Consistent with prior studies (i.e. $\alpha = 0.75 - 0.83$; $\alpha = 0.73$) [30, 31], the items demonstrated adequate reliability in our study ($\alpha = 0.969$).

4 Results

Table 1 highlights the means and standard deviations for frequency, reliabilities and descriptive statistics of the study variables. The result pointed out that male respondents were the main contributor (85.9%) and the remainder were female (14.1%), and most of them aged between 24–29 years old (30.3%) and 30–35 years old (24.2%). Likewise, the majority of the respondents have only worked less than 1 year (20.2%), 1–6 years (21.2%) and more than 9 years (32.3%) in Johor Port.

Table 1. Mean and standard deviations

Variable	Mean	SD	1	2	3	4	5
Age (years)	3.18	1.548					
Gender (1 = male, 2 = female)	1.14	0.350					
Race	1.20	0.714					
Education	2.05	1.289					
Position	3.52	2.130					
Tenure of service (years)	3.08	1.543					
Trainer directiveness			75.63	8.532	5.0418	0.56877	(0.911)
Trainee learning experiences			51.33	5.650	5.1333	0.56496	(0.942)
Trainee engagement			34.81	6.285	4.9726	0.90	(0.969)

Notes:
Estimated reliabilities (Cronbach's Alpha) are reported in parentheses. All significant tests are two-tailed.

Table 2 presents the strength of the joint relationship of trainer directiveness and trainee learning experiences with training engagement. The result showed that the highest Pearson correlation was found between trainer directiveness and trainee learning experiences ($r = 0.857$, $p < 0.01$). Yet, trainer directiveness correlate poorly with training engagement ($r = 0.353$, $p < 0.01$). Likewise, trainee learning experiences have a weak correlation with training engagement ($r = 0.351$, $p < 0.01$). However, the result indicated the relationship between trainer directiveness and training engagement is significant at 0.01, also similar with the relationship between trainee learning experiences and training engagement ($p < 0.01$). Thus, *RQ 1* clarifies trainer directiveness correlate with training engagement. Also, *RQ 2* explains trainee learning experiences correlates with training engagement.

Table 2. Correlations analysis

	Trainer directiveness	Trainee learning experiences	Trainee engagement
Trainer directiveness		0.857**	0.353**
Trainee learning experiences	0.857**		0.351**
Trainee engagement	0.353**	0.351**	

Notes:
** Correlation is significant at the 0.01 level (two-tailed).

Looking at the correlation between the relationship, the yield is positive ($r = 0.857$, 0.353 & 0.351), thus, the correlation is one direction. This was revealed when trainer uses more directive behaviors, trainee will be motivated to engage in the learning process and it affects learning engagement among them. When inferred, trainer directiveness give a positive influence in enhancing trainees' cognitive interest and motivation to learn and, subsequently, producing a higher level of engagement.

5 Discussion and Implications

Research cannot fully substantiate the trainer's advantages. Yet, it flanks scholars to examine trainer immediacy [32, 33]. In the instructional settings, a trainer should maintain a supportive and in-charge demeanor [7] to trainees. Most extant transfer studies are deficient to state that the trainer is held accountable to trainees [1]. Trainees are also often need to be directed by their trainers [6]. The directive style (e.g. encouraging trainees to talk) were mixed with trainer behaviors in conveying support (e.g. using trainees' names) and reliance on peripheral cues (e.g. use of humor) [33], hence, in this line of work, our research extends the study of trainer directiveness in influencing training engagement as an initial measure to training transfer. We discovered that trainer directiveness has a positive relationship with training engagement and likewise, trainee learning experiences correlate positively with training engagement. We propose that trainer characteristics deserve more scholarly attention as it influences training engagement through trainee learning experiences [2].

Our study provides a few practical implications. One of it is organizational need to be mindful regarding trainees are different in cognitive processing and motivational orientations. Therefore, they should conduct learner needs assessments for effective training transfer before implementation of the training as it provides valuable information to trainers on how to structure and facilitate the training contents. Training engagement is an important mechanism [27] that explains how the trainer's presentation and delivery style brings to the transfer success. And, organizations have to encourage trainers to apply various appropriate instructional styles with learner needs and contexts. The trainer should vary their styles based on follower needs and situational demands as they have different follower [34]. MacKeracher [8] enhanced two additional instructional styles (e.g. collaborating and enabling) that differ based on the amount of task complexity and resources, learner control and required expertise in the learning setting. Generally, trainers' discrepant in instructional styles and general philosophy in relation to the best training practices. A proposition to the trainers to use the directive style in instructional settings as the specialization of directiveness give a positive influence to training engagement and subsequently, provide benefits to the trainee learning.

6 Limitations and Future Research

This study has its own limitations. First, our study is confined to the exploration of training programs in Johor Port, which involve one organization and the programs are specialized to the theory-based technical segments only. Second, data were subjected to the disposition of trainees' behavior itself. We cannot deny that the information is based on trainees' favorable attitudes toward the training experiences itself as clarified before. They are often provide positive responses toward the post-training measurement although they, in fact, are less likely satisfied if convergence is shifted to more difficult objectives of transfer [1]. Likewise, we are not able to overrule that the information is not free of bias [35]. Finally, our research design specializes to the joint relationship of trainer directiveness and trainee learning experiences with training engagement only, which the rationale for the proposed relationships is theoretically grounded without describing comprehensively the strength of variables of trainer directiveness and trainee learning experiences that could reinforce our findings.

Our study provides opportunities for a subsequent research. The research found that trainees learn best when they are actively engaged [36] which then, will lead to effective training transfer. Thus, further research in this area gives a valuable return to both trainees and trainers. The trainer is one of the most important factors in determining how effective trainees perceive the training to be [37]. Although a directive style provides effects to satisfaction and transfer [6], including a more comprehensive training style measure in relation to different levels of trainee learning orientation would help to better elucidate the influence of trainer delivery styles toward training engagement. Hence, related future research from the perspective of the trainer will certainly be valuable as well. Additionally, the administration of declarative knowledge tests and evaluations of subsequent performance from the trainer and trainees standpoints might be useful as every person has different cognitive processing. And, the need for research examines the extent to which styles other than directiveness fit with different trainee learning styles (e.g. rational style) [38] or a combination of directiveness with other instructional styles to predict transfer. Researchers have shown that instructors could use different techniques to accommodate different learning styles [39]. In addition, understanding additional interactions between trainer and trainee styles are necessary to enhance training transfer in various types of trainees.

References

1. Baldwin, T.T., Ford, J.K., Blume, B.D.: The state of transfer of training research: moving toward more consumer-centric inquiry. Hum. Resour. Dev. Q. **28**(1), 17–28 (2017)
2. Burke, L.A., Hutchins, H.M.: A study of best practices in training transfer and proposed model of transfer. Hum. Resour. Dev. Q. **19**(2), 107–128 (2008)
3. Harris, T.B., Chung, W., Frye, C.L., Chiaburu, D.S.: Satisfaction guaranteed? Enhanced impact of trainer competence for autonomous trainees. Ind. Commer. Train. **46**(5) (2014, in press)
4. Friedel, C.R., Rudd, R.D.: Relationships between Students' engagement and the dissimilar cognitive styles of their undergraduate instructors. Career Tech. Educ. Res. **34**(1), 21–45 (2009)
5. Arghode, V., Wang, J.: Exploring trainers' engaging instructional practices: a collective case study. Eur. J. Train. Dev. **40**(2), 111–127 (2016)
6. Harris, T.B., Chung, W., Hutchins, H.M., Chiaburu, D.S.: Do trainer style and learner orientation predict training outcomes? J. Work. Learn. **26**(5), 331–344 (2014)

7. Wheelan, S.A.: Facilitating Training Groups: A Guide to Leadership and Verbal Intervention Skills. Prager, New York (1990)
8. MacKeracher, D.: Making Sense of Adult Learning. University of Toronto Press, Toronto (2004)
9. Ford, J.K., Kraiger, K.: The application of cognitive constructs and principles to the instructional systems model of training: implications for needs assessment, design and transfer. In: Cooper, C.I., Robertson, I.T. (eds.) International Review of Industrial & Organizational Psychology, vol. 10, pp. 1–48. Wiley, New York (1995)
10. Warr, P.B., Allan, C.: Learning strategies and occupational training. In: Cooper, C.I., Robertson, I.T. (eds.) International Review of Industrial & Organizational Psychology, pp. 254–259. Wiley, Chichester (1998)
11. Holladay, C.L., Quiñones, M.A.: Practice variability and transfer of training: the role of self-efficacy generality. J. Appl. Psychol. **88**(6), 1094–1103 (2003)
12. Goldstein, I.L., Ford, J.K.: Training in Organizations: Need Assessment, Development, and Evaluation, 4th edn. Wadsworth, Belmont (2002)
13. Towler, A.: Effects of trainer expressiveness, seductive details, and trainee goal orientation on training outcomes. Hum. Resour. Dev. Q. **20**(1), 65–84 (2009)
14. Knowles, M.S., Holton, E.F., Swanson, R.A.: The Adult Learner: The Definitive Classic in Adult Education and Human Resource Development. Gulf, Houston (1998)
15. Kolb, D.A.: Experiential Learning: Experience as the Source of Learning and Development, vol. 1. Prentice-Hall, Englewood Cliffs (1984)
16. Hayes, J., Allinson, C.W.: Matching learning styles and instructional strategy: an application of the person-environment interaction paradigm. Percept. Motors Skills **76**(1), 63–79 (1993)
17. Awoniyi, E.A., Griego, O.V., Morgan, G.A.: Person-environment fit and transfer of training. Int. J. Train. Dev. **6**(1), 25–35 (2002)
18. Pacini, R., Epstein, S.: The relation of rational and experiential information processing styles to personality, basic beliefs and the ratio-bias phenomenon. J. Pers. Soc. Psychol. **76**(6), 972–987 (1999)
19. Novak, T.P., Hoffman, D.L.: The fit of thinking style and situation: new measures of situation-specific experimental and rational cognition. J. Consum. Res. **36**(1), 56–72 (2009)
20. Epstein, S.: Emotion and Self-Theory. Guilford Press, New York (1993)
21. Zeyer, A.: Motivation to learn science and cognitive style. Eurasia J. Math. Sci. Technol. Educ. **6**(2), 121–128 (2010)
22. Stipek, D.J.: Motivation and instruction. In: Berliner, D.C., Calfee, R.C. (eds.) Handbook of Educational Psychology, 1st edn., pp. 8–113. Macmillan, New York (1996)
23. Brophy, J.E.: Motivation Students to Learn, 3rd edn. Routledge, New York (2010)
24. Ghosh, P., Satyawadi, R., Joshi, J.P., Ranjan, R., Singh, P.: Towards more effective training programmes: a study of trainer attributes. Ind. Commer. Train. **44**(4), 194–202 (2012)
25. Christophel, D.M.: The relationship among teacher immediacy behaviors, student motivation and learning. Commun. Educ. **39**(4), 323–340 (1990)
26. Towler, A.J., Dipboye, R.L.: Effects of trainer expressiveness, organization and trainee goal orientation on training outcomes. J. Appl. Psychol. **86**(4), 664–673 (2001)
27. Rangel, B., Chung, W., Harris, T.B., Carpenter, N.C., Chiaburu, D.S., Moore, J.L.: Rules of engagement: the joint influence of trainer expressiveness and trainee experiential learning style on engagement and training transfer. Int. J. Train. Dev. **19**(1), 18–31 (2015)
28. Bangert-Drowns, R.L., Pyke, C.: A taxonomy of student engagement with educational software: an exploration of literate thinking with electronic text. J. Educ. Comput. Res. **24** (3), 213–234 (2001)

29. Yanson, R.: The Relationship Between Training Design and Trainee Differences on Training Outcomes: An Experimental Investigation of the Treatment of Socialization and Training Content in the E-Learning Environment. State University of New York at Albany, New York (2012)
30. Schaufeli, W.B., Bakker, A.B.: UBES: Utrechtse Bevlogenheidschaal (UWES: Utrecht Engagement Scale). University of Utrecht, Utrecht (2003)
31. Mauno, S., Kinnunen, U., Ruokolainen, M.: Job demands and resources as antecedents of work engagement: a longitudinal study. J. Vocat. Behav. **70**(1), 149–171 (2007)
32. Gorham, J.: The relationship between verbal teacher immediacy behaviors and student learning. Commun. Educ. **37**(1), 40–53 (1988)
33. Sanders, J.A., Weiseman, R.L.: The effects of verbal and nonverbal teacher immediacy on perceived cognitive, affective and behavioral learning in the multicultural classroom. Commun. Educ. **39**(4), 341–353 (1990)
34. House, R.J., Mitchell, T.R.: Path goal theory of leadership. J. Contemp. Bus. **3**(4), 81–97 (1974)
35. Chiaburu, D.S., Sawyer, K.B., Thoroughgood, C.N.: Transferring more than learned in training: employees' and managers' (over) generalization of skills. Int. J. Sel. Assess. **18**(4), 380–393 (2010)
36. Colquitt, J.A., LePine, J.A., Noe, R.A.: Toward an integrative theory of training motivation: a meta-analytic path analysis of 20 years of research. J. Appl. Psychol. **85**, 678–707 (2000)
37. Towler, A., Arman, G., Quesnell, T., Hoffman, L.: How charismatic trainers inspire others to learn through positive affectivity. Comput. Hum. Behav. **32**, 221–228 (2014)
38. Epstein, S.: Integration of the cognitive and the psychodynamic unconscious. Am. Psychol. **49**(8), 709–724 (1994)
39. Hawk, T.F., Shah, A.J.: Using learning style instruments to enhance students learning. Decis. Sci. J. Innov. Educ. **5**, 1–19 (2007)

The Impact of Employees' Own Knowledge on the Effectiveness of Project Managers' Development Programs

Pawel Pietras(✉) and Maciej Szczepanczyk

Lodz University of Technology, Lódz, Poland
{pawel.pietras,maciej.szczepanczyk}@p.lodz.pl

Abstract. The article presents experiences from the implementation of three development programs for employees in the construction industry. Development programs were implemented in organizations culturally rooted in three different countries: Poland, Austria and France. Each of the programs had a similar goal - to increase the competence of project managers. However, each of the programs was based on different assumptions and used the experience of the organisations' employees to a different level. It resulted from cultural differences in the approach to knowledge management of the organization. The article attempts to evaluate the effectiveness of development programs in the function of using the knowledge of the employees. The research was conducted for over two years, it resulted from the duration of development programs of the surveyed organisations.

Keywords: Training · Development of managers · Project management competencies · Organization culture

1 Introduction

The main objective of each company's operation are attempts to increase its value. A higher corporate value may result from numerous factors, such as for example, a greater number of satisfied clients, better products, more professional services, qualified employees or established reputation. It is possible to achieve each of the aforementioned elements by focusing on a broadly defined staff development. Organizing an efficient system of staff development gives a chance of having high-tech products, providing services of a higher quality to a greater number of clients, which as a consequence, should lead to an improved corporate value. In a highly competitive business environment, knowledge possessed by employees becomes one of the company's key goals since having adequate knowledge allows to take decisions and actions that trigger a lower risk. Thus, it is necessary to undertake development activities, particularly those stimulating employees to develop and providing proper conditions that would enable them to create new solutions. What is especially important is to use the knowledge, both explicit and tacit, that employees already have.

The article presents results of studies carried out during three development projects for construction industry employees. Each of them was based on certain assumptions and to a different extent used experience of a company's employees. In these studies,

W. Karwowski et al. (Eds.): AHFE 2019, AISC 963, pp. 385–390, 2020.
https://doi.org/10.1007/978-3-030-20135-7_38

there was a thesis indicating that usefulness of knowledge acquired by employees depends on whether it is shared by the company staff or whether it comes from the outside. In the conclusion the article encourages its readers to think of the impact of organization's national culture on stimulating the concept of knowledge sharing.

2 Literature Review

Nowadays achievement of business targets at the level of an organization is mainly based on people who are well-prepared to work, satisfied and professional. Employees who meet these requirements are more willing to take up daring actions completing them more efficiently and effectively and being aware of the contribution they make to achievement of the organization's overall targets. Ultimately, it will result in the employees being appreciated and awarded with bonuses, however, it will also motivate them to work more diligently and with greater commitment [1].

The tools which support commitment to work are career planning and development which, if approached appropriately, may transform employees who have been useless so far into those very desirable to a given institution. Career planning is a necessary element of management in a modern organization and should be practised in all kinds of environment. It is particularly important if employees are an invaluable source of knowledge for an organization and tasks they perform involve a high level of specialization. This issue is emphasized by Strandberg [2] who indicates that some HR managers are aware of the importance of career planning especially in the case of employees who perform special tasks or provide the organization with their expert knowledge. Such employees are engaged in the process of their career planning and thus their satisfaction and motivation to work increase, and eventually they continue to work for the organization for a longer time.

The aspect is also emphasized by Riveros and Tsai [3] who indicate that employees included in development schemes are more motivated to work, hence they are also more willing to get involved in operations of the organization who cares about them.

Effective plans of employee development have an impact on their routine activities by improving their creativity and courage when adopting new, more efficient solutions which in a longer perspective leads to professional success and completion of the organization's targets [4].

They are also significant in terms of new and young employees who, because of their status, may not be aware of the applied standards and ways of work performance [12]. Moreover, they may also find it difficult to adapt themselves to their work environment. Including them in development schemes and providing them with care and support from the very beginning of their professional career may ultimately bring considerable benefits [5].

Additionally, it will also give them knowledge resources necessary for skillful, efficient and professional performance of work tasks. The knowledge provided through development schemes may comprise experience and information gained, as well as values and approaches adopted so far by the organization [6]. Moreover, it may be used in practice by employees who have a greater experience, supervisors or external experts. What is particularly important in this context is individual knowledge sharing between employees during e.g. organized workshop and training sessions or present problem solving, as well as tacit (implicit) knowledge which constitutes an organization's resources. Such activities boost the organization and improve its competitive power [7]. Therefore, the crucial aspect here is an efficient system of knowledge management which, as Kiessling [8] believes, will become the key factor of an organization's competitiveness in the market and that will have an impact on increasing the quality of human resources in companies. Organizations having broad knowledge will be able to improve their creativity and efficiency, which will enable them to reach a new level of quality [9]. It will also play a role in improving the efficiency and effectiveness of the organization's employees [10].

3 Organizations Included in the Study and Assumptions of Development Projects

The study included three enterprise operating in the construction industry. First of them, Alpha, is an enterprise of Polish origin and capital, well-known in Poland. The company was established at the beginning of the 1990's and year by year it gradually developed its current position of one of the most recognizable Polish companies in the residential and commercial construction services sector. The second enterprise, Beta, is a Polish branch office of a large group of companies operating in the infrastructure construction industry. Its headquarters are located in Austria and the company is well-recognized in the road construction sector in the country triangle including Austria, Germany and Lithuania. The third enterprise, Gamma, is a company dealing with highly-specialized construction works, cooperating with most enterprises involved in various forms of construction services. It is a part of a French construction corporate group recognized worldwide.

In the years 2016–2018 all of the three organizations conducted development schemes addressed to project managers. The overall objective of these schemes was professional development in management and leadership.

The scheme carried out by Alpha company included altogether 36 selected project managers and 86 selected engineers. Employees who were to take part in training programs were selected by managers of branches they worked in. The two development schemes formed a unified training and counselling offer for employees and were known as: Akademia Managera [Academy of Managers] and Akademia Inżyniera [Academy of Engineers]. The authors of the development schemes were employees of an external company providing training and consulting services who following of an analysis based on a questionnaire distributed among future employees, prepared schemes of several dozens of training courses addressed to the company's staff. Next, the training courses were conducted by external (as regards a specific company) coaches.

As for Beta company, the development scheme included 46 employees, mainly project managers, however, the in this group there were a few individuals taking both higher positions, e.g. project directors, as well as lower ones, such as assistants or junior project managers. The development scheme was preceded by an analysis of needs carried out as assessment center that indicated competency gaps that employees had as compared to the profile of required competency prepared by HR department.

Employees in whom the assessment center identified competency gaps (46 of 61 verified persons) were indicated as the program participants. The authors of the development scheme were employees of an external training-consultancy company which prepared programs of over a dozen of training courses supplementing the required competencies. The training courses were then conducted by external coaches.

In Gamma company, the development scheme included all employees who were project managers and construction engineers. The development scheme was prepared by the staff of the corporate group (the French mother company). The training courses were conducted by the corporate employees, both of the mother company and the Polish branch, and an external coach played the role of a consultant for internal coaches and a coach responsible for didactic efficiency of the scheme who supported external coaches during the training courses.

4 Impact of the Organizational Knowledge on Training Schemes

Each organization, not only the ones that invest in development schemes, invests in development schemes being aware of the influence they have on work efficiency of its employees. In the analyzed cases, the business reasons for taking up development schemes for employees were improving the level of project managers' competencies, which was expected to directly translate into increased economic (financial) efficiency of projects, a lower number of inadequate (expensive) risk responses and improved planning quality. All these three criteria applied for assessing the quality of work of project managers are obviously related to experience in project management, hence, transfer of organizational knowledge during training schemes is a key factor from the point of view of their efficiency and even usefulness in professional work.

Training courses are one of the key tools of knowledge transfer in an organization [11], therefore selecting them as a tool for sharing the required knowledge was the most relevant choice. However, the question should be raised in what way, if any, these development schemes may support transfer of knowledge, including tacit knowledge, within an organization. The study assumes that the time devoted by a company's in-house employees (i.e. professional staff, therefore excluding human resources employees) may constitute a certain indicator of knowledge transfer. In this study it is referred to as the involvement indicator and it was divided into three categories. The first one is called the influence indicator and it is calculated as a ratio of the number of work hours that in-house professional employees devoted to preparing the training scheme to the overall number of hours devoted to preparing the training schemes. The second indicator is referred to as the impact indicator and it is calculated as a ratio of the number of work hours that in-house professional employees spent preparing training materials and process to the overall number of hours spent preparing training materials and process. The third indicator is referred to as the real time indicator and it was calculated as the ratio of the number of hours of training conducted by professional employees to the overall number of training hours. The study was not aimed at developing any model of knowledge transfer, however, when thinking about such a model, one should assume the reference values of the said indicators the level of which would show that the expected knowledge transfer is effective. By referring to one's own experience and observations only, it may be assumed that the reference value (relevant, ensuring an effective influence on knowledge transfer) for the influence indicator should be 0.7, 0.5 for the impact indicator and 0.7 for the real transfer indicator. It results from the fact that the training programme (meaning both its contents and the training duration) should not include subject-matter issues only, and a professional employee is not required to have a relatively broad a teaching knowledge as it is the responsibility of the human resources staff and external consultants.

An interesting case is Beta company. After about 1.5 year of the development scheme, at the suggestions of the participating employees, the training schedule was changed by including training sessions developed based on the participants' own projects and prepared by the company's professional employees. As Table 1 presents, this solution significantly increased the values of the influence indicator and impact indicator. If the development scheme was, from the very beginning, based on such assumption, the values of the indicators would be comparable to those achieved by Gamma company.

Table 1. The values of the said indicators in the three analyzed organizations.

Indicator	Alpha	Beta	Gamma
Influence	0.1	0.4	0.8
Impact	0.2	0.4	0.5
Real Transfer	0.1	0.3	0.7
Involvement	**0.4**	**1.1**	**2.0**

At the end of each training cycle, the participants assessed the training by completing specially designed forms. Each of the three companies already had their own form templates, however, for the purpose of the study, all of them were asked to assess the efficiency of training from the point of view of its usefulness for a given work position.

The average assessment result (measured in the scale from 1 to 5, where 1 means not very useful and 5 means very useful) is: 3.78 for Alpha company, 4.42 for Beta company and 4.93 for Gamma company.

The assessments of particular participants were obviously significantly differentiated; however, a distinct correlation may be noticed between the values of the said indicators and the assessment of the training. The higher involvement of the company's own professional employees in preparing and, possibly, conducting the training, the better result of assessment regarding the usefulness of the acquired knowledge in a given work position taken by the employee.

5 Summary

The presented results were a part of a broader study on quality and efficiency of training schemes. They indicate a rather strong interdependency between involvement of a company's own professional employees in preparing and conducting training courses focused on improvement of competencies necessary in a given enterprise. Apart from the presented analytical data, another reflection, reaching beyond the study scope, comes to mind. It is interesting how much the cultural background of the analyzed companies influenced their development policies. The first case was an entirely (in terms of their capital and history) Polish company, where in the Polish management culture achievement of effects is cultivated, and work efficiency, in common terms, means time spent on work. The second case was a company rooted, in terms of its capital, rooted in German culture (the head office in Austria, the majority of the share capital is German) where a conservative business approach and a clear and well-structured organization are applied. It was this company that in the middle of the development scheme decided to introduce a considerable change towards increasing efficiency. The third company was culturally rooted in France where a model of a socially developed business and employees' participation in any activities is entirely common. In that organization creating development schemes by consultants "foreign" to the organization was out of the question. They could only fulfill an auxiliary function.

References

1. Hameed, A., Waheed, A.: Employee development and its affect on employee performance a conceptual framework. Int. J. Bus. Soc. Sci. **2**(13), 224–229 (2011)
2. Strandberg, C.: The role of human resource management in corporate social responsibility issue brief and roadmap. Report for Industry Canada. Burnaby, BC: Strandberg Consulting (2009)
3. Riveros, A.M., Tsai, T.S.: Career commitment and organizational commitment in for profit and non-profit sectors. Int. J. Emerg. Sci. **1**(3), 324–340 (2011)
4. Hirsh, W.: Career development in employing organization: practices and challenges from a UK perspective, Guidance for Work-force Development Conference, held by CEDEFOP (2007)
5. Witko, K., Bernes, K.B., Magnusson, K., Bardick, A.D.: Senior high school career planning: what students want. J. Educ. Enquiry, **6**(1) (2009)
6. Hooff, B.V.D., Hendrix, L.: Eargerness and Willingness to Share: The Relevance of Different Attitudes Towards Knowledge Sharing. Stadtmagistrat Innsbruck, Innsbruck (2004)
7. Verbeke, W., Belschak, F., Bagozzi, R., Wuyts, S.: Gaining access to intrafirm knowledge: an internal market perspective on knowledge sharing. J. Hum. Perform. **24**(3), 205–230 (2011)
8. Kiessling, T., Richey, G., Meng, J., Dabic, M.: Exploring knowledge management to organizational performance outcomes in a transitional economy. J. World Bus. **44**(4), 421–433 (2009)
9. Dickel, D.G., Moura, G.L.: Organizational performance evaluation in intangible criteria: a model based on knowledge management and innovation management. RAI J. Technol. Manage. Innovation **13**(3), 211–220 (2016)
10. Byukusenge, E., Munene, J., Orobia, L.: Knowledge management and business performance: mediating effect of innovation. J. Bus. Manage. Sci. **4**(4), 82–92 (2016)
11. Cancialosi, C.: 6 Key Steps to Influencing Effective Knowledge Transfer in Your Business, Forbes, 8 December 2014
12. Stankiewicz-Mróz, A.: Approach to the issues of leadership in the processes of companies acquisitions. Procedia Manuf. **3**, 793–798 (2015)

Students' Independent Learning Activity as an Effective Method of Acquiring Professional Competencies

Margarita Salnikova, Yulia Salnikova(✉), Marina Soroka, and Valentina Stolyarova

Platov South-Russian State Polytechnic University (NPI), Prosveshcheniya, 132, 346428 Novocherkassk, Russian Federation
salnikovamag@gmail.com, yuliasalnikova@gmail.com, kedrrostov40a@rambler.ru, stolval73@mail.ru

Abstract. The article represents the model of the organization of independent out-of-class students' learning activities. The main purpose of the model is to stimulate students' cognitive activity while transferring accent from the academic knowledge to making independent research and practical skills application. The model is based on the results of students' psychological characteristics diagnostics that had been carried out by authors in the period 2016–2018 in order to define the features of personal motivational sphere. The model implies an application of project-oriented educational technologies based on the interaction scheme "teacher-student" and "student-student". It is described the experience of model integration in the courses "Marketing" and "Mathematics in Management" for bachelor-students of the polytechnic university. Within the presented model, the training tasks are designed so that all members of the group are interconnected but at the same time independent in making research and solving problems.

Keywords: Students' learning activity · Out-of-class independent work · Project-oriented teaching · Small groups · Motivation level

1 Introduction

Nowadays, governments of many countries make efforts to improving modern educational standards and principles in order to comply with the innovative technologies development, rapidly changing working conditions and demands of labor market.

In particular, according to Russian Federal State Educational Standards of higher education the graduates of the polytechnic university are to master the system of general cultural and professional competences. Among these competences is the ability to self-organization and self-education, which is impossible to acquire without a high level of motivation.

Modern higher school is supposed to contribute to the intensification of the educational and cognitive activity of students, to develop their creative abilities and thinking culture, the ability to orient themselves in the constantly changing and increasing flow of scientific, technical and socio-political information. In order to

W. Karwowski et al. (Eds.): AHFE 2019, AISC 963, pp. 391–399, 2020.
https://doi.org/10.1007/978-3-030-20135-7_39

achieve these purposes, the Federal State Educational Standard provides a large amount of educational time and content for out-of-class independent work of students. By out-of-class independent work of students we mean the planned educational and research work performed by students with the teacher's methodological guidance and instructions but without his (or her) direct participation. This is complex and qualitatively different type of activity aimed at developing students' cognitive independence, increasing the effectiveness of educational material perception, establishing interdisciplinary connections, providing opportunities for the individualization of education. Independent work of students can be considered as a basis and start point for lifelong self-education and willingness to improve different types of skills in the future. In this connection, the main teachers' task is to help students organize, rationally plan and implement the results of their out-of-class independent work.

The inclusion of independent work in the curriculum (that is, its regulation as a special type of educational activity) characterizes a fundamentally new stage in the organization of the educational process. The above mentioned determines the relevance of the research topic.

2 Features of First-Year Students' Adaptation and Motivation to Study at the University

The current modernization and innovation processes in Russian higher education requires new teaching approaches. Today one of the main criteria for the professional competence, along with knowledge and skills, is the ability to independently acquire and apply knowledge. In the system of lifelong education, the development of cognitive motives and abilities of an individual is of great importance as the basis for successful learning and readiness for self-education [1].

It is the initial stage of studying at the university that plays a special role in the development of the individual's professional qualities. The first year at the university forms the students' attitudes toward studying process itself and future professional activities. The effectiveness of learning is determined not only by the students' abilities but also by how quickly and successfully their adaptation to higher education will pass, which, in turn, affects the level of learning motivation, moral satisfaction and psychological comfort.

The teacher needs to know the characteristics of the need-motivational sphere of the students in order to develop learning situations, use teaching methods and techniques aimed at increasing students' interest in the subject, the level of motivation to learn, the formation of the ability to learn. Therefore, it is necessary to diagnose the individual psychological characteristics of first-year students, which results will help to identify the degree of adaptation to the educational process and the availability of educational and professional motives for studying at the university.

3 Results of the Psychological and Pedagogical Diagnostics of First-Year Students

This section presents the results of three surveys carried out by the authors in the 2016–2017 and 2017–2018 academic years within the psychological and pedagogical diagnostics of first-year students of Innovation and Production Management Department (Platov South-Russian State Polytechnic University).

3.1 The Survey "Adaptation to Student Life"

Below are the results of the survey "Is it difficult for you to get used to student life?" (Fig. 1).

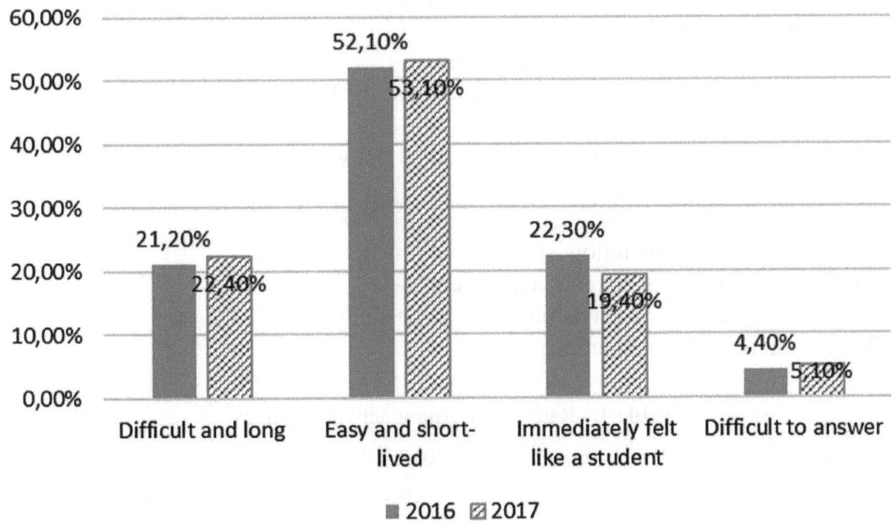

Fig. 1. Adaptation to student life.

Diagnostics showed that some students have difficulties with self-assessment and do not consider adaptation as an important issue (4.4% in 2016–2017, 5.1% in 2017–2018). For the majority of freshmen adaptation was easy and short-lived (52.1% in 2016–2017, 53.1% in 2017–2018), some respondents immediately felt like a student (22.3% in 2016–2017, 19.4% - in 2017–2018). At the same time, a considerable part of first-year students noted that they had been getting used to student life with difficulty and for a long time (21.2% in 2016–2017, 22.4% in 2017–2018).

Such a situation is determined by the novelty of both the learning activity itself and the conditions in which it occurs. First-year students do not have enough skills and abilities necessary to master the university program successfully. Consequently, they experience the psychological discomfort of transitioning from a school system to a university system.

3.2 The Survey "Motives of Everyday Learning Activities"

The study of the motives of first-year students' learning activities was carried out according to the method of Rean and Yakunin [2]. The students were asked to choose the five most significant motives for learning activities from the given list. The frequency of choosing one or another motive was determined over the entire sample. The more often a particular motive is chosen, the higher its rank, the more it dominates the motive system.
List of motives:

1. To become a first-class specialist.
2. To get a diploma.
3. To be able studying successfully in subsequent courses.
4. To pass exams for "good" and "excellent".
5. To receive a scholarship on a constant basis.
6. To acquire profound knowledge.
7. To be well prepared for the next classes.
8. Not to neglect any subject of the educational cycle.
9. To keep up with fellow students.
10. To ensure the success of future professional activities.
11. To fulfill pedagogical requirements.
12. To be respected by the teacher.
13. To be an example for fellow students.
14. To get an approval of parents and others.
15. To avoid rejection and punishment for poor study.
16. To get intellectual satisfaction.

Table 1. Rating of learning activity motives

2016–2017	2017–2018
To get a diploma	To become a first-class specialist
To acquire profound knowledge	To get a diploma
To become a first-class specialist	To ensure the success of future professional activities
To pass exams for "good" and "excellent"	To acquire profound knowledge
To ensure the success of future professional activities	To be able studying successfully in subsequent courses

The most unpopular motives of educational activity include (Table 1):

- to keep up with fellow students;
- to fulfill pedagogical requirements;
- to be an example for fellow students;
- to be well prepared for classes.

Analyzing the results of the survey, we can conclude that first-year students have adequate motivation, they strive to become highly qualified specialists, get a diploma, acquire deep and solid knowledge. At the same time, they consider that being prepared for classes and fulfillment of pedagogical requirements are not necessary conditions for achieving the goal. This is because most of them do not know how to learn and exercise self-assessment and self-control of their actions.

3.3 The Survey "Motivation to Study at the University"

The study of students' motivation to study at the university was conducted by the method of Ilyina [2]. The method includes three scales: "The acquisition of knowledge" (the desire to acquire knowledge); "Mastering a profession" (the desire to master professional knowledge and skills, to form professionally important qualities); "Getting a diploma" (the desire to acquire a diploma in the formal assimilation of knowledge, the desire to find workarounds for exams and tests). The predominance of motives on the first two scales indicates an adequate choice of a profession for a student and satisfaction with it. The results of the survey of first-year students are shown in Fig. 2.

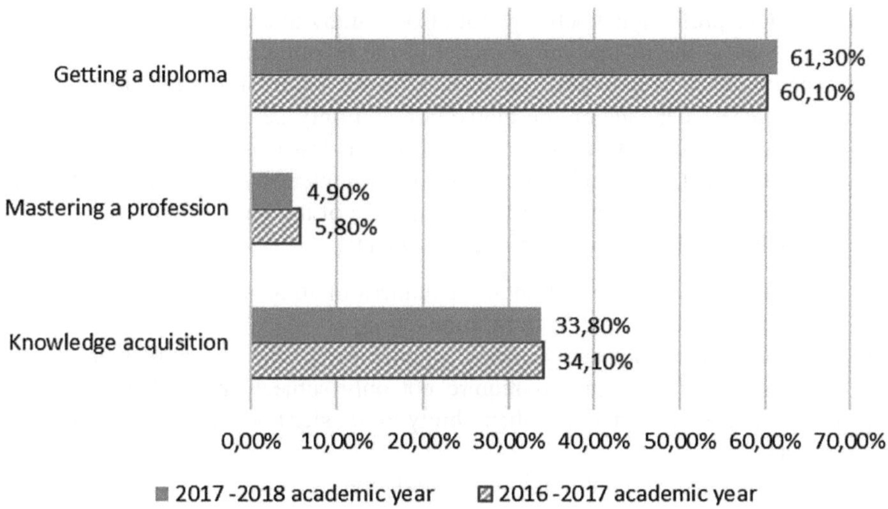

Fig. 2. Motivation to study at the university

The survey results showed that external motives of educational activity prevail: for most of the first-year students, the dominant motive for learning is to get a diploma while they have rather little idea about the chosen profession.

During the analysis of the survey conducted, the following was also revealed:

- 37.8% note that they have not enough willpower to learn without reminding by the university administration;
- 42.2% of respondents believe that they can do their best only when they are stimulated and motivated;
- 44.4% of respondents answered that it is difficult to force themselves to study subjects, which are not with the future profession;
- 68.9% of students admit that they use cheat sheets when taking exams;

- 33.3% believe that examinations must be passed with a minimum of efforts;
- 42.2% of the students surveyed believe that it is the laziness that prevents them from studying properly.

The success of teaching in higher school depends on various factors: the students' level of pre-university training, the ability to plan their educational activities, the motives for choosing an educational institution, the psychological characteristics of students, the organization of the educational process at the university, and many others. Teachers often consider students "as information processing devices" who attend lectures, read textbooks, perform assigned tasks and, when required, demonstrate their knowledge by passing tests and exams [3].

It is assumed that if the students voluntarily entered the university, they are already motivated and interested in the studied disciplines. The experience of teaching at the university shows that this is not always the case. Many first-year students are not able to learn without systematic external control. Besides, they have low motivation not only to study the subjects of the general education cycle, which they do not associate with their future professional activity, but also to study at a university in general.

The analysis of the mathematical model of the learning process (which takes into account the influence of individual parameters on the result) shows that approximately 78% of the success depends on the motivation and only 22% is related to a person's abilities. This confirms that the success of learning is less dependent on the individual abilities of the student, but largely is determined by his interest in the learning process [4].

In this connection, among the professional tasks of a higher school teacher are the tasks of developing the cognitive sphere of students:

- to assess the levels of motivation and readiness of students to learn new material;
- to take steps for increasing learning motivation;
- to reveal the importance and scientific significance of the discipline being taught;
- to select tasks which solutions require not only subject knowledge but also the general intelligence of students, their ability to transfer knowledge from one subject area to another;
- to develop creative skills through the formulation of research educational tasks [5].

4 The Model of Organization of Out-of-Class Independent Work of Students

The model of organizing out-of-class independent work of students in the educational process contains the following components:

- subject component (students and teacher);
- motivation–targeted component (choice of purpose and determination of the motive for independent activity);
- content component (the volume and structure of the content of educational material for out-of-class independent work of students);
- organizational activity component (tasks for out-of-class independent work of students);

- control component (control of out-of-class independent work of students);
- assessment component (assessment of the results of out-of-class independent work of students).

The model involves the use of design-oriented educational technologies based on the "teacher–student" and "student–student" interaction scheme. It is not only a teacher who attracts students to the educational process but the students themselves, interacting with each other, influence the motivation of the group mates.

The project method of teaching is not fundamentally new in world pedagogy. It originated in the early twentieth century in the United States, it was also called the method of problems and it was connected with the ideas of the humanistic philosophy and education developed by the American philosopher and teacher J. Dewey, as well as his follower V. X. Kilpatrick. In Russia, the ideas of project education emerged in 1905 under the guidance of the Russian pedagogue S. Shatsky.

The project method is always focused on students' independent activities, organically combined with a group approach to learning, involves a certain set of teaching and learning techniques that allow solving a particular problem as a result of independent actions of students with the mandatory presentation of these results. As a pedagogical technology, the method of projects includes a set of research and problem methods, which are creative in their essence [6].

Within studying the "Marketing" discipline, students of bachelor program "Management" are offered to do marketing research, which is aimed at developing the skills of applying theoretical knowledge to specific market situations and conditions, analyzing the market and developing an optimal strategy with taking into account various external factors and the ambiguity of market participants' behavior.

Subjects of the projects include a wide range of marketing problems within the framework of Novocherkassk: fast food market; beauty industry market; preschool education market; medical services market; advertising services market. The analysis determines the size and growth rate of the industry; market leaders; market volume; distribution of market shares; market development trends.

The process of working on a project consists of several stages.

The preparatory stage consists in choosing a project topic, setting a goal and objectives for a project, identifying sources of information, report forms and evaluation criteria, forming groups and distributing responsibilities among participants. In order to ensure proper organization of work, the distribution of roles in subgroups of 5–6 people (the choice of the responsible coordinator, designer, speaker, etc.) should be based on the results of the Belbin's "Team Roles" test [7]. This allows to determine the contribution of each participant to the team's work and the nature of interaction with the rest of the group.

Stage of independent work: project participants collect information, conduct research, analyze the results, draw conclusions, prepare project reports.

The final stage is the results presentation. The project is considered completed if each member of the group is aware of all aspects of the research, explains and answers all questions of the teacher and other students.

The motivating role of evaluating the results of educational activities is not in doubt, but it is important that the evaluation should provide a qualitative analysis of the students' learning activities, contain information about the level of competence.

Work on the project involves the independent activity and interest of students but at the same time causes difficulties, therefore the teacher should provide the consultative support on the project topic.

The considered result of out-of-class independent work of students on the project is aimed at the formation of such competencies as [8]

- ability to self-organization and self-education;
- ability to work in a team;
- acquiring the strategic analysis skills, development and implementation of an organization's strategy aimed at ensuring competitiveness;
- ability to analyze the connections between the functional strategies of companies in order to make management decisions.

Teachers often have to deal with the problem of misunderstanding by students of the need to study certain disciplines. The results of above given surveys showed that 44,4% of students have low motivation to study subjects which are not related straightly to their future profession. Particularly, it concerns "Mathematics in management". For example, first-year students often have a question: "Why would a future bachelor-manager study mathematics?". Teachers know that the learning material is absorbed deeper and stronger if students are interested in the topic being studied and maintain this cognitive interest. In order to increase the motivation to study mathematics in the study of the discipline "Mathematics in Management" students can be offered an educational research project "The connection between the oil price and ruble exchange rate." While working on the project, students solve the tasks:

- collecting information about the oil price and the ruble exchange rate to the dollar in selected periods of time and processing of the statistical data;
- creating various correlation models using the available mathematical packages and choosing the optimal models for the selected periods;
- analysis of the results.

The presented project implements the principle of unity of the fundamental and applied aspects in teaching mathematics. The theoretical material used by students in the process of working on a project creates a fundamental mathematical base for research and solving applied problems. The project establishes interdisciplinary connections between such disciplines as "Mathematics in Management", "Economics" and "Informatics". The use of interdisciplinarity in teaching mathematics allows students to demonstrate various application areas of mathematics and thereby increase the motivation to study it.

5 Conclusion

The project-oriented technology, developed by authors, is aimed at creating students' sustainable cognitive motivation, developing cognitive interests and organizational-activity skills. Mastering key competences while out-of-class independent work of students, proceeds more successfully within the framework of the proposed organizational model. Work in small groups develops personal qualities, interpersonal skills and cooperation, increases the activity of participants in the educational process united by a single goal. In a heterogeneous group, a participant with a higher level of training and motivation is interested in the success of the group. A lower-level student experiences the situation when there is a motive "not to let the group down". Thus, students go through the stage of consolidating and comprehension of all the information and received results, which is especially suitable for capable students who are not attracted to the traditional way of learning through repetitions.

The model implementation has led to the following results:

- increasing the students' motivation level and emotional involvement in the educational process;
- better understanding by students the significance and importance of the subject and its relation to the practical professional field;
- ability to use flexibly the knowledge of the subject while fulfilling variety of practical tasks;
- comprehension of the interdisciplinary connections and the ability to transfer knowledge from one subject area to another;
- increasing the general intellectual level of the students;
- development of communication and cooperation skills.

References

1. Bakshaeva, N.A., Verbickiy, A.A.: Psychology of Students' Motivation (Психология мотивации студентов). Logos, Moscow (2006)
2. Il'in, E.P.: Motivation and Motives (Мотивация и мотивы). Peter, St. Petersburg (2003)
3. Berns, R.: Self-concept Development and Education (Развитие Я-концепции и воспитание). Progress, Moscow (1986)
4. Solodova, E.A.: New Models in the Education System: The Synergetic Approach (Новые модели в системе образования: Синергетический подход). The book house "Librocom", Moscow (2012)
5. Smirnov, S.D.: Psychology and Pedagogy for Teachers of Higher School (Психоло-гия и педагогика для преподавателей высшей школы). Publishing House of Bau-man MSTU, Moscow (2014)
6. Polat, E.S., Bukharkina, M.Y., Moiseeva, M.V., Petrov, A.E.: New Pedagogical and Information Technologies in the Education System. Publishing Center "Academy", Moscow (2002)
7. The Nine Belbin Team Roles. https://www.belbin.com/about/belbin-team-roles/
8. Federal State Educational Standard of Higher Professional Education in the field of training 38.03.02 «Management» (Bachelor degree). Approved by Ministry of Education and Science of the Russian Federation, order from Yan. 12, 2016, № 7. http://fgosvo.ru/uploadfiles/fgosvob/380302.pdf

Author Index

W. Karwowski et al. (Eds.): AHFE 2019, AISC 963, pp. 401–402, 2020.
https://doi.org/10.1007/978-3-030-20135-7

MIX
Papier aus verantwortungsvollen Quellen
Paper from responsible sources
FSC® C105338

If you have any concerns about our products,
you can contact us on
ProductSafety@springernature.com

In case Publisher is established outside the EU,
the EU authorized representative is:
Springer Nature Customer Service Center GmbH
Europaplatz 3, 69115 Heidelberg, Germany

Printed by Libri Plureos GmbH
in Hamburg, Germany